MECHANICS
A Graduate Course

MECHANICS
A Graduate Course

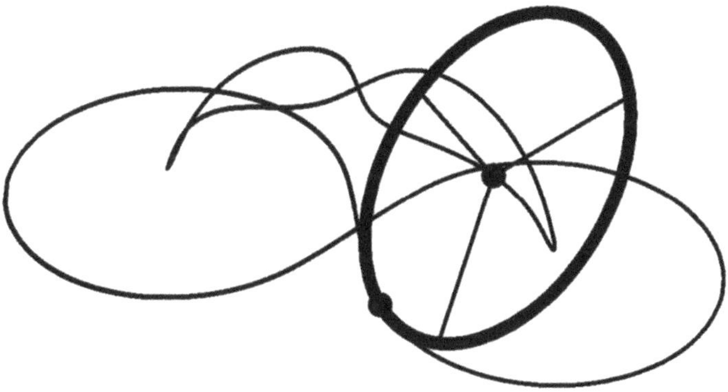

Dmitry Garanin
The City University of New York, USA

World Scientific

NEW JERSEY • LONDON • SINGAPORE • BEIJING • SHANGHAI • HONG KONG • TAIPEI • CHENNAI • TOKYO

Published by

World Scientific Publishing Co. Pte. Ltd.
5 Toh Tuck Link, Singapore 596224
USA office: 27 Warren Street, Suite 401-402, Hackensack, NJ 07601
UK office: 57 Shelton Street, Covent Garden, London WC2H 9HE

Library of Congress Control Number: 2026944379

British Library Cataloguing-in-Publication Data
A catalogue record for this book is available from the British Library.

MECHANICS
A Graduate Course

ISBN 978-981-98-2885-2 (hardcover)
ISBN 978-981-98-2959-0 (paperback)
ISBN 978-981-98-2886-9 (ebook for institutions)
ISBN 978-981-98-2887-6 (ebook for individuals)

For any available supplementary material, please visit
https://www.worldscientific.com/worldscibooks/10.1142/14757#t=suppl

Desk Editor: Muhammad Ihsan Putra

Typeset by Stallion Press
Email: enquiries@stallionpress.com

Contents

Introduction

Mechanics is a branch of physics that studies the motion of material bodies and the conditions of their equilibrium. The latter constitutes *statics*, which is particularly important in engineering. *Kinematics* examines the general properties of motion regardless of its causes, with special attention to the role of constraints. *Dynamics*, the largest and most important part of mechanics, studies motion caused by forces or interactions.

Mechanics can be categorized according to the systems studied: mechanics of *material points*, mechanics of *rigid bodies*, mechanics of *elastic bodies*, and mechanics of fluids (*hydro-* and *aerodynamics*). At the core of each area lies Newton's second law, the fundamental equation of motion. Mechanics of material points and rigid bodies are described by ordinary differential equations (ODEs). Within the mechanics of material points, we distinguish between systems of one or few bodies and many-body systems. For rigid bodies, the ODEs include variables for the positions and velocities of their centers of mass as well as angles defining their orientation. The mechanics of elastic bodies and fluids (collectively, continuum mechanics) is more complex, requiring partial differential equations. In many cases, continuum mechanics couples with thermodynamics, especially in aerodynamics. This course focuses on systems described by ODEs, including particles and rigid bodies.

Classical mechanics has two important limitations. First, it applies only when object speeds are much smaller than the speed of light ($v \ll c$); otherwise, *relativistic mechanics* must be used. Second, it applies to bodies with sufficiently large mass and/or kinetic energy; for small particles such as electrons, *quantum mechanics* is required.

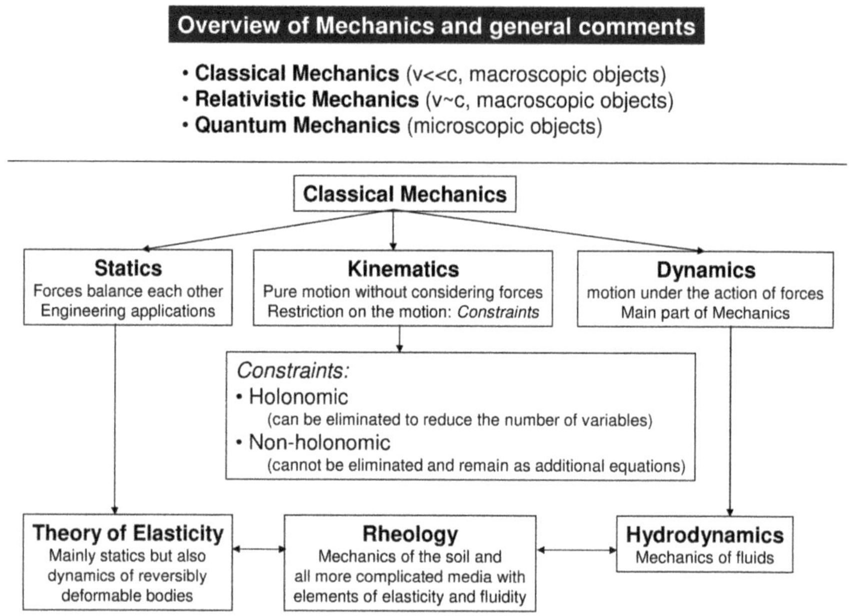

Figure 1. Overview of mechanics.

Regarding theoretical approaches, mechanics splits into three parts: Newtonian, Lagrangian, and Hamiltonian mechanics. Newtonian mechanics offers the most straightforward formulation, directly applying Newton's second law. It is efficient in most cases, especially for the consideration of particles under the influence of forces. Lagrangian mechanics is more sophisticated and based on the least action principle. It is efficient for the consideration of more general mechanical systems having constraints, in particular, mechanisms. Hamiltonian mechanics represents the most mathematically sophisticated approach and, while less practical for many applications, serves a crucial role in bridging classical mechanics to quantum mechanics.

In this course, we will consider Newtonian, Lagrangian, and Hamiltonian mechanics, as well as some additional advanced topics.

Chapter 1

Newtonian Mechanics

The basis of Newtonian mechanics is Newton's laws. Especially important is the second Newton's law, which defines the equation of motion of a particle of mass m subject to the influence of a force $\mathbf{F}$:

$$m\ddot{\mathbf{r}} = \mathbf{F}, \tag{1.0.1}$$

where $\ddot{\mathbf{r}} \equiv d^2\mathbf{r}/dt^2 \equiv \partial_t^2\mathbf{r}$ is the double time derivative of the position vector $\mathbf{r}$ of the particle, i.e., its acceleration $\mathbf{a}$. This second-order differential equation can be written as the system of two first-order differential equations

$$\begin{aligned} \dot{\mathbf{r}} &= \mathbf{v}, \\ \dot{\mathbf{v}} &= \mathbf{F}/m, \end{aligned} \tag{1.0.2}$$

where

$$\mathbf{v} = \dot{\mathbf{r}} \tag{1.0.3}$$

is the particle's velocity. The force $\mathbf{F}$ can depend on the particle's position and velocity, as well as explicitly on time: $\mathbf{F} = \mathbf{F}(\mathbf{r}, \mathbf{v}, t)$.

In the absence of the force (free particle), the solution of the above equation of motion is

$$\begin{aligned} \mathbf{v} &= \mathbf{v}_0, \\ \mathbf{r} &= \mathbf{r}_0 + \mathbf{v}_0 t, \end{aligned} \tag{1.0.4}$$

where $\mathbf{v}_0$ and $\mathbf{r}_0$ are integration constants of the ordinary differential equations (ODE) above, being physically the initial conditions: velocity and position at $t = 0$. This is Newton's first law, now having only a historical meaning. Philosophically, it is kept as an axiom stating the existence of inertial frames, in which there is no acceleration in the absence of forces.

Newton's second law is a differential equation for a generally three-component vector variable $\mathbf{r} \equiv (x, y, z)$. Correspondingly, Eq. (1.0.1) can be written in components as a system of three equations:

$$m\ddot{x} = F_x,$$
$$m\ddot{y} = F_y,$$
$$m\ddot{z} = F_z. \tag{1.0.5}$$

In general, these three equations are not independent and may be coupled via the force depending on all three position and/or velocity components.

In general, Newton's equations of motion require a numerical solution. Among many numerical methods for solving systems of ODEs, one should mention the still less-known Taylor method (http://taylorcenter.org/Gofen/TaylorMethod.htm) based on the Taylor expansion. It provides outstanding accuracy. On the site http://taylorcenter.org/Exploratorium/, you can find numerical implementations and visualizations of mechanical problems included in this book and beyond. A specialty of this software is the stereo rendering of trajectories in three dimensions that can be viewed through red-blue glasses as a real-time animation.

Systems of N particles are described by a generally coupled system of ODEs consisting of second Newton's second law for each particle

$$m_i\ddot{\mathbf{r}}_i = \mathbf{F}_i, \quad i = 1, \ldots, N. \tag{1.0.6}$$

Each force $\mathbf{F}_i$ can be represented as the sum of the external force $\mathbf{F}_i^{\text{ext}}$ and inter-particle interaction forces $\mathbf{f}_{ij}$,

$$\mathbf{F}_i = \mathbf{F}_i^{\text{ext}} + \sum_j \mathbf{f}_{ij}. \tag{1.0.7}$$

According to Newton's third law, interaction forces are anti-symmetric,

$$\mathbf{f}_{ij} = -\mathbf{f}_{ji}. \tag{1.0.8}$$

1.1 Mechanics of a single particle

In this section, simple examples of equations of motion for a single particle will be considered with the help of basic mathematical methods.

1.1.1 *Motion with a constant force*

For $\mathbf{F} = \text{const.}$ in Eq. (1.0.1), one readily finds the solution

$$\mathbf{v} = \mathbf{v}_0 + \mathbf{a}t,$$
$$\mathbf{r} = \mathbf{r}_0 + \mathbf{v}_0 t + \frac{1}{2}\mathbf{a}t^2, \tag{1.1.1}$$

where $\mathbf{a} = \mathbf{F}/m$ is constant acceleration. The validity of this solution can be checked by differentiation over time. The trajectory $\mathbf{r}(t)$ (a parabola) is confined to the plane, the orientation of which is specified by the two vectors $\mathbf{v}_0$ and $\mathbf{a}$. Thus, this motion is effectively two-dimensional. It is convenient to choose the coordinate system such that the trajectory is in xy-plane while $z = 0$.

1.1.2 *Motion with a viscous damping*

If a body is moving in a viscous fluid or in the air at a small enough speed, it is experiencing a drag force linear in the velocity. For a particle with symmetric shape, the drag force is opposite to its velocity,

$$\mathbf{F}_v = -\alpha \mathbf{v}. \tag{1.1.2}$$

Newton's second law with a drag force is usually written explicitly as

$$m\ddot{\mathbf{r}} + \alpha\dot{\mathbf{r}} = \mathbf{F}, \tag{1.1.3}$$

where $\mathbf{F}$ means all other forces. This equation can be rewritten as

$$\ddot{\mathbf{r}} + \Gamma\dot{\mathbf{r}} = \mathbf{F}/m, \tag{1.1.4}$$

where $\Gamma \equiv \alpha/m$ is a characteristic relaxation rate measured in s^{-1}. If $\mathbf{F}$ is the explicit function of only the time, one can rewrite this equation via the velocity,

$$\dot{\mathbf{v}} + \Gamma\mathbf{v} = \mathbf{F}/m. \tag{1.1.5}$$

After solving this first-order ODE, one can find $\mathbf{r}(t)$ by simple integration, $\mathbf{r}(t) = \int dt\mathbf{v}(t)$.

Let us start with the uniform ODE with $\mathbf{F} = 0$, i.e.,

$$\dot{\mathbf{v}} + \Gamma\mathbf{v} = 0. \tag{1.1.6}$$

In this case, the motion is one-dimensional along a line in $3d$ space. It is convenient to choose the coordinates, so that the motion is along the x-axis, while $y = z = 0$. Using, for brevity, the notation $v_x = v$, one obtains the equation

$$\dot{v} + \Gamma v = 0, \tag{1.1.7}$$

which can be solved as follows:

$$\begin{aligned} \frac{dv}{v} &= -\Gamma dt, \\ \int \frac{dv}{v} &= -\Gamma \int dt, \\ \ln|v| &= -\Gamma t + \text{const.}, \end{aligned} \tag{1.1.8}$$

which finally yields

$$v = v_0 e^{-\Gamma t}. \tag{1.1.9}$$

Returning to the general vector form, one obtains

$$\mathbf{v} = \mathbf{v}_0 e^{-\Gamma t}. \tag{1.1.10}$$

Now, integrating this equation yields

$$\mathbf{r} = \mathbf{r}_0 + \frac{\mathbf{v}_0}{\Gamma}\left(1 - e^{-\Gamma t}\right). \tag{1.1.11}$$

This method of solution works for a class of nonlinear first-order ODEs.

For linear homogeneous ODEs, one can use a more powerful method based on searching the solution in the exponential form such as

$$\mathbf{v} \propto e^{\lambda t}. \tag{1.1.12}$$

Substitution into Eq. (1.1.6) yields the algebraic equation for λ:

$$\lambda + \Gamma = 0, \tag{1.1.13}$$

having the solution $\lambda = -\Gamma$. Thus, the solution of the ODE has the form

$$\mathbf{v} = \mathbf{C}e^{-\Gamma t} = \mathbf{v}_0 e^{-\Gamma t}, \tag{1.1.14}$$

which coincides with Eq. (1.1.10).

In the presence of a constant force, e.g., gravity, Eq. (1.1.5) can be solved by a small modification of the method above. The problem has a non-trivial asymptotic solution that can be found by setting $\dot{\mathbf{v}} = 0$. This yields

$$\mathbf{v}_\infty = \mathbf{F}/(m\Gamma), \tag{1.1.15}$$

the stationary velocity at large times. Then one can rewrite Eq. (1.1.5) in terms of the new variable $\mathbf{u} \equiv \mathbf{v} - \mathbf{v}_\infty$ as

$$\dot{\mathbf{u}} + \Gamma \mathbf{u} = 0. \tag{1.1.16}$$

This is mathematically the same equation as before and it can be solved in a similar way. Finally, one obtains

$$\mathbf{v} = \mathbf{v}_\infty + (\mathbf{v}_0 - \mathbf{v}_\infty)\, e^{-\Gamma t}. \tag{1.1.17}$$

Integration of this formula yields

$$\mathbf{r} = \mathbf{r}_0 + \mathbf{v}_\infty t + \frac{\mathbf{v}_0 - \mathbf{v}_\infty}{\Gamma}\left(1 - e^{-\Gamma t}\right). \tag{1.1.18}$$

Finally, Eq. (1.1.5) can be solved in quadratures for any time-dependent force $\mathbf{F}(t)$ by the method of variation of constants. This method allows us to find the solution of a non-uniform ODE (or system of ODEs), including ODE with coefficients depending on time if the solution of the corresponding uniform ODE is known. Using this method, one searches for the solution of Eq. (1.1.5) in the form

$$\mathbf{v}(t) = \mathbf{C}(t) e^{-\Gamma t}. \tag{1.1.19}$$

Substituting this into Eq. (1.1.5), one obtains the equation for the "variable constant" $\mathbf{C}(t)$:

$$\dot{\mathbf{C}}(t) e^{-\Gamma t} = \mathbf{F}(t)/m. \tag{1.1.20}$$

Solving for $\dot{\mathbf{C}}(t)$ and integrating the result, one obtains

$$\mathbf{C}(t) = \int dt' e^{\Gamma t'} \frac{\mathbf{F}(t')}{m}. \tag{1.1.21}$$

Substituting this into Eq. (1.1.19), one obtains

$$\mathbf{v}(t) = e^{-\Gamma t} \int dt' e^{\Gamma t'} \frac{\mathbf{F}(t')}{m}. \tag{1.1.22}$$

In this expression, the indefinite integral contains an integration constant. One can work it out by rewriting the result in terms of a definite integral as

$$\mathbf{v}(t) = \mathbf{v}_0 e^{-\Gamma t} + e^{-\Gamma t} \int_0^t dt' e^{\Gamma t'} \frac{\mathbf{F}(t')}{m}. \tag{1.1.23}$$

To check this formula, one can differentiate it over t, obtaining Eq. (1.1.5).

1.1.3 *Harmonic oscillator*

The most elementary example of a harmonic oscillator is a body of mass m attached to a spring with spring constant k. Adding the viscous damping and generalizing Eq. (1.1.3), one writes down the equation

$$m\ddot{\mathbf{r}} + \alpha\dot{\mathbf{r}} + k\mathbf{r} = \mathbf{F}, \tag{1.1.24}$$

where $\mathbf{F}$ is an external force, as before. In the following, we will consider the case of the body performing a linear motion along the x-axis. Dividing by the mass, one obtains the equation

$$\ddot{x} + 2\Gamma\dot{x} + \omega_0^2 x = f, \tag{1.1.25}$$

where

$$\Gamma \equiv \frac{\alpha}{2m}, \quad \omega_0 \equiv \sqrt{\frac{k}{m}}, \quad f \equiv \frac{F}{m}. \tag{1.1.26}$$

Here, we have defined Γ in a way different from above for the sake of simplicity of the formulas; ω_0 is the frequency of oscillations in the absence of damping.

Solution of the homogeneous equation ($f = 0$), in accordance with the general method, can be searched in the form

$$x(t) = Ce^{i\Omega t}. \tag{1.1.27}$$

The imaginary i has been inserted in anticipation of an oscillating motion of the system. Substituting this into Eq. (1.1.25), one obtains the quadratic equation

$$-\Omega^2 + 2i\Gamma\Omega + \omega_0^2 = 0, \tag{1.1.28}$$

having the solution

$$\Omega_\pm = i\Gamma \pm \tilde{\omega}_0, \quad \tilde{\omega}_0 \equiv \sqrt{\omega_0^2 - \Gamma^2}. \tag{1.1.29}$$

Thus, the solution of the ODE has the form

$$x(t) = x_+(t) + x_-(t) = C_+e^{i\Omega_+ t} + C_-e^{i\Omega_- t}, \tag{1.1.30}$$

where $C_\pm$ are integration constants. Using the relation

$$e^{i\varphi} \equiv \cos\varphi + i\sin\varphi, \tag{1.1.31}$$

one can rewrite the result in explicitly real form

$$x(t) = C_1 e^{-\Gamma t}\cos\tilde{\omega}_0 t + C_2 e^{-\Gamma t}\sin\tilde{\omega}_0 t, \tag{1.1.32}$$

where $C_{1,2}$ is another set of integration constants. The latter can be found from the initial conditions

$$x(0) = x_0, \quad \dot{x}(0) = v_0, \tag{1.1.33}$$

i.e.,

$$x(0) = C_1 = x_0 \tag{1.1.34}$$

and

$$\dot{x}(0) = -\Gamma C_1 + \tilde{\omega}_0 C_2 = v_0. \tag{1.1.35}$$

One finds

$$C_1 = x_0, \quad C_2 = \frac{v_0 + \Gamma x_0}{\tilde{\omega}_0}. \tag{1.1.36}$$

Thus,

$$x(t) = x_0 e^{-\Gamma t} \cos \tilde{\omega}_0 t + \frac{v_0 + \Gamma x_0}{\tilde{\omega}_0} e^{-\Gamma t} \sin \tilde{\omega}_0 t. \tag{1.1.37}$$

Let us look at this solution. According to Eq. (1.1.29), in the absence of damping, $\Gamma = 0$, the body is oscillating with the frequency ω_0. Damping reduces the oscillation frequency that turns to zero at $\Gamma = \omega_0$. In the strong-damping limit $\Gamma > \omega_0$, the motion of the body is aperiodic.

Let us now consider the motion of the harmonic oscillator under the influence of an external force. Using the method of variation of constants in Eq. (1.1.30), one searches for the solution in the form

$$x(t) = C_+(t) e^{i\Omega_+ t} + C_-(t) e^{i\Omega_- t}. \tag{1.1.38}$$

The "variable constants" satisfy the system of equations

$$\begin{aligned} \dot{C}_+(t) x_+(t) + \dot{C}_-(t) x_-(t) &= 0, \\ \dot{C}_+(t) \dot{x}_+(t) + \dot{C}_-(t) \dot{x}_-(t) &= f(t). \end{aligned} \tag{1.1.39}$$

Its solution is

$$\dot{C}_+(t) = \frac{\begin{vmatrix} 0 & x_-(t) \\ f(t) & \dot{x}_-(t) \end{vmatrix}}{\begin{vmatrix} x_+(t) & x_-(t) \\ \dot{x}_+(t) & \dot{x}_-(t) \end{vmatrix}} = \frac{-f(t) x_-(t)}{x_+(t)\dot{x}_-(t) - \dot{x}_+(t) x_-(t)}. \tag{1.1.40}$$

Using

$$x_+(t)\dot{x}_-(t) - \dot{x}_+(t) x_-(t) = i\left(\Omega_- - \Omega_+\right) e^{i(\Omega_- + \Omega_+)t} = -2i\tilde{\omega}_0 e^{-2\Gamma t}, \tag{1.1.41}$$

one obtains

$$\dot{C}_+(t) = -\frac{i}{2}\frac{f(t)}{\tilde{\omega}_0}e^{i\Omega_- t+2\Gamma t} = -\frac{i}{2}\frac{f(t)}{\tilde{\omega}_0}e^{-i\Omega_+ t} \tag{1.1.42}$$

and, similarly,

$$\dot{C}_-(t) = \frac{i}{2}\frac{f(t)}{\tilde{\omega}_0}e^{i\Omega_+ t+2\Gamma t} = \frac{i}{2}\frac{f(t)}{\tilde{\omega}_0}e^{-i\Omega_- t}. \tag{1.1.43}$$

Integrating these two formulas and substituting the result in Eq. (1.1.38), one obtains

$$x(t) = x_{\text{free}}(t) + x_{\text{forced}}(t), \tag{1.1.44}$$

where $x_{\text{free}}(t)$ is the solution of the uniform ODE describing the free oscillator and given by Eq. (1.1.37), whereas

$$x_{\text{forced}}(t) = -\frac{i}{2\tilde{\omega}_0}e^{i\Omega_+ t}\int_0^t dt' f(t')e^{-i\Omega_+ t'} + \frac{i}{2\tilde{\omega}_0}e^{i\Omega_- t}\int_0^t dt' f(t')e^{-i\Omega_- t'} \tag{1.1.45}$$

is the response to the external force. The latter can be rewritten in the explicitly real form as

$$x_{\text{forced}}(t) = \frac{1}{\tilde{\omega}_0}\int_0^t dt' f(t')e^{-\Gamma(t-t')}\sin\left[\tilde{\omega}_0(t-t')\right]. \tag{1.1.46}$$

One can check that $x_{\text{forced}}(0) = \dot{x}_{\text{forced}}(0) = 0$, i.e., the forced solution is independent of the initial conditions and does not change the form of $x_{\text{free}}(t)$.

1.1.4 *Charged particle in a magnetic field*

A moving particle having the charge q experiences the so-called Lorentz force from the magnetic field $\mathbf{B}$:

$$F_L = q\left[\mathbf{v} \times \mathbf{B}\right]. \tag{1.1.47}$$

Since this force depends on the velocity, it is convenient to write the equation of motion in terms of the velocity, as was done in the case of the viscous drag force,

$$m\dot{\mathbf{v}} = q\left[\mathbf{v} \times \mathbf{B}\right]. \tag{1.1.48}$$

In the basic case of uniform magnetic field that will be considered below, z-axis will be chosen along $\mathbf{B}$, so that $B_z = B$ and $B_x = B_y = 0$. The equation of motion in components has the form

$$\begin{aligned} \dot{v}_x &= \omega_c v_y, \\ \dot{v}_y &= -\omega_c v_x, \\ \dot{v}_z &= 0, \end{aligned} \tag{1.1.49}$$

where

$$\omega_c \equiv \frac{qB}{m} \tag{1.1.50}$$

is the cyclotron frequency. Differentiating the first equation over time and substituting the second equation, one obtains the second-order ODE for v_x:

$$\ddot{v}_x + \omega_c^2 v_x = 0. \tag{1.1.51}$$

This is the harmonic-oscillator equation considered in the previous section. Thus, v_x will be oscillating in time. The same equation can be obtained for v_y, thus v_y will be oscillating, too. On the other hand, motion in the direction of the field is a free motion,

$$\dot{v}_z = 0, \quad v_z = \text{const.} = v_{z0}, \quad z = z_0 + v_{z0}t. \tag{1.1.52}$$

The solution of Eq. (1.1.51) can be obtained from Eq. (1.1.32):

$$v_x(t) = C_{x1}\cos\omega_c t + C_{x2}\sin\omega_c t. \tag{1.1.53}$$

Integration constants obtained from the initial conditions have the form

$$C_{x1} = v_x(0) = v_{x0}, \tag{1.1.54}$$

$$C_{x2} = \dot{v}_x(0)/\omega_c = v_y(0) = v_{y0}, \tag{1.1.55}$$

thus the final solution reads

$$v_x(t) = v_{x0}\cos\omega_c t + v_{y0}\sin\omega_c t. \tag{1.1.56}$$

Similarly, one obtains

$$v_y(t) = v_{y0}\cos\omega_c t - v_{x0}\sin\omega_c t. \tag{1.1.57}$$

It is easy to show that

$$v_x^2(t) + v_y^2(t) = v_{x0}^2 + v_{y0}^2 = \text{const.} \tag{1.1.58}$$

Thus, the vector (v_x, v_y) is rotating at the constant rate ω_c, while the kinetic energy $E_k = m\mathbf{v}^2/2$ is conserved. The time dependences of x and y can be obtained by integration of the results for v_x and v_y. One obtains

$$x = x_0 + \frac{v_{x0}}{\omega_c}\sin\omega_c t - \frac{v_{y0}}{\omega_c}\cos\omega_c t,$$
$$y = y_0 + \frac{v_{y0}}{\omega_c}\sin\omega_c t + \frac{v_{x0}}{\omega_c}\cos\omega_c t. \tag{1.1.59}$$

This trajectory is circular with the center at an arbitrary (x_0, y_0). The radius of the circle R can be found by the calculation similar to the above:

$$R^2 = (x - x_0)^2 + (y - y_0)^2 = \frac{v_{x0}^2 + v_{y0}^2}{\omega_c^2}. \tag{1.1.60}$$

Thus, the cyclotron radius is given by

$$R = \frac{v_\perp}{\omega_c} = \frac{mv_\perp}{qB}, \tag{1.1.61}$$

where $v_\perp = \sqrt{v_x^2 + v_y^2}$.

1.1.5 *Phase portraits*

For systems with one degree of freedom, such as one particle in one dimension, it is useful to make the so-called phase portraits showing the trajectories in the *phase space* of the position and velocity. To find the trajectories, one needs to solve the equation of motion. If the energy is conserved, one can just plot the energy landscape, where the trajectories can be read out as lines of the constant energy. To find the direction of motion along these lines, one needs to use the equations of motion.

Consider, as an example, the mathematical pendulum having the energy

$$E = \frac{mv^2}{2} + mgl\,(1 - \cos\phi), \tag{1.1.62}$$

where m is the mass attached to the end of a massless rod of length l and g is the gravity acceleration. The angle ϕ if defined with respect to the reference direction down. The velocity $\mathbf{v}$ is directed perpendicularly to the rod and can be expressed via the time derivative of the angle (the angular velocity) as $v = l\dot{\phi}$. The energy landscape in terms of ϕ and $\dot{\phi}$ is shown in Fig. 1.1.1. There are two regimes of the motion. The first regime is finite motion around the energy minimum $E_{\min} = 0$, which corresponds to oscillations of the pendulum. The second regime is the regime of infinite motion in which the pendulum is rotating in one direction crossing the point of the vertical

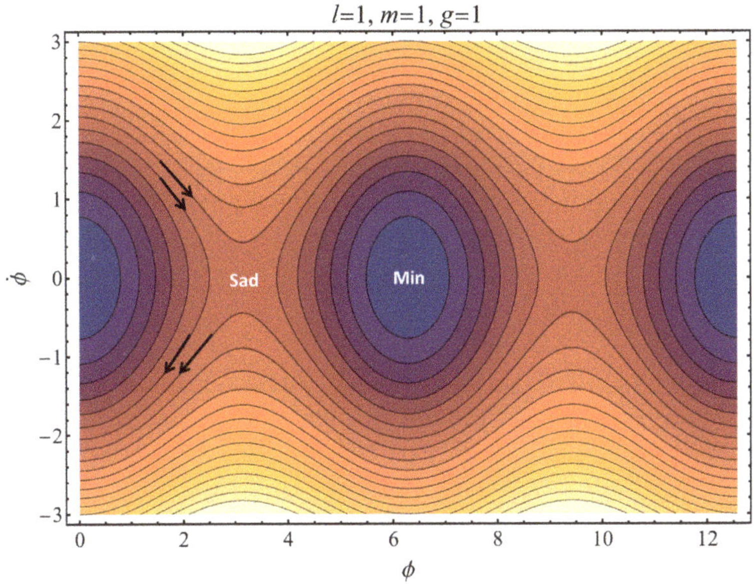

Figure 1.1.1. Phase portrait of the mathematical pendulum. “Sad” and “Min” stand for the saddle point and the minimum of the energy. The direction of motion along the lines of constant energy is shown by the arrows.

orientation. The points $\phi = \pi + 2\pi n$, $\dot{\phi} = 0$ with $n = 0, \pm 1, \pm 2, \ldots$ are the saddle points of the energy landscape.

1.2 Momentum

Momentum of a particle is defined as

$$\mathbf{p} = m\mathbf{v}. \tag{1.2.1}$$

It can be used to write Newton's second law in the form

$$\dot{\mathbf{p}} = \mathbf{F}. \tag{1.2.2}$$

In the absence of forces, momentum as well as the velocity are conserved. The above is not essentially new.

Significance of momentum becomes apparent if one considers a system of interacting particles and defines the total momentum as

$$\mathbf{P} = \sum_i \mathbf{p}_i. \tag{1.2.3}$$

Differentiating it over time, one obtains

$$\dot{\mathbf{P}} = \sum_i \dot{\mathbf{p}}_i = \sum_i \mathbf{F}_i. \tag{1.2.4}$$

Separating the forces into external and internal, Eq. (1.0.7), and using Newton's third law, Eq. (1.0.8), one can see that only external forces change the momentum of the system,

$$\dot{\mathbf{P}} = \sum_i \mathbf{F}_i^{\text{ext}}. \tag{1.2.5}$$

In an isolated system, there are only interaction forces between the particles that are changing momenta of individual particles, whereas the total momentum is conserved. Conservation of the total momentum is important, in particular, in collisions. During collisions, usually systems can be considered as effectively isolated because collisions occur during a very short time, so that external forces cannot change momenta significantly during collisions. On the contrary, internal forces during collisions are large and important.

1.3 Angular momentum, torque

The angular momentum of a particle can be defined with respect to any "origin" point $\mathbf{r}_O$. For this, we define the relative position $\mathbf{r}'$ via

$$\mathbf{r} = \mathbf{r}' + \mathbf{r}_O, \tag{1.3.1}$$

where $\mathbf{r}$ is the "absolute" position vector of the particle defined with respect to a particular frame or coordinate system. The relative angular momentum is defined by

$$\mathbf{l}' = [\mathbf{r}' \times \mathbf{p}] = [(\mathbf{r} - \mathbf{r}_O) \times \mathbf{p}] = \mathbf{l} - [\mathbf{r}_O \times \mathbf{p}], \tag{1.3.2}$$

where $\mathbf{l} = [\mathbf{r} \times \mathbf{p}]$ is the "absolute" angular momentum. (Although $\mathbf{r}_O$ can depend on time, defining the angular momentum with respect to it does not mean switching to the moving frame as we still use $\mathbf{p}$ and not $\mathbf{p}'$, the momentum in the moving frame.)

The time derivative of the angular momentum is given by

$$\dot{\mathbf{l}}' = [(\dot{\mathbf{r}} - \dot{\mathbf{r}}_O) \times \mathbf{p}] + [(\mathbf{r} - \mathbf{r}_O) \times \dot{\mathbf{p}}]. \tag{1.3.3}$$

With $\dot{\mathbf{p}} = \mathbf{F}$ and $[\dot{\mathbf{r}} \times \mathbf{p}] = [\mathbf{v} \times m\mathbf{v}] = \mathbf{0}$, this equation becomes

$$\dot{\mathbf{l}}' = [\mathbf{p} \times \mathbf{v}_O] + \boldsymbol{\tau}', \tag{1.3.4}$$

where $\mathbf{v}_O = \dot{\mathbf{r}}_O$ is the velocity of the origin and

$$\boldsymbol{\tau}' = [(\mathbf{r} - \mathbf{r}_O) \times \mathbf{F}] \tag{1.3.5}$$

is the relative torque that depends on the origin. The first term in the rhs of Eq. (1.3.4) is the kinematic term due to the dynamical change of the origin point. If the force is central, i.e., directed everywhere away from or toward a particular central point, one can choose a frame having the origin at this point. Then, if the origin is not moving, the torque is zero and the angular momentum is conserved.

The total relative angular momentum of a system of particles is defined in a similar way by

$$\mathbf{L}' = \sum_i [(\mathbf{r}_i - \mathbf{r}_O) \times \mathbf{p}_i] = \mathbf{L} - [\mathbf{r}_O \times \mathbf{P}], \tag{1.3.6}$$

where $\mathbf{P}$ is the total momentum defined by Eq. (1.2.3) and $\mathbf{L} = \sum_i [\mathbf{r}_i \times \mathbf{p}_i]$ is the "absolute" angular momentum. The latter is the sum of the relative angular momentum and the contribution due to the motion of the system as a whole. If $\mathbf{P} = \mathbf{0}$, the relative angular momentum does not depend on the position of the origin $\mathbf{r}_O$ and coincides with the absolute angular momentum, $\mathbf{L}' = \mathbf{L}$.

Differentiation over time yields

$$\dot{\mathbf{L}}' = \sum_i [\mathbf{p}_i \times \mathbf{v}_O] + \sum_i [(\mathbf{r}_i - \mathbf{r}_O) \times \mathbf{F}_i]. \tag{1.3.7}$$

Separating the forces into external and internal ones, Eq. (1.0.7), for the internal torque, renaming i and j and using Newton's third law, one can write

$$\boldsymbol{\tau}'_{\text{int}} = \sum_{ij} [(\mathbf{r}_i - \mathbf{r}_O) \times \mathbf{f}_{ij}] = \sum_{ij} [(\mathbf{r}_j - \mathbf{r}_O) \times \mathbf{f}_{ji}] = -\sum_{ij} [(\mathbf{r}_j - \mathbf{r}_O) \times \mathbf{f}_{ij}]. \tag{1.3.8}$$

One can symmetrize $\boldsymbol{\tau}'_{\text{int}}$ by taking half of each expression:

$$\boldsymbol{\tau}'_{\text{int}} = \frac{1}{2} \sum_{ij} [(\mathbf{r}_i - \mathbf{r}_j) \times \mathbf{f}_{ij}]. \tag{1.3.9}$$

In nature, interaction forces between particles are directed along the line connecting the particles. Thus the cross-product in the formula above disappears, and therefore, the total relative torque is due to the external forces only:

$$\boldsymbol{\tau}' = \sum_i \left[(\mathbf{r}_i - \mathbf{r}_O) \times \mathbf{F}_i^{\text{ext}}\right] = \boldsymbol{\tau} - \mathbf{r}_O \times \sum_i \mathbf{F}_i^{\text{ext}}, \tag{1.3.10}$$

where $\boldsymbol{\tau} = \sum_i \left[\mathbf{r}_i \times \mathbf{F}_i^{\text{ext}}\right]$ is the total absolute torque. If the sum of all forces is zero, the relative torque does not depend on the origin and coincides with the absolute torque, $\boldsymbol{\tau}' = \boldsymbol{\tau}$.

Now, the equation of motion for the relative angular momentum above becomes

$$\dot{\mathbf{L}}' = [\mathbf{P} \times \mathbf{v}_O] + \boldsymbol{\tau}'. \tag{1.3.11}$$

In the treatment of particular problems, primes are usually dropped. For the rotation around the center of mass or a fixed point, the kinematic term in the rhs of this equation does not arise.

1.4 Work, energy and potential forces

Infinitesimal work done by the force $\mathbf{F}$ on a particle undergoing displacement $d\mathbf{r}$ is defined by

$$\delta A = \mathbf{F} \cdot d\mathbf{r}. \tag{1.4.1}$$

In this formula, δA is used instead of dA since work is not a function and the above is not its differential. Power W is work done per unit of time,

$$W = \frac{\delta A}{dt} = \mathbf{F} \cdot \frac{d\mathbf{r}}{dt} = \mathbf{F} \cdot \mathbf{v}. \tag{1.4.2}$$

If $\mathbf{F}$ is acting on a free particle, the latter will accelerate or decelerate, and the work on the particle will change its kinetic energy defined by

$$E_k = \frac{m\mathbf{v}^2}{2}. \tag{1.4.3}$$

Indeed, the infinitesimal change of the energy can be related to the infinitesimal work as

$$dE_k = m\mathbf{v} \cdot d\mathbf{v} = m\frac{d\mathbf{v}}{dt} \cdot \mathbf{v}dt = \mathbf{F} \cdot d\mathbf{r} = \delta A. \tag{1.4.4}$$

Potential forces are forces that can be expressed via gradients of position-dependent potential energy U:

$$\mathbf{F} = -\nabla U(\mathbf{r}) = -\frac{\partial U}{\partial \mathbf{r}}. \tag{1.4.5}$$

Work done by potential forces is independent of the trajectory and depends only on the initial and final positions,

$$A_{12} = \int_{\mathbf{r}_1}^{\mathbf{r}_2} \mathbf{F} \cdot d\mathbf{r} = -\int_{\mathbf{r}_1}^{\mathbf{r}_2} \frac{\partial U}{\partial \mathbf{r}} \cdot d\mathbf{r} = U(\mathbf{r}_1) - U(\mathbf{r}_2). \tag{1.4.6}$$

The work on a closed path is zero for potential forces,

$$\oint \mathbf{F} \cdot d\mathbf{r} = 0 \tag{1.4.7}$$

because the change of U is zero. Equation (1.4.6) can be rewritten as the definition of potential energy via the force,

$$U(\mathbf{r}) = U(\mathbf{r}_0) - \int_{\mathbf{r}_0}^{\mathbf{r}} \mathbf{F} \cdot d\mathbf{r}, \tag{1.4.8}$$

where $U(\mathbf{r}_0)$ is the (arbitrary) value of potential energy at the reference point $\mathbf{r}_0$. Potential energy is a secondary or auxiliary physical quantity, not directly measurable and introduced via the force as a primary quantity, and it is defined up to an arbitrary additive constant. Using Eq. (1.4.8), one first has to be sure that the integral does not depend on the path. To check whether $\mathbf{F}$ is a potential force, one can use the Stokes theorem to express the closed-path line integral as the flux through the surface spanned by the closed path,

$$\oint \mathbf{F} \cdot d\mathbf{r} = \int_S \operatorname{rot} \mathbf{F} \cdot d\mathbf{S}. \tag{1.4.9}$$

If

$$\operatorname{rot} \mathbf{F} \equiv \operatorname{curl} \mathbf{F} \equiv \nabla \times \mathbf{F} = \begin{vmatrix} \mathbf{e}_x & \mathbf{e}_y & \mathbf{e}_z \\ \partial_x & \partial_y & \partial_z \\ F_x & F_y & F_z \end{vmatrix} = 0, \tag{1.4.10}$$

$\mathbf{F}$ is a potential force.

Not all forces are potential. For instance, friction forces are not. An example of potential forces is gravity near the Earth's surface,

$$U = mg\mathbf{r} \cdot \mathbf{e}_z + \text{const.}, \tag{1.4.11}$$

where $\mathbf{e}_z$ is the unit vector directed up. The general gravitational force between two masses has the potential energy

$$U = G\frac{mM}{|\mathbf{r}|}, \tag{1.4.12}$$

where the origin of the coordinate system is put at the big mass M (e.g., the sun) and $\mathbf{r}$ points to the position of the small mass m (e.g., the Earth). The gravitational force acting on m has the form

$$\mathbf{F} = -\frac{\partial U}{\partial \mathbf{r}} = -G\frac{mM}{r^2}\frac{\mathbf{r}}{r}, \tag{1.4.13}$$

where the relation

$$\frac{\partial |\mathbf{r}|}{\partial \mathbf{r}} = \frac{\mathbf{r}}{|\mathbf{r}|} \tag{1.4.14}$$

was used. This is an example of a radial force that conserves the angular momentum.

A special kind of force is Lorentz force, Eq. (1.1.47). As it is not doing any work, there is no potential energy associated with it. For the same reason, this force does not lead to dissipation of energy as friction forces do.

The total energy of the particle

$$E = \frac{m\mathbf{v}^2}{2} + U(\mathbf{r}) \tag{1.4.15}$$

can be shown to be dynamically conserved (i.e., to be an *integral of motion*) if there are no other forces such as friction. Indeed,

$$\dot{E} = m\dot{\mathbf{v}} \cdot \mathbf{v} + \frac{\partial U}{\partial \mathbf{r}} \cdot \dot{\mathbf{r}} = (m\dot{\mathbf{v}} - \mathbf{F}) \cdot \mathbf{v} = 0 \tag{1.4.16}$$

via Newton's second law. Due to the conservation of the total energy, systems with potential forces are called conservative. In the presence of the viscous friction force, Eq. (1.1.2), one obtains

$$\dot{E} = m\dot{\mathbf{v}} \cdot \mathbf{v} + \frac{\partial U}{\partial \mathbf{r}} \cdot \dot{\mathbf{r}} = \left(-\alpha\mathbf{v} - \frac{\partial U}{\partial \mathbf{r}} + \frac{\partial U}{\partial \mathbf{r}}\right) \cdot \mathbf{v} = -\alpha\mathbf{v}^2, \tag{1.4.17}$$

which describes dissipation of energy.

Let us obtain the total energy E as the first integral of Newton's second law. Dot-multiplying it by $\mathbf{v}$ and manipulating the expressions, one obtains

$$0 = \left(m\dot{\mathbf{v}} + \frac{\partial U}{\partial \mathbf{r}}\right) \cdot \mathbf{v} = \frac{m}{2}\frac{d\mathbf{v}^2}{dt} + \frac{\partial U}{\partial \mathbf{r}} \cdot \dot{\mathbf{r}} = \frac{d}{dt}\left(\frac{m\mathbf{v}^2}{2} + U(\mathbf{r})\right). \tag{1.4.18}$$

Integrating this, one obtains Eq. (1.4.15) with $E = \text{const.}$ as an integral of motion.

The total energy of a system of interacting particles has the form

$$E = \sum_i \frac{m_i \mathbf{v}_i}{2} + U\left(\{\mathbf{r}_i\}\right), \tag{1.4.19}$$

where the potential energy depends, in general, on the positions of all particles. In the absence of non-conservative forces, conservation of this many-body energy follows from Newton's second law as above. In most cases, potential energy includes one-particle and two-particle terms,

$$U\left(\{\mathbf{r}_i\}\right) = \sum_i U_0\left(\mathbf{r}_i\right) + \frac{1}{2}\sum_{ij} V\left(|\mathbf{r}_i - \mathbf{r}_j|\right). \tag{1.4.20}$$

The factor 1/2 in the interaction term is inserted to compensate for double counting of interacting pairs ij and ji, so that the interaction energy between each two particles is just V. Note that the interaction energy is translationally invariant (does not change under transformation $\mathbf{r}_i \Rightarrow \mathbf{r}_i + \mathbf{a}$). Using Eq. (1.4.14), one obtains the corresponding force,

$$\mathbf{F}_i = -\frac{\partial U}{\partial \mathbf{r}_i} = -\frac{\partial U_0}{\partial \mathbf{r}_i} + \sum_j \mathbf{f}_{ij}, \tag{1.4.21}$$

where interaction forces are given by

$$\mathbf{f}_{ij} = -V'\left(|\mathbf{r}_i - \mathbf{r}_j|\right)\frac{\mathbf{r}_i - \mathbf{r}_j}{|\mathbf{r}_i - \mathbf{r}_j|}, \tag{1.4.22}$$

where V' is the derivative of the function over its argument. One can see that the third Newton's law $\mathbf{f}_{ji} = -\mathbf{f}_{ij}$ is obeyed. As it was said in the comment to Eq. (1.3.9), interaction forces are directed along the line connecting the two particles.

1.5 Center of mass, reduced mass

For any system of point masses, one can define the center of mass (CM) or center of inertia as

$$\mathbf{R} = \frac{1}{M}\sum_i m_i \mathbf{r}_i, \tag{1.5.1}$$

where $M = \sum_i m_i$ is the total mass. For a solid body, CM is defined by a corresponding integral. The velocity of CM is related to the total momentum,

$$\mathbf{V} = \frac{1}{M} \sum_i m_i \mathbf{v}_i = \frac{\mathbf{P}}{M}. \tag{1.5.2}$$

This formula evokes an image of a system considered as one body of mass M moving with the velocity $\mathbf{V}$. Dynamics of CM is due to the external forces,

$$M\dot{\mathbf{V}} = \dot{\mathbf{P}} = \sum_i \mathbf{F}_i^{\text{ext}}, \tag{1.5.3}$$

where Eq. (1.2.5) was used. CM of an isolated system is moving with a constant velocity.

If one defines the angular momentum in Eq. (1.3.6) with respect to the CM, $\mathbf{r}_O = \mathbf{R}$, then in Eq. (1.3.11), $\mathbf{v}_O = \mathbf{V} = \mathbf{P}/M$ and the kinematic term vanishes. The kinetic energy of a system of particles can be rewritten by expressing the particles' coordinates as $\mathbf{r}_i = \mathbf{r}'_i + \mathbf{r}_O = \mathbf{r}'_i + \mathbf{R}$, where $\mathbf{r}'_i$ are coordinates with respect to the CM. Using $\mathbf{v}_i = \mathbf{v}'_i + \mathbf{V}$, one obtains

$$E_k = \frac{1}{2} \sum_i m_i \left(\mathbf{v}'_i + \mathbf{V}\right)^2 = \frac{1}{2} \sum_i m_i \mathbf{v}_i'^2 + \frac{M\mathbf{V}^2}{2} + \mathbf{P}' \cdot \mathbf{V} = E'_k + \frac{M\mathbf{V}^2}{2}, \tag{1.5.4}$$

where we used the obvious relation $\mathbf{P}' = 0$. Thus, the kinetic energy consists of the internal (relative) kinetic energy and the kinetic energy due to the motion of the system as the whole.

An isolated system of two interacting masses can be described as one so-called *reduced mass* moving around the CM of the system. Choosing a coordinate system with the origin at the CM, one obtains the constraint on the positions of the bodies

$$m_1 \mathbf{r}_1 + m_2 \mathbf{r}_2 = 0. \tag{1.5.5}$$

As the single dynamical variable, one can choose the position-difference vector

$$\mathbf{r} \equiv \mathbf{r}_1 - \mathbf{r}_2. \tag{1.5.6}$$

Using these two equations, one can express the individual positions as

$$\mathbf{r}_1 = \frac{m_2}{m_1 + m_2} \mathbf{r}, \quad \mathbf{r}_2 = -\frac{m_1}{m_1 + m_2} \mathbf{r}. \tag{1.5.7}$$

The total energy of the system

$$E = \frac{m_1 \mathbf{v}_1^2}{2} + \frac{m_2 \mathbf{v}_2^2}{2} + U\left(|\mathbf{r}_1 - \mathbf{r}_2|\right) \tag{1.5.8}$$

becomes

$$E = \frac{m\mathbf{v}^2}{2} + U\left(|\mathbf{r}|\right), \tag{1.5.9}$$

where $\mathbf{v} = \dot{\mathbf{r}}$ and

$$m \equiv \frac{m_1 m_2}{m_1 + m_2} \tag{1.5.10}$$

is the reduced mass. The equation of motion of the system can be obtained from either of the two equations for the individual bodies, e.g.,

$$m_1 \ddot{\mathbf{r}}_1 = m\ddot{\mathbf{r}} = -\frac{\partial U}{d\mathbf{r}_1} = -\frac{\partial U}{d\mathbf{r}}. \tag{1.5.11}$$

The resulting equation of motion for the reduced mass

$$m\ddot{\mathbf{r}} = -\frac{\partial U}{d\mathbf{r}} \tag{1.5.12}$$

can be obtained from the equation of motion for the second body as well.

1.6 One-dimensional conservative motion

For one-dimensional conservative systems, conservation of the total energy can be used to reduce the second-order differential equation to a first-order differential equation that can be straightforwardly integrated. The basic form of a one-dimensional system, a particle, has the energy

$$E = \frac{mv^2}{2} + U(x). \tag{1.6.1}$$

In addition, there are systems that become one-dimensional because of a constraint. For instance, a mass on a light rod (pendulum) making a motion along a circle in the xy-plane can be described by a single dynamical variable, the angle φ, which makes it effectively one-dimensional. Solving the above equation for $v = \dot{x}$, one arrives at the first-order ODE:

$$\dot{x} = \sqrt{\frac{2\left[E - U(x)\right]}{m}}, \tag{1.6.2}$$

where the integral of motion E plays the role of an integration constant. Further integration of Eq. (1.6.2) is done as follows:

$$t = \int dt = \sqrt{\frac{m}{2}} \int \frac{dx}{\sqrt{E - U(x)}}. \tag{1.6.3}$$

This defines $x(t)$ implicitly. Another integration constant comes from the indefinite integral.

Motion of a mechanical system can take place only in the region $E > U$, where kinetic energy is positive. Regions $E < U$ are inaccessible for the system and are called *barrier regions*. The system hits barriers and changes direction of its motion at *turning points*, where $E = U$, and thus, $v = 0$. Positions of turning points depend on the energy. If the motion of the system is limited by left and right turning points x_1 and x_2 being the two roots of the equation $E = U(x)$, the particle performs oscillations between these points with the period

$$T(E) = \sqrt{2m} \int\limits_{x_1}^{x_2} \frac{dx}{\sqrt{E - U(x)}}. \tag{1.6.4}$$

In general, the period depends on the energy E. If E is close to the minimum of $U(x)$ where the latter is parabolic,

$$U(x) \cong U_0 + \frac{k}{2}(x - x_0)^2, \tag{1.6.5}$$

the system becomes a harmonic oscillator with the period independent of the energy. The motion of the harmonic oscillator has been determined above by solving its equation of motion. Within the present approach, setting $U_0 = x_0 = 0$, one finds $x_{1,2} = \pm x_m$, where $x_m = \sqrt{2E/k}$. Then one can integrate Eq. (1.6.3) as follows:

$$t = \sqrt{\frac{m}{k}} \int \frac{dx}{\sqrt{x_m^2 - x^2}} = \frac{1}{\omega_0}\left(\arcsin\frac{x}{x_m} + \varphi_0\right), \tag{1.6.6}$$

where $\omega_0 \equiv \sqrt{k/m}$ is the oscillator frequency and φ_0 is an integration constant. Resolving this formula for x, one finally obtains

$$x(t) = x_m \sin(\omega_0 t - \varphi_0), \tag{1.6.7}$$

which is a form of Eq. (1.1.32).

Square potential energy near the minimum is a borderline. If $U(x)$ is subsquare (grows slower than x^2), the frequency of oscillations decreases with the amplitude. For a supersquare $U(x)$, oscillation frequency increases with the amplitude. For instance, for the potential energy $U(x) = A|x|^n$, turning points are $x_{1,2} = \pm x_m$, where $x_m = (E/A)^{1/n}$. The period given by Eq. (1.6.4) becomes

$$T(E) = 2\sqrt{\frac{2m}{A}} \int_0^{x_m} \frac{dx}{\sqrt{x_m^n - x^n}}. \tag{1.6.8}$$

Changing integration variable to $y \equiv x/x_m$, one obtains

$$T(E) = 2\sqrt{\frac{2m}{A}} x_m^{1-n/2} \int_0^1 \frac{dy}{\sqrt{1-y^n}} = 2\sqrt{\frac{2m}{A}} \left(\frac{E}{A}\right)^{1/n-1/2} \int_0^1 \frac{dy}{\sqrt{1-y^n}}, \tag{1.6.9}$$

where the integral is just a number. For the frequency, one has

$$\omega_0(E) \propto 1/T \propto E^{(n-2)/(2n)}, \tag{1.6.10}$$

which illustrates the above statement.

Consider a particle in a washboard potential

$$U(x) = U_0 [1 - \cos(ax)]. \tag{1.6.11}$$

Near the minima at $ax = 2\pi n$, $n = 0, \pm 1, \pm 2$, etc., one has $U(x) \cong kx^2/2$ with $k = U_0 a^2$ and behaves similar to the harmonic oscillator. The whole problem is mathematically equivalent to that of pendulum. At $ax = \pi + 2\pi n$, potential energy has maxima, $U_{\max} = 2U_0$. Let us calculate the period of oscillations, considering the region around the minimum at $x = 0$. Equation (1.6.4) becomes

$$T = 2\sqrt{\frac{2m}{U_0}} \int_0^{x_m} \frac{dx}{\sqrt{\cos(ax) - \cos(ax_m)}}, \tag{1.6.12}$$

where x_m is the right turning point satisfying $E = U_0 [1 - \cos(ax_m)]$. The integral can be transformed into the form similar to those above:

$$T = 2\sqrt{\frac{m}{U_0}} \int_0^{x_m} \frac{dx}{\sqrt{\sin^2(ax_m/2) - \sin^2(ax/2)}}. \tag{1.6.13}$$

Next, one can employ the variable change

$$\begin{aligned} \sin(ax/2) &= \sin(ax_m/2)\sin\xi, \\ (a/2)\cos(ax/2)\,dx &= \sin(ax_m/2)\cos\xi\, d\xi, \end{aligned} \tag{1.6.14}$$

which yields

$$T = \frac{4}{a}\sqrt{\frac{m}{U_0}} \int_0^{\pi/2} \frac{1}{\cos(ax/2)} \frac{\cos\xi d\xi}{\sqrt{1-\sin^2\xi}} = \frac{4}{a}\sqrt{\frac{m}{U_0}} \int_0^{\pi/2} \frac{d\xi}{\cos(ax/2)}. \quad (1.6.15)$$

Now, eliminating x in the integrand, one obtains

$$T = \frac{4}{a}\sqrt{\frac{m}{U_0}} K\left[\sin^2(ax_m/2)\right],$$

where $K(m)$ is the elliptic integral of the first kind (here m is not the mass!),

$$K(m) = \int_0^{\pi/2} \frac{d\xi}{\sqrt{1 - m\sin^2\xi}}. \quad (1.6.16)$$

Let us consider limiting cases of this formula. In the case of small oscillations near the minimum, $ax_m \ll 1$, and one can use

$$K(m) \cong \int_0^{\pi/2} d\xi \left(1 + \frac{1}{2} m \sin^2\xi\right) = \frac{\pi}{2}\left(1 + \frac{1}{4}m\right). \quad (1.6.17)$$

Then the period becomes

$$T = T_0\left[1 + \frac{1}{16}(ax_m)^2 + \cdots\right], \quad T_0 = \frac{2\pi}{a}\sqrt{\frac{m}{U_0}}. \quad (1.6.18)$$

If turning points approach the maxima of $U(x)$, then $m = \sin^2(ax_m/2) \to 1$. At this point, the elliptic integral logarithmically diverges and so does the period. This is physically understandable because the particle is spending a lot of time near turning points when the latter approach the maxima.

If the energy of the particle exceeds the maxima of the potential energy, its motion becomes unbounded and unidirectional, so that there is no period any longer. Different regimes can be conveniently represented on the two-dimensional phase space of the particle (x, v). Small oscillations near minima make elliptic trajectories in the phase space. Unbounded motion makes infinite curved trajectories. In the limit of a very high energy, trajectories become straight lines $v = \sqrt{2E/m} = \text{const}$. Special trajectories are those corresponding to $E = U_{\max}$ and separate bound and unbound trajectories. They are called separatrices.

1.7 Systems with constraints and special coordinates

Some mechanical systems include *constraints* that are restricting motion of its parts. Simplest examples are mass on the incline, two masses connected by a light rod and a mathematical pendulum (mass on light rod, the other point of the rod fixed at the pivot point or fulcrum). Constraints that can be eliminated leading to a decrease in the number of dynamical variables of the system (its degrees of freedom) are called *holonomic*. More rare *non-holonomic* constraints cannot be eliminated and they have to be added to the equations of motion via reaction forces. A cylinder rolling on a plane without slipping is a system with a holonomic constraint, as the rotation angle of the cylinder can be expressed via displacement of its CM or vice versa. However, a disk or a sphere rolling on a plane without slipping can make a more complicated motion than back and forth, so that the constraint cannot be eliminated and is non-holonomic. Removing friction makes the sphere on a plane a holonomic system, while the disc remains non-holonomic. Finally, any rigid body can be considered as a collection of point masses with lots of constraints, although this point of view is not practically significant. In some important cases, constraints can be resolved by choosing a special coordinate system.

1.7.1 *Polar coordinate system: Pendulum*

Consider, as an example, a mathematical pendulum (simply pendulum), a mass m on a light rod of length l, moving in the xy-plane. The constraint has the simple form $\sqrt{x^2+y^2} \equiv r = l$. Employing the constraint, it is convenient to change to polar coordinates with the center at the fulcrum

$$x = r\cos\varphi, \quad y = r\sin\varphi. \tag{1.7.1}$$

For the position vector, one obtains the expression

$$\mathbf{r} = l\,(\mathbf{e}_x\cos\varphi + \mathbf{e}_y\sin\varphi). \tag{1.7.2}$$

One can see that, under the constraint, the two-component vector $\mathbf{r} = (x, y)$ has been reduced to a single variable φ.

The velocity of the pendulum has the form

$$\mathbf{v} = \dot{\mathbf{r}} = l\,(-\mathbf{e}_x\sin\varphi + \mathbf{e}_y\cos\varphi)\,\dot{\varphi}. \tag{1.7.3}$$

One can check $\mathbf{v}\cdot\mathbf{r} = 0$, i.e., both vectors are perpendicular. This means that velocity is tangential with respect to the circular trajectory the pendulum is making. The kinetic energy reads

$$E_k = \frac{m\mathbf{v}^2}{2} = \frac{ml^2\omega^2}{2}, \tag{1.7.4}$$

where

$$\omega \equiv \dot{\varphi} \tag{1.7.5}$$

is the angular velocity.

The acceleration is given by

$$\dot{\mathbf{v}} = l\,(-\mathbf{e}_x \sin\varphi + \mathbf{e}_y \cos\varphi)\,\ddot{\varphi} - l\,(\mathbf{e}_x \cos\varphi + \mathbf{e}_y \sin\varphi)\,\dot{\varphi}^2. \tag{1.7.6}$$

One can see that the first term in $\dot{\mathbf{v}}$ is tangential and collinear with velocity, while the second term is collinear with $\mathbf{r}$ and directed toward the center (fulcrum). This is the centripetal acceleration. Introducing the angular velocity, one can write the centripetal acceleration in the form

$$\mathbf{a}_c = -\omega^2\mathbf{r}. \tag{1.7.7}$$

It is convenient to project everything on the (local) direction of the circular trajectory and the perpendicular (radial) direction. For this, one can introduce orthogonal unit vectors

$$\begin{aligned} \mathbf{e}_r &= \mathbf{r}/r = \mathbf{e}_x \cos\varphi + \mathbf{e}_y \sin\varphi, \\ \mathbf{e}_\varphi &= -\mathbf{e}_x \sin\varphi + \mathbf{e}_y \cos\varphi, \end{aligned} \tag{1.7.8}$$

which satisfy

$$\frac{\partial \mathbf{e}_r}{\partial \varphi} = \mathbf{e}_\varphi, \quad \frac{\partial \mathbf{e}_\varphi}{\partial \varphi} = -\mathbf{e}_r. \tag{1.7.9}$$

Note also the expression for the gradient in the polar coordinate system

$$\frac{\partial}{\partial \mathbf{r}} = \mathbf{e}_r \frac{\partial}{\partial r} + \mathbf{e}_\varphi \frac{1}{r}\frac{\partial}{\partial \varphi}. \tag{1.7.10}$$

Equations (1.7.8) can be used to rewrite the above formulas as

$$\begin{aligned} \mathbf{r} &= l\mathbf{e}_r, \\ \mathbf{v} &= l\omega\mathbf{e}_\varphi, \\ \dot{\mathbf{v}} &= l\dot{\omega}\mathbf{e}_\varphi - l\omega^2\mathbf{e}_r. \end{aligned} \tag{1.7.11}$$

These results can also be obtained as

$$\mathbf{v} = \dot{\mathbf{r}} = l\dot{\mathbf{e}}_r = l\frac{\partial \mathbf{e}_r}{\partial \varphi}\dot{\varphi} = l\omega\mathbf{e}_\varphi, \tag{1.7.12}$$

etc. In components, one has

$$\begin{aligned} v_r &= 0, \quad v_\varphi = l\omega, \\ \dot{v}_r &= -l\omega^2, \quad \dot{v}_\varphi = l\dot{\omega}. \end{aligned} \tag{1.7.13}$$

Note that $\dot{v}_r \neq \partial_t v_r = 0$.

Constraint causes an additional normal reaction force

$$\mathbf{N} = f_N \mathbf{e}_r, \tag{1.7.14}$$

which helps keep the body on the trajectory. Let us project the equation of motion

$$m\dot{\mathbf{v}} = \mathbf{F} + \mathbf{N}, \tag{1.7.15}$$

where $\mathbf{F}$ is the external force, onto $\mathbf{e}_r$ and $\mathbf{e}_\varphi$. One obtains

$$\begin{aligned} m\dot{\mathbf{v}} \cdot \mathbf{e}_r &= -ml\omega^2 = (\mathbf{F} + \mathbf{N}) \cdot \mathbf{e}_r = F_r + f_N, \\ m\dot{\mathbf{v}} \cdot \mathbf{e}_\varphi &= ml\dot{\omega} = (\mathbf{F} + \mathbf{N}) \cdot \mathbf{e}_\varphi = F_\varphi. \end{aligned} \tag{1.7.16}$$

Here, the second equation yields the equation of motion for the pendulum,

$$\dot{\omega} = F_\varphi / (ml). \tag{1.7.17}$$

The first equation yields the reaction force,

$$f_N = -ml\omega^2 - F_r. \tag{1.7.18}$$

The latter may be important in the problem of mechanical stability of a real-life rod but is irrelevant in the idealized consideration of constraints. Thus, one can just project Newton's second law onto the direction of the trajectory enforced by the constraint, which leads to Eq. (1.7.17).

If $\mathbf{F}$ is a potential force,

$$F_\varphi = \mathbf{F} \cdot \mathbf{e}_\varphi = -\frac{\partial U}{\partial \mathbf{r}} \cdot \mathbf{e}_\varphi. \tag{1.7.19}$$

Using Eq. (1.7.10), one obtains

$$F_\varphi = -\frac{1}{r}\frac{\partial U}{\partial \varphi} = -\frac{1}{l}\frac{\partial U}{\partial \varphi}. \tag{1.7.20}$$

For the gravity force acting on the pendulum,

$$\mathbf{F} = mg\mathbf{e}_x, \quad U(\mathbf{r}) = -mg\mathbf{r} \cdot \mathbf{e}_x, \tag{1.7.21}$$

one obtains

$$F_\varphi = \mathbf{F} \cdot \mathbf{e}_\varphi = mg\mathbf{e}_x \cdot \mathbf{e}_\varphi = -mg\sin\varphi. \tag{1.7.22}$$

Potential energy expressed via φ has the form

$$U(\varphi) = -mg\mathbf{r}\cdot\mathbf{e}_x = -mgl\,(\mathbf{e}_x\cos\varphi + \mathbf{e}_y\sin\varphi)\cdot\mathbf{e}_x = -mgl\cos\varphi, \tag{1.7.23}$$

so that the total kinetic energy becomes

$$E = \frac{ml^2\omega^2}{2} - mgl\cos\varphi. \tag{1.7.24}$$

From Eq. (1.7.20), one obtains

$$F_\varphi = -\frac{1}{l}\frac{\partial U}{\partial\varphi} = -mg\sin\varphi, \tag{1.7.25}$$

same as above. Thus, the equation of motion for the pendulum, Eq. (1.7.17), takes the final form

$$\ddot{\varphi} + \omega_0^2\sin\varphi = 0, \quad \omega_0 = \sqrt{\frac{g}{l}}. \tag{1.7.26}$$

Here ω_0 is the frequency of pendulum's oscillations in the limit of a small amplitude, where $\sin\varphi \cong \varphi$. Note that because of the constraint the problem has become non-linear.

Polar coordinate system can also be used without the constraint $r = l$. All formulas can be easily generalized for this case. One important application of this formalism is planetary motion.

1.7.2 *Spherical coordinate system: Spherical pendulum*

Consider a mass m on a light rod of length a, fixed at a fulcrum at the other side, the system now being able to move on a sphere of radius a. To resolve the constraint, one can choose the spherical coordinate system

$$x = r\sin\theta\cos\varphi, \quad y = r\sin\theta\sin\varphi, \quad z = r\cos\theta, \tag{1.7.27}$$

where θ is azimuthal angle and φ is polar angle. In this system, the constraint has the simple form $r = a$. One can introduce the local orthogonal frame defined by

$$\begin{aligned}
\mathbf{e}_r &= \mathbf{r}/r = \mathbf{e}_x\sin\theta\cos\varphi + \mathbf{e}_y\sin\theta\sin\varphi + \mathbf{e}_z\cos\theta,\\
\mathbf{e}_\theta &= \mathbf{e}_x\cos\theta\cos\varphi + \mathbf{e}_y\cos\theta\sin\varphi - \mathbf{e}_z\sin\theta,\\
\mathbf{e}_\varphi &= -\mathbf{e}_x\sin\varphi + \mathbf{e}_y\cos\varphi.
\end{aligned} \tag{1.7.28}$$

These vectors satisfy

$$\frac{\partial \mathbf{e}_r}{\partial \theta} = \mathbf{e}_\theta, \quad \frac{\partial \mathbf{e}_r}{\partial \varphi} = \mathbf{e}_\varphi \sin\theta, \tag{1.7.29}$$

$$\frac{\partial \mathbf{e}_\theta}{\partial \theta} = -\mathbf{e}_r, \quad \frac{\partial \mathbf{e}_\theta}{\partial \varphi} = \mathbf{e}_\varphi \cos\theta, \tag{1.7.30}$$

$$\frac{\partial \mathbf{e}_\varphi}{\partial \theta} = 0, \quad \frac{\partial \mathbf{e}_\varphi}{\partial \varphi} = -\mathbf{e}_x \cos\varphi - \mathbf{e}_y \sin\varphi = -\mathbf{e}_r \sin\theta - \mathbf{e}_\theta \cos\theta, \tag{1.7.31}$$

as well as

$$\mathbf{e}_r \times \mathbf{e}_\theta = \mathbf{e}_\varphi \tag{1.7.32}$$

plus cyclic permutations. The gradient in the spherical system has the form

$$\frac{\partial}{\partial \mathbf{r}} = \mathbf{e}_r \frac{\partial}{\partial r} + \mathbf{e}_\theta \frac{1}{r}\frac{\partial}{\partial \theta} + \mathbf{e}_\varphi \frac{1}{r\sin\theta}\frac{\partial}{\partial \varphi}. \tag{1.7.33}$$

In the following, all vectors will be projected onto the local frame. The velocity of the spherical pendulum can be obtained by differentiating the first line of Eq. (1.7.28) using Eq. (1.7.29):

$$\frac{\mathbf{v}}{a} = \frac{\dot{\mathbf{r}}}{a} = \frac{\partial \mathbf{e}_r}{\partial \theta}\dot{\theta} + \frac{\partial \mathbf{e}_r}{\partial \varphi}\dot{\varphi} = \mathbf{e}_\theta \dot{\theta} + \mathbf{e}_\varphi \sin\theta \dot{\varphi}. \tag{1.7.34}$$

Now, total energy of the particle on a sphere takes the form

$$E = \frac{ma^2}{2}\left(\dot{\theta}^2 + \sin^2\theta \dot{\varphi}^2\right) + U(\theta, \varphi). \tag{1.7.35}$$

In Lagrangian mechanics considered in the following chapter, the Lagrange function $\mathcal{L} = E_k - U$ (similar to the total energy above) is used to obtain equations of motion by differentiation, a straightforward procedure. In Newtonian mechanics, one has to set up the vector equation of motion, Newton's second law, and project it onto the axes. This leads to the same equations as Lagrange equations. However, it is more difficult to obtain them since one has to work out the acceleration.

As was done for the polar coordinate system above, to obtain the equation of motion in the spherical system, it is sufficient to project components of Newton's second law onto the local frame discarding radial terms. One obtains two equations:

$$(m\dot{\mathbf{v}} - \mathbf{F}) \cdot \mathbf{e}_\theta = 0, \quad (m\dot{\mathbf{v}} - \mathbf{F}) \cdot \mathbf{e}_\varphi = 0. \tag{1.7.36}$$

For the acceleration, discarding terms containing $\mathbf{e}_r$, one obtains

$$\begin{aligned}
\frac{\dot{\mathbf{v}}}{a} &= \dot{\mathbf{e}}_\theta\dot{\theta} + \mathbf{e}_\theta\ddot{\theta} + \dot{\mathbf{e}}_\varphi \sin\theta\dot{\varphi} + \mathbf{e}_\varphi \sin\theta\ddot{\varphi} + \mathbf{e}_\varphi \cos\theta\dot{\theta}\dot{\varphi} \\
&= \left(\frac{\partial \mathbf{e}_\theta}{\partial\theta}\dot{\theta} + \frac{\partial \mathbf{e}_\theta}{\partial\varphi}\dot{\varphi}\right)\dot{\theta} + \mathbf{e}_\theta\ddot{\theta} + \left(\frac{\partial \mathbf{e}_\varphi}{\partial\theta}\dot{\theta} + \frac{\partial \mathbf{e}_\varphi}{\partial\varphi}\dot{\varphi}\right)\sin\theta\dot{\varphi} \\
&\quad + \mathbf{e}_\varphi \sin\theta\ddot{\varphi} + \mathbf{e}_\varphi \cos\theta\dot{\theta}\dot{\varphi} \\
&= \mathbf{e}_\varphi \cos\theta\dot{\varphi}\dot{\theta} + \mathbf{e}_\theta\ddot{\theta} - \mathbf{e}_\theta \cos\theta\dot{\varphi}\sin\theta\dot{\varphi} + \mathbf{e}_\varphi \sin\theta\ddot{\varphi} + \mathbf{e}_\varphi \cos\theta\dot{\theta}\dot{\varphi}
\end{aligned} \tag{1.7.37}$$

or, finally,

$$\frac{\dot{\mathbf{v}}}{a} = \mathbf{e}_\theta\left(\ddot{\theta} - \sin\theta\cos\theta\dot{\varphi}^2\right) + \mathbf{e}_\varphi\left(\sin\theta\ddot{\varphi} + 2\cos\theta\dot{\theta}\dot{\varphi}\right). \tag{1.7.38}$$

Thus, the equations of motion of a body confined to the sphere, Eq. (1.7.36), become

$$\ddot{\theta} - \sin\theta\cos\theta\dot{\varphi}^2 = \frac{1}{ma}F_\theta, \tag{1.7.39}$$

$$\sin\theta\ddot{\varphi} + 2\cos\theta\dot{\theta}\dot{\varphi} = \frac{1}{ma}F_\varphi, \tag{1.7.40}$$

where $F_\theta \equiv \mathbf{F}\cdot\mathbf{e}_\theta$, etc. The kinetic part of these equations is strongly nonlinear which is the price for the reduction of the number of dynamic variables by the constraint.

To obtain the components of the force, one has to express the potential energy via θ and φ and use Eq. (1.7.33). This yields

$$F_\theta = -\frac{1}{r}\frac{\partial U}{\partial\theta}, \quad F_\varphi = -\frac{1}{r\sin\theta}\frac{\partial U}{\partial\varphi}, \tag{1.7.41}$$

with $r \Rightarrow l$. If gravity force is applied,

$$U(\mathbf{r}) = mg\mathbf{r}\cdot\mathbf{e}_z = mga\cos\theta, \tag{1.7.42}$$

then

$$F_\theta = mg\sin\theta, \quad F_\varphi = 0. \tag{1.7.43}$$

Consider angular momentum $\mathbf{l} = \mathbf{r}\times m\mathbf{v}$ of a body confined to a sphere. Expressing it in the spherical coordinate system and using Eq. (1.7.32), one obtains

$$\mathbf{l} = a\mathbf{e}_r \times ma\left(\mathbf{e}_\theta\dot{\theta} + \mathbf{e}_\varphi \sin\theta\dot{\varphi}\right) = ma^2\left(\mathbf{e}_\varphi\dot{\theta} - \mathbf{e}_\theta \sin\theta\dot{\varphi}\right). \tag{1.7.44}$$

The torque is defined by

$$\boldsymbol{\tau} = \mathbf{r} \times \mathbf{F} = a\mathbf{e}_r \times (\mathbf{e}_\theta F_\theta + \mathbf{e}_\varphi F_\varphi) = a\mathbf{e}_\varphi F_\theta - a\mathbf{e}_\theta F_\varphi. \tag{1.7.45}$$

In the case of gravity force directed along z-axis, as defined above, the component of the torque $\boldsymbol{\tau} = \mathbf{r} \times \mathbf{F}$ along z-axis is zero: $\tau_z = \boldsymbol{\tau} \cdot \mathbf{e}_z = a\mathbf{e}_\varphi F_\theta \cdot \mathbf{e}_z = 0$. This means that z-component of the angular momentum is conserved. Using Eq. (1.7.28), one obtains

$$l_z = \mathbf{l} \cdot \mathbf{e}_z = -ma^2 \sin\theta\dot{\varphi}\mathbf{e}_\theta \cdot \mathbf{e}_z = ma^2 \sin^2\theta\dot{\varphi} = \text{const.} \tag{1.7.46}$$

A similar result can be obtained from Eq. (1.7.40) in the case $F_\varphi = 0$ by multiplying by the integrating factor $\sin\theta$ as follows:

$$\left(\sin\theta\ddot{\varphi} + 2\cos\theta\dot{\theta}\dot{\varphi}\right)\sin\theta = \partial_t \sin^2\theta\dot{\varphi} = 0. \tag{1.7.47}$$

Integrating this, one obtains Eq. (1.7.46). Eliminating $\dot{\varphi}$ in Eq. (1.7.39), one obtains an autonomous equation of motion for θ:

$$\ddot{\theta} = \frac{l_z^2}{(ma^2)^2}\frac{\cos\theta}{\sin^3\theta} + \frac{F_\theta}{ma}. \tag{1.7.48}$$

In the axially-symmetric case, $U = U(\theta)$, one has $l_z = \text{const.}$, and the energy, Eq. (1.7.35), can be expressed as

$$E = \frac{ma^2}{2}\left(\dot{\theta}^2 + \frac{l_z^2}{(ma^2)^2}\frac{1}{\sin^2\theta}\right) + U(\theta). \tag{1.7.49}$$

This is an effectively one-dimensional motion that can be integrated after resolving for $\dot{\theta}$. Equation (1.7.49) itself can be obtained as an integral of motion of Eq. (1.7.48) multiplying by the integrating factor $\dot{\theta}$, similar to what was done in the main text. Moreover, one can obtain Eq. (1.7.35) by integrating Eqs. (1.7.39) and (1.7.40).

1.8 Motion in a central field

Consider motion of a body in a central field, $U = U(r)$, so that $\mathbf{F} = -\nabla U = -(dU/dr)\,\mathbf{e}_r$ is directed radially. In this case, the angular momentum is conserved, $\mathbf{l} = \mathbf{r} \times m\mathbf{v} = \text{const.}$, see discussion after Eq. (1.3.4). Since both $\mathbf{r}$ and $\mathbf{v}$ are perpendicular to $\mathbf{l}$, the body is moving in the plane perpendicular to $\mathbf{l}$. It is convenient to use polar coordinate system to describe this motion. Using

$$\mathbf{r} = \mathbf{e}_r r, \quad \mathbf{v} = \mathbf{e}_\varphi \dot{\varphi} r + \mathbf{e}_r \dot{r} \tag{1.8.1}$$

(see Sec. 1.7.1), one obtains

$$l = mr^2\dot{\varphi} = \text{const.} \tag{1.8.2}$$

As $r > 0$, one can see that φ is changing monotonically and $\dot{\varphi}$ is not changing the sign. This equation can be rewritten in terms of the sectorial velocity $\dot{S}$ as

$$\dot{S} \equiv \frac{1}{2} r^2 \dot{\varphi} = \frac{l}{2m} = \text{const.} \tag{1.8.3}$$

This is Kepler's second law.

Substituting Eq. (1.8.2) into the energy, one obtains

$$E = \frac{m\mathbf{v}^2}{2} + U(r) = \frac{m}{2}\left(\dot{r}^2 + r^2\dot{\varphi}^2\right) + U(r) = \frac{m}{2}\left(\dot{r}^2 + \frac{l^2}{m^2 r^2}\right) + U(r) \tag{1.8.4}$$

or

$$E = \frac{m\dot{r}^2}{2} + U_{\text{eff}}(r), \tag{1.8.5}$$

where

$$U_{\text{eff}}(r) = U(r) + \frac{l^2}{2mr^2}. \tag{1.8.6}$$

The last term here is the so-called centrifugal energy. Then one can proceed as in the case of one-dimensional motion. First, one resolves the formula above for $\dot{r}$:

$$\dot{r} = \sqrt{\frac{2}{m}\left[E - U_{\text{eff}}(r)\right]}. \tag{1.8.7}$$

Integrating this, one obtains the dependence $r(t)$ implicitly:

$$t = \sqrt{\frac{m}{2}} \int \frac{dr}{\sqrt{E - U_{\text{eff}}(r)}}. \tag{1.8.8}$$

The trajectory of the body is defined in the polar system by the dependence $r(\varphi)$. For the inverse function $\varphi(r)$ using Eqs. (1.8.2) and (1.8.7), one obtains

$$\frac{d\varphi}{dr} = \frac{\dot{\varphi}}{\dot{r}} = \sqrt{\frac{m}{2}} \frac{l/(mr^2)}{\sqrt{E - U_{\text{eff}}(r)}}, \tag{1.8.9}$$

and thus,

$$\varphi = \int \frac{dr}{r^2} \frac{l}{\sqrt{2m\left[E - U_{\text{eff}}(r)\right]}}. \tag{1.8.10}$$

Due to the centrifugal energy, the body cannot fall into the center for typical attractive forces ($U \propto 1/r$) because the positive centrifugal energy is growing faster. As in the case of one-dimensional motion, the body is moving within the classically accessible region $U_{\text{eff}}(r) < E$. If the radial motion is between the two turning points, $r_1 < r < r_2$, this is a bound state and the body is orbiting around the center. The state with only one turning point is a *scattering* state. The body is coming from infinity and then goes away after scattering on attracting or repelling center. If one chooses a turning point as the origin for φ, the traejectory $r(\varphi)$ will be symmetric with respect to this turning point, $r(\varphi) = r(-\varphi)$.

The period of motion in bound states is defined as time needed for the body to go from one turning point to the other and back:

$$T = \sqrt{2m} \int_{r_1}^{r_2} \frac{dr}{\sqrt{E - U_{\text{eff}}(r)}}. \tag{1.8.11}$$

The rotation angle corresponding to the period of motion is given by

$$\Delta\varphi = l\sqrt{\frac{2}{m}} \int_{r_1}^{r_2} \frac{dr}{r^2} \frac{1}{\sqrt{E - U_{\text{eff}}(r)}}. \tag{1.8.12}$$

Closed orbits correspond to $\Delta\varphi = 2\pi m/n$, where m and n are natural numbers. Closed orbit is a special case, since for an arbitrary $U(r)$, the rotating angle $\Delta\varphi$ is arbitrary. However, there are two important particular cases, in which orbits are closed. First, this is gravitation, $U \propto 1/r$. Second, this is harmonic oscillator, $U \propto r^2$.

If two turning points are close to each other, $r_1 \cong r_2 \cong R$, then the orbit is nearly circular. In this case, in the integral for $\Delta\varphi$, one can approximate

$1/r^2 \Rightarrow 1/R^2$. After that, the integral becomes the same as for the period and one obtains the relation $T = mR^2\Delta\varphi/l$. With $\Delta\varphi = 2\pi$ and $l = mrv$, one obtains $T = 2\pi R/v$, as it should be. This is the case of the rotation of the Earth around the Sun.

1.8.1 *Kepler's problem*

Consider the important case $U \propto 1/r$ that corresponds to gravitational interaction between stars and planets, as well as to the Coulomb interaction of charged particles. The effective energy of Eq. (1.8.6) has the form

$$U_{\text{eff}}(r) = -\frac{\alpha}{r} + \frac{l^2}{2mr^2}, \quad \alpha \equiv GMm, \tag{1.8.13}$$

where G is gravitational constant, M and m are masses of the two gravitating bodies. If $M \gg m$ (the Sun and the Earth), one can consider the light body m rotating around the heavy body M put in the center. This will be our main case. If the masses are comparable, they will be both moving, but the problem can be reduced to a one-body problem using the reduced mass (see Sec. 1.5). This effective energy has the minimum $U_{\text{eff}}(r_0) = E_{\min}$ with

$$r_0 = \frac{l^2}{\alpha m}, \quad E_{\min} = -\frac{\alpha^2 m}{2l^2}. \tag{1.8.14}$$

For $E_{\min} < E < 0$, there are two turning points and the motion is bounded (orbiting). For $E > 0$, there is only one turning point and the motion is unbounded (scattering). The trajectory can be found from Eq. (1.8.10). With the new variable $u = 1/r$, $du = -dr/r^2$, one obtains

$$\begin{aligned}\varphi &= -l\int\frac{du}{\sqrt{2m\left(E+\alpha u-\frac{l^2}{2m}u^2\right)}} = -\int\frac{du}{\sqrt{\frac{2mE}{l^2}+\frac{2m\alpha}{l^2}u-u^2}}\\ &= -\int\frac{du}{\sqrt{\frac{2mE}{l^2}+\frac{m^2\alpha^2}{l^4}-\left(u-\frac{m\alpha}{l^2}\right)^2}} = -\arccos\frac{u-\frac{m\alpha}{l^2}}{\sqrt{\frac{2mE}{l^2}+\frac{m^2\alpha^2}{l^4}}}.\end{aligned} \tag{1.8.15}$$

This can be rewritten as

$$\frac{r_0}{r} = 1 + \epsilon\cos\varphi, \tag{1.8.16}$$

where

$$\epsilon \equiv \sqrt{1 + \frac{E}{|E_{\min}|}}. \tag{1.8.17}$$

For $E < 0$, this trajectory is an ellipse (Kepler's first law) with the *eccentricity* ϵ and axes

$$a = \frac{r_0}{1 - \epsilon^2} = \frac{\alpha}{2|E|}, \quad b = \frac{r_0}{\sqrt{1 - \epsilon^2}} = \frac{l}{\sqrt{2m|E|}}. \tag{1.8.18}$$

Note that eccentricity is defined as

$$\epsilon \equiv \sqrt{1 - b^2/a^2}. \tag{1.8.19}$$

In the case $E = E_{\min}$, one has $\epsilon = 0$, and the ellipse degenerates into a circle $r = r_0 = \frac{l^2}{m\alpha} = \frac{m^2 r^2 v^2}{m\alpha}$. From this, one obtains $r = \alpha/(mv^2)$.

The area of the ellipse is

$$S = \pi ab = \frac{\pi \alpha l}{\sqrt{(2|E|)^3 m}}. \tag{1.8.20}$$

This area is being covered during the period of motion. With the help of Eq. (1.8.3), one obtains $S = \dot{S}T$ and

$$T = \frac{S}{\dot{S}} = \frac{\pi \alpha l}{\sqrt{(2|E|)^3 m}} \frac{2m}{l} = \pi \alpha \sqrt{\frac{m}{2|E|^3}}, \tag{1.8.21}$$

depending only on the energy and diverging for $E \to 0$. One can express the period via the linear size of the orbit a using Eq. (1.8.18). The result is

$$T = 2\pi \sqrt{\frac{m}{\alpha}} a^{3/2} = \frac{2\pi}{\sqrt{GM}} a^{3/2}. \tag{1.8.22}$$

The obtained relation $T^2/a^3 = \text{const.}$ is Kepler's third law. Note that the constant does not depend on m, i.e., it is the same for all planets of the same star.

Time dependence of the motion can be conveniently represented in parametric form. For this, one can rewrite Eq. (1.8.8) as

$$t=\sqrt{\frac{m}{2|E|}}\int\frac{rdr}{\sqrt{-r^2+\frac{\alpha}{|E|}r-\frac{l^2}{2m|E|}}}=\sqrt{\frac{ma}{\alpha}}\int\frac{rdr}{\sqrt{-\left(r-a\right)^2+a^2\epsilon^2}}. \tag{1.8.23}$$

With the substitution

$$r=a\left(1-\epsilon\cos\xi\right), \tag{1.8.24}$$

the integral can be calculated as

$$\begin{aligned} t&=\sqrt{\frac{ma}{\alpha}}\int\frac{a\left(1-\epsilon\cos\xi\right)a\epsilon\sin\xi d\xi}{\sqrt{a^2\epsilon^2\left(1-\cos^2\xi\right)}}=\sqrt{\frac{ma^3}{\alpha}}\int\left(1-\epsilon\cos\xi\right)d\xi\\ &=\sqrt{\frac{m}{\alpha}}a^{3/2}\left(\xi-\epsilon\sin\xi\right). \end{aligned} \tag{1.8.25}$$

These two formulas provide the dependence $r(t)$ parametrically. The period of motion corresponds to changing of ξ by 2π: r returns to the same value, while t increases by T given by Eq. (1.8.22).

The time dependence of the angle φ is now defined by Eq. (1.8.16). Moreover, one can express Descartes coordinates (x,y) parametrically too. First, one finds

$$x=r\cos\varphi=\frac{r_0-r}{\epsilon}=\frac{a\left(1-\epsilon^2\right)-a\left(1-\epsilon\cos\xi\right)}{\epsilon}=a\left(\cos\xi-\epsilon\right). \tag{1.8.26}$$

Then one finds

$$y=\sqrt{r^2-x^2}=a\sqrt{1-\epsilon^2}\sin\xi. \tag{1.8.27}$$

Cases of unbound motion, $E>0$, and repelling potential $U(r)=\alpha/r$ (unbound motion for any energy) can be considered in a similar way.

1.8.2 *Scattering by a central field*

Consider an unbound motion of a body or a particle in a central field. The particle is coming from infinity, experiences the action of the central force, and then goes to infinity again, but, having changed its direction of motion. Changing of the direction of motion of a body as the result of its interaction with other bodies is called *scattering*. As was mentioned above, the trajectory is symmetric, $r(\varphi)=r(-\varphi)$, if one chooses the turning point

r_1 as the origin of φ. Thus, the change of direction (deflection angle) χ as the result of scattering can be obtained from Eq. (1.8.10) as

$$\chi = \pi - 2\varphi_0, \tag{1.8.28}$$

where

$$\varphi_0 = \int_{r_1}^{\infty} \frac{dr}{r^2} \frac{l}{\sqrt{2m\left[E - U_{\text{eff}}(r)\right]}} \tag{1.8.29}$$

is a half of the angle between the initial and final parts of the trajectory. In the following, it is convenient, instead of integrals of motion E and l, to use the speed at infinity v_∞ and the target distance ρ. The latter is the minimal distance from the center corresponding to the straight trajectory in the absence of the central force. Using

$$E = \frac{mv_\infty^2}{2}, \quad l = \rho m v_\infty, \tag{1.8.30}$$

one can rewrite Eq. (1.8.29) in the form

$$\chi = \pi - 2\int_{r_1}^{\infty} \frac{dr}{r^2} \frac{\rho}{\sqrt{1 - \frac{2U_{\text{eff}}(r)}{mv_\infty^2}}}, \quad \frac{2U_{\text{eff}}(r)}{mv_\infty^2} = \frac{\rho^2}{r^2} + \frac{2U(r)}{mv_\infty^2}. \tag{1.8.31}$$

In experiments, there is usually a beam of particles with different ρ that are being scattered. It is convenient to introduce the differential scattering cross-section $d\sigma$ defined by

$$d\sigma(\chi) = dN/n, \tag{1.8.32}$$

where dN is the number of particles scattered during a unit of time within the angular interval $d\chi$ around χ and n is the number of particles crossing a unit of area of the beam during a unit of time. It is assumed that n is uniform. One can see that $d\sigma$ has the unit of area. Since $\chi = \chi(\rho)$, particles scattered within $d\chi$ are those within the target distance interval $d\rho = |d\rho/d\chi|\, d\chi$ that corresponds to $d\chi$. The number of such particles scattered during a unit of time is $dN = 2\pi\rho d\rho n$. Thus, finally, one obtains

$$d\sigma(\chi) = 2\pi\rho(\chi)\left|\frac{d\rho}{d\chi}\right| d\chi, \tag{1.8.33}$$

where $\rho(\chi)$ and $d\rho/d\chi$ can be found from Eq. (1.8.31).

If the particle is being scattered on another particle that is at rest before the collision, one can change into the center-of-mass frame and consider scattering of a particle with the reduced mass on the force center located at the CM, using the formulas above. After that, one has to change back into the laboratory frame that results in the redefinition of the angles and transformation of the results that requires some algebra (see *Mechanics* by Landau and Lifshitz).

Consider scattering of a charged particle by the Coulomb field, $U(r) = \alpha/r$, as in celestial mechanics. Calculating the integral in Eq. (1.8.31) as was done in Eq. (1.8.15), one arrives at

$$\chi = 2\arcsin\frac{1}{\sqrt{1+\rho^2/\rho_0^2}}, \tag{1.8.34}$$

where

$$\rho_0 \equiv \frac{\alpha}{mv_\infty^2} \tag{1.8.35}$$

is the characteristic distance. Resolving for ρ, one obtains

$$\rho^2 = \rho_0^2 \cot^2\frac{\chi}{2}. \tag{1.8.36}$$

Now, Eq. (1.8.33) yields

$$d\sigma = \pi\rho_0^2 \frac{\cos(\chi/2)}{\sin^3(\chi/2)} d\chi. \tag{1.8.37}$$

Instead of scattering within $d\chi$, one can consider scattering within the infinitesimal body angle

$$d\Omega = 2\pi \sin\chi d\chi. \tag{1.8.38}$$

This yields the Rutherford formula

$$d\sigma = \left(\frac{\alpha}{2mv_\infty^2}\right)^2 \frac{d\Omega}{\sin^4(\chi/2)}. \tag{1.8.39}$$

In some cases, one can define the total scattering cross-section

$$\sigma = \int d\sigma = \int d\chi \frac{d\sigma}{d\chi}. \tag{1.8.40}$$

With the help of Eq. (1.8.33), it can be written as

$$\sigma = 2\pi \int \rho d\rho. \tag{1.8.41}$$

The image behind this formula is a scattering center in the form of a circle of some finite radius a and the area $\sigma = \pi a^2$. If the target distance ρ satisfies $\rho < a$, the particle will hit the target and be scattered. Otherwise, it will not be scattered at all. For all interactions that decrease gradually at infinity, particles with *all* ρ will be scattered. Correspondingly, the integral in Eq. (1.8.40) diverges at small χ. Only if the interaction has a cut-off, the total scattering cross-section is finite. The simplest example is elastic scattering on a rigid sphere of radius a. Here, one immediately finds

$$\rho = a \sin \varphi_0 = a \sin \frac{\pi - \chi}{2} = a \cos \frac{\chi}{2}. \tag{1.8.42}$$

Substitution into Eq. (1.8.33) yields the differential cross-section

$$d\sigma = \frac{\pi a^2}{2} \sin \chi d\chi. \tag{1.8.43}$$

Integrating over χ yields

$$\sigma = \int d\sigma = \frac{\pi a^2}{2} \int_0^{\pi} \sin \chi d\chi = \pi a^2, \tag{1.8.44}$$

as it should be. However, for interactions with the cut-off, the above calculation is only a consistency test because the result for σ is obvious and can be written immediately.

One can also define total scattering cross-section for specific events, for instance, for falling onto the attracting center $U = -\alpha/r^2$. In this case, the effective potential energy of Eq. (1.8.6) reads

$$U_{\text{eff}}(r) = -\frac{\alpha}{r^2} + \frac{l^2}{2mr^2} = \left(-\alpha + \frac{l^2}{2m}\right) \frac{1}{r^2}. \tag{1.8.45}$$

Absorbed will be all particles for which the coefficient in $U_{\text{eff}}(r)$ is negative:

$$\frac{l^2}{2m} = \frac{(\rho m v_\infty)^2}{2m} < \alpha, \tag{1.8.46}$$

otherwise particles will escape to infinity. In terms of the target distance, this condition has the form

$$\rho^2 < \rho_0^2 \equiv \frac{2\alpha}{m v_\infty^2}. \tag{1.8.47}$$

Total scattering cross-section then becomes

$$\sigma = \pi\rho_0^2 = \frac{2\pi\alpha}{mv_\infty^2}. \tag{1.8.48}$$

Although in this case, one can find σ easily, finding the differential cross-section requires a more serious calculation.

1.8.3 *Small-angle scattering*

If the target distance ρ or particle's speed is large, the trajectory is nearly a straight line, say, along x-axis. Passing the force center, the particle acquires a perpendicular momentum p_y as the result. Now, the small scattering angle χ can be found as

$$\chi \cong \frac{p_y}{mv_\infty} \ll 1. \tag{1.8.49}$$

The momentum p_y can be found, using $dt = dx/v_\infty$ as

$$p_y = \int_{-\infty}^{\infty} dt F_y = -\frac{1}{v_\infty}\int_{-\infty}^{\infty} dx \left|\frac{\partial U(x,y)}{\partial y}\right|_{y=\rho}. \tag{1.8.50}$$

The final result becomes

$$\chi = -\frac{1}{mv_\infty^2}\int_{-\infty}^{\infty} dx\, \frac{\partial U(x,y)}{\partial y}\bigg|_{y=\rho}. \tag{1.8.51}$$

This formula can be generalized for any general potential energy $U(x, y, z)$.

As an example, consider scattering by the central field with $U(r) = \alpha/r$. The formula above yields

$$\begin{aligned}\chi &= -\frac{\alpha}{mv_\infty^2}\int_{-\infty}^{\infty} dx\, \frac{\partial}{\partial y}\frac{1}{\sqrt{x^2+y^2}}\bigg|_{y=\rho} = \frac{\alpha}{mv_\infty^2}\int_{-\infty}^{\infty} dx \frac{x}{(\rho^2+x^2)^{3/2}} \\ &= \frac{\alpha}{mv_\infty^2}\int_0^{\infty} \frac{du}{(\rho^2+u)^{3/2}} = -\frac{2\alpha}{mv_\infty^2}\frac{1}{\sqrt{\rho^2+u}}\bigg|_0^{\infty} = \frac{2\alpha}{mv_\infty^2\rho} = \frac{2\rho_0}{\rho},\end{aligned} \tag{1.8.52}$$

which coincides with Eq. (1.8.34) in the limit $\rho \gg \rho_0$.

1.9 Problems on Newtonian Mechanics

1.9.1 *Overdamped harmonic oscillator*

Using the general solution of the equation of motion for the free harmonic oscillator

$$\ddot{x} + 2\Gamma\dot{x} + \omega_0^2 x = 0, \tag{1.9.1}$$

obtain the explicitly real solution in the overdamped case $\Gamma < \omega_0$.

Solution: The general solution has the form

$$x(t) = C_+ e^{i\Omega_+ t} + C_- e^{i\Omega_- t}, \tag{1.9.2}$$

where

$$\Omega_\pm = i\Gamma \pm \tilde{\omega}_0, \quad \tilde{\omega}_0 \equiv \sqrt{\omega_0^2 - \Gamma^2} = i\sqrt{\Gamma^2 - \omega_0^2}. \tag{1.9.3}$$

Thus, the explicitly real solution has the form

$$x(t) = C_+ \exp\left[\left(-\Gamma - \sqrt{\Gamma^2 - \omega_0^2}\right)t\right] + C_- \exp\left[\left(-\Gamma + \sqrt{\Gamma^2 - \omega_0^2}\right)t\right]. \tag{1.9.4}$$

1.9.2 *Power absorbed by harmonic oscillator*

A harmonic force is applied to a harmonic oscillator at rest at $t = 0$:

$$\ddot{x} + 2\Gamma\dot{x} + \omega_0^2 x = f(t) = f_0 \sin(\omega t). \tag{1.9.5}$$

Calculate the average absorbed power using

$$P_{\text{abs}} = \frac{1}{t}\int_0^t dt' \dot{x}(t') f(t') \tag{1.9.6}$$

at all times that are multiples of the period $T = 2\pi/\omega$. Consider short-time and long-time limits for a weakly damped oscillator, $\Gamma \ll \omega_0$. Use the solution of the equation of motion from the book keeping only resonant terms near resonance. Analyze the solution at long times and at $\Gamma = 0$.

Solution: Since at $t = 0$, the oscillator was at rest, there is no free solution. The forced solution is given by

$$x(t) = -\frac{f_0}{4\omega_0}\frac{e^{i\omega t} - e^{(-\Gamma + i\omega_0)t}}{\Gamma - i\omega_0 + i\omega} + \text{c.c.}, \tag{1.9.7}$$

where in the case of weak damping, we set $\tilde{\omega}_0 \Rightarrow \omega_0$. It is convenient to first integrate the absorbed power by parts to simplify subsequent integration:

$$P_{\text{abs}} = \frac{1}{t}x(t)f(t) - \frac{1}{t}\int_0^t dt' x(t')\dot{f}(t'). \tag{1.9.8}$$

First,

$$\begin{aligned}\frac{1}{t}x(t)f(t) &= -\frac{f_0^2}{4\omega_0 t}\frac{e^{i\omega t} - e^{(-\Gamma+i\omega_0)t}}{\Gamma - i\omega_0 + i\omega}\frac{e^{i\omega t} - e^{-i\omega t}}{2i} + \text{c.c.}\\ &= i\frac{f_0^2}{8\omega_0 t}\frac{e^{2i\omega t} - 1 - e^{(-\Gamma+i\omega_0+i\omega)t} + e^{(-\Gamma+i\omega_0-i\omega)t}}{\Gamma - i\omega_0 + i\omega} + \text{c.c.}\end{aligned} \tag{1.9.9}$$

Near resonance, $\omega_0 + \omega \cong 2\omega$, then $e^{2i\omega t} = 1$ because $t = n2\pi/\omega$. Thus, one obtains

$$\frac{1}{t}x(t)f(t) = i\frac{f_0^2}{8\omega_0 t}\frac{e^{-\Gamma t}\left(-1 + e^{(i\omega_0 - i\omega)t}\right)}{\Gamma - i\omega_0 + i\omega} + \text{c.c.} \tag{1.9.10}$$

and further

$$\begin{aligned}\frac{1}{t}x(t)f(t) &= i\frac{f_0^2}{8\omega_0 t}\\ &\times \frac{e^{-\Gamma t}\left\{-1 + \cos\left[(\omega - \omega_0)t\right] + i\sin\left[(\omega - \omega_0)t\right]\right\}(\Gamma - i\omega_0 + i\omega)}{(\omega - \omega_0)^2 + \Gamma^2} + \text{c.c.}\end{aligned} \tag{1.9.11}$$

and

$$\frac{1}{t}x(t)f(t) = \frac{f_0^2}{4\omega_0 t}\frac{\left\{-1 + \cos\left[(\omega - \omega_0)t\right]\right\}(\omega_0 - \omega) - \Gamma\sin\left[(\omega - \omega_0)t\right]}{(\omega - \omega_0)^2 + \Gamma^2}. \tag{1.9.12}$$

This term vanishes at resonance, and thus, it can be neglected.

Further, using

$$\dot{f}(t) = \omega f_0 \cos(\omega t) = \frac{\omega}{2} f_0 \left(e^{i\omega t} + e^{-i\omega t}\right), \tag{1.9.13}$$

one calculates

$$\begin{aligned}
P_{\text{abs}} &= -\frac{1}{t}\int_0^t dt' x(t')\dot{f}(t') \\
&= \frac{f_0^2\omega}{8\omega_0}\frac{1}{t}\int_0^t dt' \left[\frac{e^{i\omega t'} - e^{(-\Gamma+i\omega_0)t'}}{\Gamma - i\omega_0 + i\omega}\left(e^{i\omega t'} + e^{-i\omega t'}\right) + \text{c.c.}\right] \\
&= \frac{f_0^2\omega}{8\omega_0}\frac{1}{t}\int_0^t dt' \left[\frac{e^{2i\omega t'} + 1 - e^{(-\Gamma+i\omega_0+i\omega)t'} - e^{(-\Gamma+i\omega_0-i\omega)t'}}{\Gamma - i\omega_0 + i\omega} + \text{c.c.}\right] \\
&= \frac{f_0^2\omega}{8\omega_0 t}\frac{1}{\Gamma - i\omega_0 + i\omega}\left[\frac{e^{2i\omega t} - 1}{2i\omega} + t + \frac{e^{(-\Gamma+i\omega_0+i\omega)t} - 1}{\Gamma - i\omega_0 - i\omega}\right. \\
&\quad \left. + \frac{e^{(-\Gamma+i\omega_0-i\omega)t} - 1}{\Gamma - i\omega_0 + i\omega}\right] + \text{c.c.}
\end{aligned} \tag{1.9.14}$$

Here, the less-singular term containing $\omega + \omega_0$ has to be dropped in the resonance approximation. The first term vanishes since t is a multiple of the period. What remains can be split into two parts:

$$P_{\text{abs}} = P_{\text{abs}}^{(1)} + P_{\text{abs}}^{(2)}, \tag{1.9.15}$$

where

$$P_{\text{abs}}^{(1)} = \frac{f_0^2}{8}\frac{1}{\Gamma - i\omega_0 + i\omega} + \text{c.c.} = \frac{f_0^2}{4}\frac{\Gamma}{(\omega - \omega_0)^2 + \Gamma^2} \tag{1.9.16}$$

and

$$P_{\text{abs}}^{(2)} = \frac{f_0^2}{8t}\frac{e^{(-\Gamma+i\omega_0-i\omega)t} - 1}{(\Gamma - i\omega_0 + i\omega)^2} + \text{c.c.}, \tag{1.9.17}$$

and we set $\omega \Rightarrow \omega_0$ in the prefactor near resonance. The first term, $P_{\text{abs}}^{(1)}$, does not depend on time and is the stationary absorption at large times, $t \gtrsim 1/\Gamma$. It peaks at resonance, $\omega = \omega_0$. The integral absorption is

$$\int_{-\infty}^{\infty} d\omega P_{\text{abs}}^{(1)}(\omega) = \frac{f_0^2}{4}\int_{-\infty}^{\infty} d\omega \frac{\Gamma}{(\omega - \omega_0)^2 + \Gamma^2} = \frac{f_0^2}{4}\pi. \tag{1.9.18}$$

In the limit $\Gamma \to 0$, Eq. (1.9.16) becomes a δ-function:

$$P_{\text{abs}}^{(1)} = \frac{f_0^2}{4} \pi \delta\left(\omega - \omega_0\right). \tag{1.9.19}$$

Since, by definition of the δ-function,

$$\int_{-\infty}^{\infty} d\omega \delta\left(\omega - \omega_0\right) = 1, \tag{1.9.20}$$

approximating the peak by δ-function preserves integral absorption.

It can be shown that $P_{\text{abs}}^{(2)}$ is important at short times $t \lesssim 1/\Gamma$. Thus, to simplify the calculation, we set $\Gamma \Rightarrow 0$ to obtain

$$P_{\text{abs}}^{(2)} = \frac{f_0^2}{4t} \frac{1 - \cos\left[\left(\omega - \omega_0\right) t\right]}{\left(\omega - \omega_0\right)^2}. \tag{1.9.21}$$

One can scale this contribution as

$$P_{\text{abs}}^{(2)} = \frac{f_0^2}{4} t \frac{1 - \cos\left[\left(\omega - \omega_0\right) t\right]}{\left[\left(\omega - \omega_0\right) t\right]^2}, \tag{1.9.22}$$

from which one can see that the peak width is $\Delta\omega \propto 1/t$ and the peak height is $\propto t$. Thus, this peak is very broad at the beginning and becomes narrower in the course of time. The integral absorption is given by

$$\int_{-\infty}^{\infty} d\omega P_{\text{abs}}^{(2)}(\omega) = \frac{f_0^2}{4} \int_{-\infty}^{\infty} dx \frac{1 - \cos x}{x^2} = \frac{f_0^2}{4} \sqrt{2\pi} \tag{1.9.23}$$

and is time-independent.

The general picture is the following: At short times $t \lesssim 1/\Gamma$, there is a peak ($P_{\text{abs}}^{(2)}$) with time-dependent broadening. At long times $t \gtrsim 1/\Gamma$, the width of the peak $P_{\text{abs}}^{(2)}$ falls below that of $P_{\text{abs}}^{(1)}$, so that the latter remains. To prove that there is only one peak at any time rather than two peaks of different widths, one has to use the complete expression for $P_{\text{abs}}^{(2)}$ that contains Γ.

1.9.3 *Energy loss by a free weakly damped harmonic oscillator*

Consider a free weakly damped harmonic oscillator, $\Gamma \ll \omega_0$. Using the general solution for the free oscillator, calculate the time dependence of its energy and the relative energy loss during one period. Calculate the energy loss per period perturbatively using the solution for a free undamped oscillator and the formula for the time derivative of the total energy due to viscous drag.

Solution: The general solution has the form

$$x(t) \cong C_1 e^{-\Gamma t} \cos \omega_0 t + C_2 e^{-\Gamma t} \sin \omega_0 t, \tag{1.9.24}$$

where we have set $\tilde{\omega}_0 \Rightarrow \omega_0$ in the weak-damping limit. The total energy of the oscillator is given by

$$E = \frac{m\dot{x}^2}{2} + \frac{kx^2}{2} = \frac{m}{2}\left(\dot{x}^2 + \omega_0^2 x^2\right). \tag{1.9.25}$$

In the weak-damping limit,

$$\dot{x}(t) \cong -C_1 e^{-\Gamma t}\omega_0 \sin \omega_0 t + C_2 e^{-\Gamma t}\omega_0 \cos \omega_0 t. \tag{1.9.26}$$

Substituting x and $\dot{x}$ into the energy, one obtains

$$E \cong \frac{m\omega_0^2}{2} e^{-2\Gamma t}\left[(C_1 \cos \omega_0 t + C_2 \sin \omega_0 t)^2 + (-C_1 \sin \omega_0 t + C_2 \cos \omega_0 t)^2\right] \tag{1.9.27}$$

and, finally,

$$E \cong \frac{m\omega_0^2}{2} e^{-2\Gamma t}\left(C_1^2 + C_2^2\right). \tag{1.9.28}$$

The relative energy loss is defined by

$$\eta \equiv \frac{\Delta E}{E} = \frac{e^{-2\Gamma t} - e^{-2\Gamma(t+T)}}{e^{-2\Gamma t}} = 1 - e^{-2\Gamma T} \cong 2\Gamma T = \frac{4\pi\Gamma}{\omega_0} \ll 1. \tag{1.9.29}$$

Let us now consider the perturbative solution of the problem. For the undamped motion, one has

$$x(t) = C_1 \cos \omega_0 t + C_2 \sin \omega_0 t \tag{1.9.30}$$

and

$$v(t) = \dot{x}(t) = -C_1\omega_0 \sin \omega_0 t + C_2\omega_0 \cos \omega_0 t. \tag{1.9.31}$$

The time derivative of the total energy is given by

$$\dot{E} = -\alpha v^2 = -2m\Gamma v^2 = 2m\Gamma v^2\omega_0^2\left(-C_1 \sin \omega_0 t + C_2 \cos \omega_0 t\right)^2. \tag{1.9.32}$$

Integrating this, one obtains

$$\Delta E = -\int_0^T \dot{E}dt = m\Gamma v^2\omega_0^2\left(C_1^2 + C_2^2\right)T. \tag{1.9.33}$$

Now,

$$\eta \equiv \frac{\Delta E}{E} = \frac{m\Gamma v^2\omega_0^2\left(C_1^2 + C_2^2\right)T}{\frac{m\omega_0^2}{2}\left(C_1^2 + C_2^2\right)} = 2\Gamma T = \frac{4\pi\Gamma}{\omega_0}, \tag{1.9.34}$$

same as above.

1.9.4 *Motion of a charged particle in mutually perpendicular electric and magnetic fields*

Integrate equation of motion of a charged particle in mutually perpendicular electric and magnetic fields:

$$m\dot{\mathbf{v}} = q\left[\mathbf{v} \times \mathbf{B}\right] + q\mathbf{E}, \tag{1.9.35}$$

$\mathbf{E} \perp \mathbf{B}$. For simplicity, consider the case of the velocity perpendicular to $\mathbf{B}$.

Solution: Choose z-axis along $\mathbf{B}$ and x-axis along $\mathbf{E}$, then the particle will be moving in xy-plane. The equation of motion in components has the form

$$\begin{aligned} \dot{v}_x &= \omega_c v_y + a, \\ \dot{v}_y &= -\omega_c v_x, \end{aligned} \tag{1.9.36}$$

where $a \equiv qE/m$. In terms of the new variable,

$$\tilde{v}_y = v_y + a/\omega_c, \tag{1.9.37}$$

the system of equations becomes uniform:

$$\begin{aligned} \dot{v}_x &= \omega_c \tilde{v}_y, \\ \dot{\tilde{v}}_y &= -\omega_c v_x. \end{aligned} \tag{1.9.38}$$

Its solution can be found in the main text:

$$\begin{aligned} v_x(t) &= v_{x0}\cos\omega_c t + \tilde{v}_{y0}\sin\omega_c t, \\ \tilde{v}_y(t) &= \tilde{v}_{y0}\cos\omega_c t - v_{x0}\sin\omega_c t. \end{aligned} \tag{1.9.39}$$

Returning to the original variables, one obtains

$$\begin{aligned} v_x(t) &= v_{x0}\cos\omega_c t + (v_{y0} + a/\omega_c)\sin\omega_c t, \\ v_y(t) &= a/\omega_c + (v_{y0} + a/\omega_c)\cos\omega_c t - v_{x0}\sin\omega_c t. \end{aligned} \tag{1.9.40}$$

One can see that electric field creates a drift (average velocity) in the y-directions, i.e., in the direction of $\mathbf{E}\times\mathbf{B}$. The drift velocity in the natural form reads

$$\bar{\mathbf{v}} = \frac{qE}{m\omega_c}\frac{\mathbf{B}\times\mathbf{E}}{BE} = \frac{\mathbf{B}\times\mathbf{E}}{B^2}. \tag{1.9.41}$$

Integrating the velocity, one obtains the trajectory having the form of a cycloid.

1.9.5 *Period of the perturbed harmonic oscillator*

Find the dependence of the frequency of an anharmonic oscillator with

$$U(x) = \frac{kx^2}{2} + \alpha x^4 \tag{1.9.42}$$

on the energy E, considering the quartic term as perturbation. What is the applicability condition of the perturbative method?

Solution: Use the formula for the period

$$T(E) = 2\sqrt{2m}\int_0^{x_m}\frac{dx}{\sqrt{E-U(x)}}, \tag{1.9.43}$$

where the turning point x_m satisfies the equation

$$\frac{kx^2}{2} + \alpha x^4 = E. \tag{1.9.44}$$

It is convenient to scale x with x_m to do the perturbation theory in the integrand only. Choosing the new variable $y \equiv x/x_m$, one rewrites the period as

$$T = 4\sqrt{\frac{m}{k}}\int_0^1 \frac{dy}{\sqrt{1+\frac{2\alpha}{k}x_m^2 - y^2 - \frac{2\alpha}{k}x_m^2 y^4}}. \tag{1.9.45}$$

With $\sqrt{k/m} = \omega_0$ and

$$\frac{2\alpha}{k}x_m^2 \equiv \beta, \tag{1.9.46}$$

the period can be rewritten as

$$T = \frac{4}{\omega_0} \int\limits_0^1 \frac{dy}{\sqrt{1 - y^2 + \beta(1 - y^4)}}. \tag{1.9.47}$$

Expanding the integrand up to the first order in $\beta \ll 1$, one obtains

$$T = \frac{4}{\omega_0} \left(\int\limits_0^1 \frac{dy}{\sqrt{1-y^2}} - \frac{\beta}{2} \int\limits_0^1 dy \frac{1+y^2}{\sqrt{1-y^2}} \right) = \frac{4}{\omega_0} \left(\frac{\pi}{2} - \frac{\beta}{2} \frac{3\pi}{4} \right), \tag{1.9.48}$$

and then

$$T(E) = T_0 \left(1 - \frac{3}{4}\beta \right) = T_0 \left(1 - \frac{3\alpha}{2k} x_m^2 \right). \tag{1.9.49}$$

As the correction term already contains α, it is sufficient to use the lowest-order result x_m^2 obtained by neglecting α in Eq. (1.9.44), $x_m^2 = 2E/k$. The final result for the period reads

$$T(E) = T_0 \left(1 - \frac{3}{4}\beta \right) = T_0 \left(1 - \frac{3\alpha}{k^2} E \right). \tag{1.9.50}$$

It is recommended to write down the result for the frequency

$$\omega_0(E) = \omega_0 \left(1 + \frac{3\alpha}{k^2} E \right) \tag{1.9.51}$$

because its qualitative dependence is easier to memorize: Supersquare $U(x)$ — $\alpha > 0$ — frequency increases with energy.

1.9.6 *Runaway in the tilted washboard potential with damping*

Consider a particle of mass m and viscous damping α moving in the washboard potential $U(x) = U_0 [1 - \cos(ax)]$. (a) In the limit of small α, considering damping as a perturbation, find the total energy loss for the particle sliding from one top of the potential to the neighboring one (e.g., from $ax = -\pi$ to $ax = \pi$). (b) Find the critical value of the applied force F at which unbound motion sets in.

Solution:
(a) The energy change per unit of time due to the viscous damping is given by $\dot{E} = -\alpha v^2$. In the limit of small α, one can integrate this over the time

of the motion, considering the motion as unaffected by damping. In this lowest-order approximation, the total energy loss reads

$$\Delta E = \alpha \int_{0}^{t_{\max}} v^2 dt = \alpha \int_{-x_m}^{x_m} v dx. \tag{1.9.52}$$

The unperturbed total energy is given by

$$E = U_0 \left[1 - \cos(ax_m)\right] = \frac{mv^2}{2} + U_0 \left[1 - \cos(ax)\right]. \tag{1.9.53}$$

From here, one obtains

$$v = \sqrt{\frac{2U_0}{m} \left[\cos(ax) - \cos(ax_m)\right]}. \tag{1.9.54}$$

Substituting this into the formula above yields

$$\Delta E = 2\alpha \sqrt{\frac{2U_0}{m}} \int_{0}^{x_m} \sqrt{\cos(ax) - \cos(ax_m)}\, dx. \tag{1.9.55}$$

Working out this integral is similar to the calculation of the period of motion in the washboard potential. First, one writes

$$\Delta E = 4\alpha \sqrt{\frac{U_0}{m}} \int_{0}^{x_m} \sqrt{\sin^2 (ax_m/2) - \sin^2 (ax/2)}\, dx. \tag{1.9.56}$$

Then, substitution

$$\begin{aligned} \sin (ax/2) &= \sin (ax_m/2) \sin \xi, \\ (a/2) \cos (ax/2)\, dx &= \sin (ax_m/2) \cos \xi d\xi \end{aligned} \tag{1.9.57}$$

yields

$$\Delta E = 4\alpha \sqrt{\frac{U_0}{m}} \int_{0}^{\pi/2} \sin (ax_m/2) \sqrt{1 - \sin^2 \xi}\, \frac{\sin (ax_m/2) \cos \xi}{(a/2) \cos (ax/2)} d\xi, \tag{1.9.58}$$

and then

$$\Delta E = \frac{8\alpha}{a} \sqrt{\frac{U_0}{m}} \sin^2 \left(\frac{ax_m}{2}\right) \int_{0}^{\pi/2} \frac{\cos^2 \xi}{\sqrt{1 - \sin^2 (ax_m/2) \sin^2 \xi}} d\xi. \tag{1.9.59}$$

Fortunately, we have $\sin(ax_m/2) = 1$ and the integral simplifies to

$$\Delta E = \frac{8\alpha}{a}\sqrt{\frac{U_0}{m}} \int_0^{\pi/2} \cos\xi d\xi = \frac{8\alpha}{a}\sqrt{\frac{U_0}{m}}. \tag{1.9.60}$$

(b) In the absence of the applied force, the particle sliding from the top of the barrier will not overcome another barrier and turn back, gradually losing energy and ending up at the energy minimum. This is the bound regime. The unbound or runaway regime sets in if the energy gained by sliding in the direction of the applied force exceeds the energy lost due to the dissipation, i.e.,

$$F\frac{2\pi}{a} > \Delta E. \tag{1.9.61}$$

With ΔE above, this yields the criterion for the runaway regime:

$$F > \frac{4\alpha}{\pi}\sqrt{\frac{U_0}{m}}. \tag{1.9.62}$$

1.9.7 *Two connected masses on perpendicular rods*

Two masses, m_1 amd m_2 can slide without friction along two perpendicular rods and are connected by a light rod of length l. There is no gravity. Find the period of the motion of the system, assuming the initial state $\theta_2 = 0$ and $v_1 = v_0 < 0$:

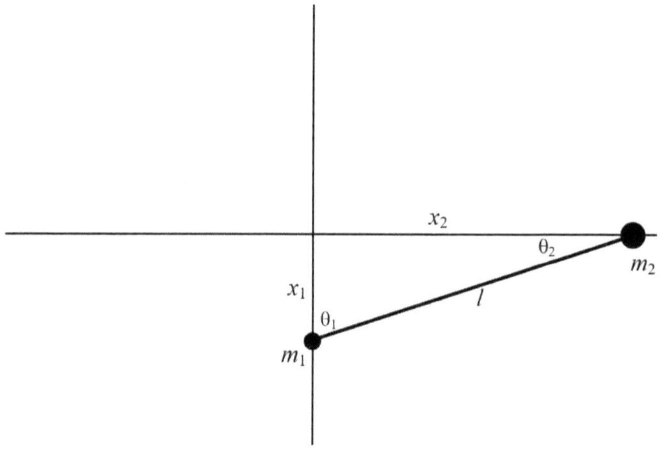

Solution: Let us consider θ_2 as the dynamic variable, then

$$x_1 = l \sin\theta_2, \quad x_2 = l\cos\theta_2. \tag{1.9.63}$$

For $v_1(0) < 0$ and $\theta_2(0) = 0$, the angle θ_2 will increase with time. The kinetic energy has the form

$$E_k = \frac{m_1 \dot{x}_1^2}{2} + \frac{m_2 \dot{x}_2^2}{2} = \frac{1}{2}\left(m_1 \cos^2\theta_2 + m_2 \sin^2\theta_2\right)\dot{\theta}_2^2. \tag{1.9.64}$$

Since E_k is conserved, $E_k = E = \text{const.}$, one can find the first integral of the equations of motion as

$$\dot{\theta}_2 = \sqrt{\frac{2E}{m_1\cos^2\theta_2 + m_2\sin^2\theta_2}} = \sqrt{\frac{2E}{m_1}}\frac{1}{\sqrt{1 + \frac{m_2 - m_1}{m_1}\sin^2\theta_2}}. \tag{1.9.65}$$

Integrating this equation, one obtains

$$\int_0^{\theta_2} d\theta \sqrt{1 + \frac{m_2 - m_1}{m_1}\sin^2\theta} = \sqrt{\frac{2E}{m_1}}t. \tag{1.9.66}$$

Changing of θ_2 from 0 to $\pi/4$ corresponds to one quarter of the period, thus the period is equal to

$$T = 4\sqrt{\frac{m_1}{2E}}\int_0^{\pi/2} d\theta\sqrt{1 - \frac{m_1 - m_2}{m_1}\sin^2\theta} = 4\sqrt{\frac{m_1}{2E}}E\left(\frac{m_1 - m_2}{m_1}\right), \tag{1.9.67}$$

where $E(m)$ is elliptic integral of the second kind. The result above is valid for $m_1 > m_2$. In the opposite case, one has to interchange m_1 and m_2. For $m_1 = m_2 = m$, the result simplifies to

$$T = \pi\sqrt{\frac{2m}{E}} = \frac{2\pi}{v_0}. \tag{1.9.68}$$

1.9.8 *Mass on a conical surface*

Write down dynamical variables (energy, angular momentum) and equation of motion for a mass moving on the surface of a cone with angle θ. Add gravity force in the direction of the cone's apex.

Solution: Placing the origin of the coordinate system at the cone's apex, one can use spherical coordinates with $\theta =$ const. and r changing together with φ. Here, to take into account the constraint, one has to project all vectors on the direction of motion $\mathbf{e}_r$, $\mathbf{e}_\varphi$, ignoring acceleration and forces in the direction $\mathbf{e}_\theta$. Writing $\mathbf{r} = r\mathbf{e}_r$ and using formulas in the main text, one obtains the velocity

$$\mathbf{v} = \dot{\mathbf{r}} = \dot{r}\mathbf{e}_r + r\frac{\partial \mathbf{e}_r}{\partial \varphi}\dot{\varphi} = \mathbf{e}_r\dot{r} + \mathbf{e}_\varphi r\sin\theta\dot{\varphi}. \tag{1.9.69}$$

Now, kinetic energy has the form

$$E_k = \frac{m\mathbf{v}^2}{2} = \frac{m}{2}\left(\dot{r}^2 + r^2\sin^2\theta\dot{\varphi}^2\right). \tag{1.9.70}$$

Angular momentum has the form

$$\mathbf{l} = \mathbf{r}\times m\mathbf{v} = mr\mathbf{e}_r \times (\mathbf{e}_r\dot{r} + \mathbf{e}_\varphi r\sin\theta\dot{\varphi}) = -mr^2\sin\theta\dot{\varphi}\mathbf{e}_\theta. \tag{1.9.71}$$

Its projection onto the symmetry axis z reads

$$l_z = \mathbf{l}\cdot\mathbf{e}_z = mr^2\sin^2\theta\dot{\varphi}. \tag{1.9.72}$$

Acceleration has the form

$$\begin{aligned}\dot{\mathbf{v}} &= \mathbf{e}_r\ddot{r} + \frac{\partial \mathbf{e}_r}{\partial \varphi}\dot{\varphi}\dot{r} + \mathbf{e}_\varphi\dot{r}\sin\theta\dot{\varphi} + \mathbf{e}_\varphi r\sin\theta\ddot{\varphi} + \frac{\partial \mathbf{e}_\varphi}{\partial \varphi}\dot{\varphi}r\sin\theta\dot{\varphi}\\ &= \mathbf{e}_r\ddot{r} + \mathbf{e}_\varphi r\sin\theta\ddot{\varphi} + \mathbf{e}_\varphi 2\sin\theta\dot{\varphi}\dot{r} - \mathbf{e}_r r\sin^2\theta\dot{\varphi}^2\end{aligned} \tag{1.9.73}$$

or, finally,

$$\dot{\mathbf{v}} = \mathbf{e}_r\left(\ddot{r} - r\sin^2\theta\dot{\varphi}^2\right) + \mathbf{e}_\varphi\left(r\sin\theta\ddot{\varphi} + 2\sin\theta\dot{\varphi}\dot{r}\right). \tag{1.9.74}$$

Thus, the equation of motion of body confined to the cone is

$$\ddot{r} - r\sin^2\theta\dot{\varphi}^2 = \frac{1}{ma}F_r, \tag{1.9.75}$$

$$r\sin\theta\ddot{\varphi} + 2\sin\theta\dot{\varphi}\dot{r} = \frac{1}{ma}F_\varphi. \tag{1.9.76}$$

The lhs of the second equation is proportional to $\dot{l}_z$. In the case, $F_\varphi = 0$ one has $l_z =$ const. and the first equation becomes autonomous:

$$\ddot{r} = \frac{l_z^2}{m^2r^3\sin^2\theta} + \frac{1}{ma}F_r. \tag{1.9.77}$$

In the case of gravity force directed toward the apex of the cone,

$$\mathbf{F} = -mg\mathbf{e}_z, \tag{1.9.78}$$

potential energy has the form

$$U = mgz = mgr\cos\theta. \tag{1.9.79}$$

Thus, ignoring θ-component, one obtains

$$\mathbf{F} = -\frac{\partial U}{\partial \mathbf{r}} = -\mathbf{e}_r\frac{\partial U}{\partial r} - \mathbf{e}_\varphi\frac{1}{r\sin\theta}\frac{\partial U}{\partial \varphi} = -\mathbf{e}_r mg\cos\theta, \tag{1.9.80}$$

i.e.,

$$F_r = -mg\cos\theta, \quad F_\varphi = 0. \tag{1.9.81}$$

1.9.9 *Bead on an elliptic trajectory*

A bead considered as a point mass m is moving along an elliptic ring in the absence of external forces and friction. Using the equation of the ellipse

$$\frac{x^2}{a^2} + \frac{y^2}{b^2} = 1, \tag{1.9.82}$$

find the period and frequency of the motion as function of the energy. Consider limiting cases.

Solution: The period can be defined as

$$T = 4\int_0^a \frac{dt}{dx}dx = 4\int_0^a \frac{dx}{\dot{x}}, \tag{1.9.83}$$

while $\dot{x}$ can be found from the energy

$$E = \frac{m\mathbf{v}^2}{2} = \frac{m}{2}\left(\dot{x}^2 + \dot{y}^2\right) \tag{1.9.84}$$

and the ellipse equation. Differentiating the latter, one obtains

$$\frac{x\dot{x}}{a^2} + \frac{y\dot{y}}{b^2} = 0. \tag{1.9.85}$$

Eliminating $\dot{y}$, one obtains

$$E = \frac{m}{2}\dot{x}^2\left[1 + \left(\frac{b}{a}\right)^4\frac{x^2}{y^2}\right]. \tag{1.9.86}$$

Here, y can be eliminated using the ellipse equation that yields

$$E = \frac{m}{2}\dot{x}^2\left[1+\left(\frac{b}{a}\right)^4 \frac{x^2}{b^2\left(1-x^2/a^2\right)}\right] = \frac{m}{2}\dot{x}^2\left[1+\left(\frac{b}{a}\right)^2 \frac{x^2/a^2}{\left(1-x^2/a^2\right)}\right]$$
$$= \frac{m}{2}\dot{x}^2\frac{1-u^2\epsilon^2}{1-u^2}, \tag{1.9.87}$$

where

$$u \equiv x/a, \quad \epsilon \equiv \sqrt{1-b^2/a^2}. \tag{1.9.88}$$

Resolving for $\dot{x}$, one obtains

$$T = 4a\int_0^1 \frac{du}{\dot{x}} = \frac{4a}{v}\int_0^1 du\sqrt{\frac{1-u^2\epsilon^2}{1-u^2}}, \tag{1.9.89}$$

where

$$v = \sqrt{\frac{2E}{m}} \tag{1.9.90}$$

is the constant speed of the bead. Note that this solution works for $b < a$. With the substitution

$$u = \sin\xi, \quad \frac{du}{\sqrt{1-u^2}} = d\xi, \tag{1.9.91}$$

the integral reduces to the complete second-kind elliptic integral

$$E(m) \equiv \int_0^{\pi/2} d\xi\sqrt{1-m\sin^2\xi}, \tag{1.9.92}$$

having particular cases

$$E(0) = \frac{\pi}{2}, \quad E(1) = 1. \tag{1.9.93}$$

Thus,

$$T = \frac{4a}{v}E\left(1-\frac{b^2}{a^2}\right), \quad b \leq a. \tag{1.9.94}$$

In the other case, the result has the form

$$T = \frac{4b}{v} E\left(1 - \frac{a^2}{b^2}\right), \quad a \leq b. \tag{1.9.95}$$

The limiting cases are the circle, $a = b$, where $T = 2\pi a/v$, and a piece of a straight line, $b = 0$, where $T = 4a/v$.

1.9.10 *Non-closed orbits*

Orbits are closed only in the cases $U(r) \propto r^2$ (spatial harmonic oscillator) and $U(r) \propto 1/r$ (gravity, Coulomb interaction). Consider the case of gravity with an additional small potential energy $\delta U(r) = -\beta/r^n$. Treating δU as a perturbation, calculate the additional angle $\delta\varphi$ by which the body is rotating during one unperturbed cycle.

Solution: The rotation angle corresponding to the period of motion is given by

$$\Delta\varphi = 2\pi + \delta\varphi = 2\int_{r_1}^{r_2} dr \frac{l/r^2}{\sqrt{2m\left[E - \frac{l^2}{2mr^2} + \frac{\alpha}{r} - \delta U(r)\right]}}, \tag{1.9.96}$$

where $\delta U = 0$ corresponds to $\delta\varphi = 0$. To find $\delta\varphi$, one can try to expand the integral in powers of δU. However, differentiation of the integrand with respect to δU yields a diverging integral. In fact, this divergence is fictitious, since one also has to expand turning points r_1 and r_2 in powers of δU, and if everything is properly taken into account, divergencies should cancel. This procedure is, of course, pretty cumbersome. Fortunately, one can use a trick to avoid these complications and make the calculation elegant, writing the formula above in the form

$$\Delta\varphi = 2\pi + \delta\varphi = -\frac{\partial}{\partial l} 2\int_{r_1}^{r_2} dr \sqrt{2m\left[E - \frac{l^2}{2mr^2} + \frac{\alpha}{r} - \delta U(r)\right]}. \tag{1.9.97}$$

Now, the expansion up to the linear term in δU does not lead to divergence and there is no contribution from differentiation of the turning points. Moreover, the resulting expression

$$\delta\varphi = \frac{\partial}{\partial l}\int_{r_1}^{r_2} dr \frac{2m\delta U(r)}{\sqrt{2m\left[E - \frac{l^2}{2mr^2} + \frac{\alpha}{r}\right]}} \tag{1.9.98}$$

is similar to the initial unperturbed formula for $\Delta\varphi$. Here, one can change from integration over r to integration over φ using the unperturbed trajectory $r(\varphi)$. With the help of

$$\frac{d\varphi}{dr} = \frac{l}{r^2}\frac{1}{\sqrt{2m\left[E - \frac{l^2}{2mr^2} + \frac{\alpha}{r}\right]}}, \tag{1.9.99}$$

one obtains

$$\delta\varphi = \frac{\partial}{\partial l}\left[\frac{2m}{l}\int_0^\pi d\varphi r^2(\varphi)\delta U(r(\varphi))\right]. \tag{1.9.100}$$

For our particular form of δU, this expression becomes

$$\delta\varphi = \frac{\partial}{\partial l}\left[\frac{2m\beta}{l}\int_0^\pi \frac{d\varphi}{r^{n-2}(\varphi)}\right] = \frac{\partial}{\partial l}\left[\frac{2m\beta}{lr_0^{n-2}}\int_0^\pi d\varphi\,(1+\epsilon\cos\varphi)^{n-2}\right], \tag{1.9.101}$$

where $r_0 = l^2/(\alpha m)$ and

$$\epsilon \equiv \sqrt{1 + \frac{E}{|E_{\min}|}} \tag{1.9.102}$$

is independent of l. Thus, the result has the form

$$\delta\varphi = \frac{\partial}{\partial l}\left[\frac{2m\beta\,(\alpha m)^{n-2}}{l^{2n-3}}\int_0^\pi d\varphi\,(1+\epsilon\cos\varphi)^{n-2}\right], \tag{1.9.103}$$

and, finally,

$$\delta\varphi = -\frac{(2n-3)\,2m\beta\,(\alpha m)^{n-2}}{l^{2n-2}}\int_0^\pi d\varphi\,(1+\epsilon\cos\varphi)^{n-2}. \tag{1.9.104}$$

In particular, one obtains

$$\delta\varphi = -\frac{2\pi m\beta}{l^2}, \quad n = 2 \tag{1.9.105}$$

and

$$\delta\varphi = -\frac{6\pi m^2\alpha\beta}{l^4}, \quad n = 3. \tag{1.9.106}$$

1.9.11 *Small-angle scattering*

Find scattering angle χ for the central potential $U(r) = \beta/(a^2 + r^2)$ in the limit of a fast-moving particle. What is the maximal value of χ? Plot differential scattering cross-section.

Solution: The fast-moving particle will be scattered at small angles χ for any target distance ρ because the potential energy is not diverging at $r \to 0$. One can use the small-scattering angle formula

$$\chi = -\frac{1}{mv_\infty^2} \int_{-\infty}^{\infty} dx \left.\frac{\partial U(x,y)}{\partial y}\right|_{y=\rho}, \tag{1.9.107}$$

which becomes

$$\chi = \frac{2\beta\rho}{mv_\infty^2} \int_{-\infty}^{\infty} \frac{dx}{(a^2+\rho^2+x^2)^2} = \frac{\pi\beta}{mv_\infty^2} \frac{\rho}{(a^2+\rho^2)^{3/2}}. \tag{1.9.108}$$

One can see that

$$\chi \propto \begin{cases} \rho, & \rho \ll a, \\ 1/\rho^2, & \rho \gg a. \end{cases} \tag{1.9.109}$$

The maximum of χ is defined by

$$\frac{\partial \chi}{\partial \rho} = \frac{\pi\beta}{mv_\infty^2} \frac{a^2 - 2\rho^2}{(a^2+\rho^2)^{5/2}} = 0, \tag{1.9.110}$$

which yields maximal scattering at $\rho = a/\sqrt{2}$,

$$\chi_{\max} = \frac{\pi\beta}{mv_\infty^2} \frac{2}{3^{3/2} a^2}. \tag{1.9.111}$$

Differential scattering cross-section is defined by

$$d\sigma = 2\pi\rho(\chi) \left|\frac{d\rho(\chi)}{d\chi}\right| d\chi. \tag{1.9.112}$$

Thus, one has to plot the function $d\sigma/d\chi$ vs χ. This can be conveniently done in the parametric form using ρ as a parameter:

$$\begin{aligned} \frac{d\sigma}{d\chi} &= 2\pi\rho \left(\frac{\partial \chi}{\partial \rho}\right)^{-1} = \frac{2mv_\infty^2}{\beta} \frac{\rho\left(a^2+\rho^2\right)^{5/2}}{|a^2 - 2\rho^2|}, \\ \chi &= \frac{\pi\beta}{mv_\infty^2} \frac{\rho}{(a^2+\rho^2)^{3/2}}. \end{aligned} \tag{1.9.113}$$

For plotting, it is convenient to use $\tilde{\chi}$ defined by

$$\tilde{\chi} \equiv \left(\frac{\pi\beta}{mv_\infty^2}\right)^2 \chi = \frac{\rho}{(a^2+\rho^2)^{3/2}}. \tag{1.9.114}$$

In terms of $\tilde{\chi}$, parametric representation of differential scattering cross-section becomes

$$\begin{aligned} \frac{d\sigma}{d\tilde{\chi}} &= \frac{2\pi\rho\left(a^2+\rho^2\right)^{5/2}}{|a^2-2\rho^2|}, \\ \tilde{\chi} &= \frac{\rho}{(a^2+\rho^2)^{3/2}}. \end{aligned} \tag{1.9.115}$$

Chapter 2

Lagrangian Mechanics

2.1 The least-action principle and Lagrange equations

Newtonian mechanics is fully sufficient practically. However, it is desirable to find a way to obtain equations of motion from some scalar generating function. For conservative systems, the complete information about the system is contained in the total energy $E = E_k + U$, expressed as a function of coordinates and velocities, or angles and their time derivatives for some systems considered above. However, there is no way to obtain equations of motion from the energy function directly. It turns out that the generating function of equations of motion is the Lagrange function or simply Lagrangian, in most cases,

$$\mathcal{L} = E_k - U. \tag{2.1.1}$$

Lagrange formalism is built upon the so-called *least-action principle*, also called the *Hamilton principle*. According to this principle, which can be put into the foundation of mechanics, the actual dynamics of the system, i.e., the actual time dependence of its *generalized coordinates* $\{q_i(t)\}$ minimizes the action on the way from state 1 to state 2,

$$\mathcal{S} = \int_{t_1}^{t_2} dt \mathcal{L}\left(q(t), \dot{q}(t), t\right), \quad q \equiv \{q_i\}. \tag{2.1.2}$$

Here, q, $\dot{q}$ may be coordinates and velocities, angles of the polar or spherical coordinate systems and their time derivatives, or whatever other dynamical variables. Varying $\mathcal{S}$ with respect to an arbitrary small deviation $\delta q(t)$

from the actual solution minimizing the action, one obtains

$$0 = \delta \mathcal{S} = \int_{t_1}^{t_2} dt \left[\mathcal{L}\left(q + \delta q, \dot{q} + \delta \dot{q}, t\right) - \mathcal{L}\left(q, \dot{q}, t\right)\right]$$

$$= \int_{t_1}^{t_2} dt \sum_i \left[\frac{\partial \mathcal{L}}{\partial q_i}\delta q_i + \frac{\partial \mathcal{L}}{\partial \dot{q}_i}\delta \dot{q}_i\right]$$

$$= \sum_i \frac{\partial \mathcal{L}}{\partial \dot{q}_i}\delta q_i \Big|_{t_1}^{t_2} + \int_{t_1}^{t_2} dt \sum_i \left[\frac{\partial \mathcal{L}}{\partial q_i} - \frac{d}{dt}\frac{\partial \mathcal{L}}{\partial \dot{q}_i}\right]\delta q_i. \tag{2.1.3}$$

The first term here vanishes because of the boundary conditions fixing $q_i(t)$, i.e., $\delta q_i(t_1) = \delta q_i(t_2) = 0$. The second term should vanish for different $\delta q_i(t)$, which requires that Lagrange equations

$$\frac{d}{dt}\frac{\partial \mathcal{L}}{\partial \dot{q}_i} - \frac{\partial \mathcal{L}}{\partial q_i} = 0 \tag{2.1.4}$$

for all i should be satisfied. Introducing vectors

$$\mathbf{q} = \{q_i\}, \quad \dot{\mathbf{q}} = \{\dot{q}_i\}, \tag{2.1.5}$$

one can write Lagrange equations in the vector form

$$\frac{d}{dt}\frac{\partial \mathcal{L}}{\partial \dot{\mathbf{q}}} - \frac{\partial \mathcal{L}}{\partial \mathbf{q}} = 0. \tag{2.1.6}$$

One can see that using Lagrange function of Eq. (2.1.1) and the above equation yields Newtonian equations of motion. For instance, for a point mass in a potential field,

$$\mathcal{L} = \frac{m\mathbf{v}^2}{2} - U(\mathbf{r}). \tag{2.1.7}$$

Identifying $\mathbf{r} = \mathbf{q}$, $\mathbf{v} = \dot{\mathbf{q}}$, from Eq. (2.1.6), one obtains

$$m\dot{\mathbf{v}} + \frac{\partial U}{\partial \mathbf{r}} = 0, \tag{2.1.8}$$

Newton's second law for potential forces.

Lagrange formalism is especially convenient for systems with holonomic constraints because in it, holonomic constraints are hidden and never appear explicitly. As an example, the Lagrange function of a pendulum considered in Newtonian mechanics above has the form

$$\mathcal{L} = \frac{ml^2\dot{\varphi}^2}{2} + mgl\cos\varphi, \tag{2.1.9}$$

where $\varphi = q$ and $\dot{\varphi} = \dot{q}$. Lagrange equation in this case reads

$$ml^2\ddot{\varphi} + mgl\sin\varphi = 0 \tag{2.1.10}$$

or

$$\ddot{\varphi} + \omega_0^2\sin\varphi = 0, \quad \omega_0^2 = g/l, \tag{2.1.11}$$

i.e., the Newtonian equation of motion after eliminating the holonomic constraint by using a polar coordinate system. Thus, in the case of holonomic constraints, using Lagrange formalism is a shorter way to equations of motion, although writing down the kinetic energy requires some habit and, in more complicated cases, a special calculation.

Lagrange formalism is also convenient in the case of using special coordinate systems even in the absence of constraints. As an example, one can consider a pendulum, in which the mass is attached to a spring and can move along the light rod. Kinetic energy in polar coordinates has the form

$$E_k = \frac{m}{2}\left(\dot{r}^2 + r^2\dot{\varphi}^2\right). \tag{2.1.12}$$

Potential energy is the sum of gravity energy and the energy of the spring,

$$U = -mgr\cos\varphi + \frac{k}{2}(r-l)^2, \tag{2.1.13}$$

where k is spring constant and l is the equilibrium position of the mass on the spring. It is convenient instead of r to use $x \equiv r - l$ as a dynamical variable. In terms of x and φ, the Lagrange function becomes

$$\mathcal{L} = \frac{m}{2}\left(\dot{x}^2 + (l+x)^2\dot{\varphi}^2\right) + mg(l+x)\cos\varphi - \frac{k}{2}x^2. \tag{2.1.14}$$

Lagrange equations are

$$\frac{d}{dt}\frac{\partial\mathcal{L}}{\partial\dot{\varphi}} - \frac{\partial\mathcal{L}}{\partial\varphi} = m\frac{d}{dt}(l+x)^2\dot{\varphi} + mg(l+x)\sin\varphi = 0 \tag{2.1.15}$$

and

$$\frac{d}{dt}\frac{\partial\mathcal{L}}{\partial\dot{x}} - \frac{\partial\mathcal{L}}{\partial x} = m\ddot{x} - m(l+x)\dot{\varphi}^2 - mg\cos\varphi + kx = 0. \tag{2.1.16}$$

They simplify to

$$(l+x)\,\ddot{\varphi}+2\dot{x}\dot{\varphi}+g\sin\varphi=0, \tag{2.1.17}$$

$$\ddot{x}-(l+x)\,\dot{\varphi}^2-g\cos\varphi+\frac{k}{m}x=0. \tag{2.1.18}$$

In the linear approximation in x and φ (small oscillations), these equations decouple:

$$\ddot{\varphi}+\frac{g}{l}\varphi=0, \quad \ddot{x}-g+\frac{k}{m}x=0. \tag{2.1.19}$$

One can see that multiplying the Lagrange function by a constant does not change Lagrange equations of motion. Thus, the choice of Lagrange function in Eq. (2.1.1) is not unique. Moreover, one can add a full time derivative of a function of coordinates and time $\partial_t f(q,t)$ to the Lagrange function,

$$\mathcal{L}'=\mathcal{L}+\frac{d}{dt}f(q,t). \tag{2.1.20}$$

This adds an invariable contribution to the action,

$$\mathcal{S}'=\mathcal{S}+f(q(t_2),t_2)-f(q(t_1),t_1), \tag{2.1.21}$$

which does not change equations of motion. Proof of irrelevance of this addition to the Lagrangian using Lagrange equations is a bit more cumbersome:

$$\begin{aligned}\frac{d}{dt}\frac{\partial}{\partial\dot{q}}\frac{d}{dt}f(q,t)-\frac{\partial}{\partial q}\frac{d}{dt}f(q,t)&=\frac{d}{dt}\frac{\partial}{\partial\dot{q}}\left[\frac{\partial f}{\partial t}+\frac{\partial f}{\partial q}\dot{q}\right]-\frac{\partial}{\partial q}\left[\frac{\partial f}{\partial t}+\frac{\partial f}{\partial q}\dot{q}\right]\\ =\frac{d}{dt}\frac{\partial f}{\partial q}-\frac{\partial^2 f}{\partial q\partial t}-\frac{\partial^2 f}{\partial q^2}\dot{q}&=\frac{\partial^2 f}{\partial q\partial t}+\frac{\partial^2 f}{\partial q^2}\dot{q}-\frac{\partial^2 f}{\partial q\partial t}-\frac{\partial^2 f}{\partial q^2}\dot{q}=0.\end{aligned} \tag{2.1.22}$$

In the presence of non-holonomic constraints, Lagrange formalism loses its elegance. The general form of a system of different non-holonomic constraints labeled by the index α is

$$\sum_i c_{\alpha i}(q)\dot{q}_i=0, \quad \alpha=1,2,\ldots \tag{2.1.23}$$

If these constraints can be integrated, i.e., represented as constraints on q, they are holonomic constraints. In general, the constraints above cannot be integrated. In this case, the way to proceed is first to note that the constraints above can be reformulated as constraints on possible variations of q, i.e.,

$$\sum_i c_{\alpha i}(q)\delta q_i = 0, \quad \alpha = 1, 2, \ldots \tag{2.1.24}$$

Constrained variation of the action can be performed with the help of the method of Lagrange multipliers. One obtains

$$0 = \delta\mathcal{S}' = \delta\mathcal{S} + \sum_\alpha \lambda_\alpha \int_{t_1}^{t_2} dt \sum_i c_{\alpha i}(q)\delta q_i, \tag{2.1.25}$$

where λ_α are Lagrange multipliers. Using Eq. (2.1.3), one arrives at constrained Lagrange equations of motion

$$\frac{d}{dt}\frac{\partial\mathcal{L}}{\partial\dot{q}_i} - \frac{\partial\mathcal{L}}{\partial q_i} - \sum_\alpha \lambda_\alpha c_{\alpha i}(q) = 0, \tag{2.1.26}$$

which have to be solved together with Eq. (2.1.23). The bottom line is that in the case of non-holonomic constraint, the Newtonian approach is better than Lagrangian because reaction forces within the former are physically more transparent than Lagrange multipliers in the latter.

2.2 Lagrangian of a particle in electromagnetic field

The Lorentz force curving trajectories of charged particles in a magnetic field is a special kind of force that is conservative but not potential. One can construct the Lagrangian for a particle in electromagnetic field that results in correct Newtonian equations of motion

$$m\dot{\mathbf{v}} = q\mathbf{E} + q\,[\mathbf{v}\times\mathbf{B}], \tag{2.2.1}$$

considered above. Here, the electric and magnetic fields can be expressed via the scalar and vector potentials as follows:

$$\mathbf{E} = -\frac{\partial\mathbf{A}}{\partial t} - \nabla\phi, \quad \mathbf{B} = \nabla\times\mathbf{A}. \tag{2.2.2}$$

Let us show that Lagrangian can be written as

$$\mathcal{L} = \frac{m\mathbf{v}^2}{2} + q\mathbf{v}\cdot\mathbf{A} - q\phi. \tag{2.2.3}$$

Indeed, the Lagrange equation becomes

$$0 = \frac{d}{dt}\frac{\partial\mathcal{L}}{\partial\mathbf{v}} - \frac{\partial\mathcal{L}}{\partial\mathbf{r}} = \frac{d}{dt}(m\mathbf{v} + q\mathbf{A}) + q\nabla\phi - q\nabla(\mathbf{v}\cdot\mathbf{A}). \tag{2.2.4}$$

Using the formula for the gradient of a dot-product

$$\nabla (\mathbf{A} \cdot \mathbf{B}) = (\mathbf{A}\nabla)\mathbf{B} + (\mathbf{B}\nabla)\mathbf{A} + \mathbf{A} \times (\nabla \times \mathbf{B}) + \mathbf{B} \times (\nabla \times \mathbf{A}), \quad (2.2.5)$$

one obtains

$$0 = m\dot{\mathbf{v}} + q\dot{\mathbf{A}} + q\nabla\phi - q(\mathbf{v}\nabla)\mathbf{A} - q\mathbf{v} \times (\nabla \times \mathbf{A}). \quad (2.2.6)$$

Finally, using the formula

$$\dot{\mathbf{A}} = \frac{d}{dt}\mathbf{A}(\mathbf{r}(t), t) = \frac{\partial \mathbf{A}}{\partial t} + (\mathbf{v}\nabla)\mathbf{A} \quad (2.2.7)$$

and Eq. (2.2.2), one arrives at Eq. (2.2.1).

Gauge transformation

$$\phi \Rightarrow \phi' \equiv \phi - \frac{\partial \chi}{\partial t}, \quad \mathbf{A} \Rightarrow \mathbf{A}' \equiv \mathbf{A} + \nabla\chi \quad (2.2.8)$$

leaves both $\mathbf{E}$ and $\mathbf{B}$ unchanged, i.e., the same $\mathbf{E}$ and $\mathbf{B}$ follow from Eq. (2.2.2) if the 4-potential $(\phi', \mathbf{A}')$ is used instead of $(\phi, \mathbf{A})$. Thus, the Lagrangian also has to be gauge-invariant. Indeed, from Eq. (2.2.3), one obtains

$$\mathcal{L}' = \frac{m\mathbf{v}^2}{2} + q\mathbf{v} \cdot (\mathbf{A} + \nabla\chi) - q\left(\phi - \frac{\partial \chi}{\partial t}\right) = \mathcal{L} + \frac{d}{dt}(q\chi), \quad (2.2.9)$$

where the full time derivative can be discarded.

2.3 Transformations and conservation laws

2.3.1 *Galilean transformation*

A free body is moving with a constant velocity by inertia if the frame, in which the motion is considered, is not moving itself with acceleration. Such a frame is called the *inertial frame*. Any frame, moving with a constant velocity with respect to the inertial frame, is also an inertial frame. Thus, there is an infinite number of inertial frames. In each of these aforementioned intertial frames, a free body will be moving with a constant velocity.

Transformation between inertial frames is *Galilean transformation*:

$$\begin{aligned} \mathbf{v} &= \mathbf{v}' + \mathbf{V}, \quad \mathbf{V} = \text{const.}, \\ \mathbf{r} &= \mathbf{r}' + \mathbf{V}t + \text{const.} \end{aligned} \quad (2.3.1)$$

Since all inertial frames are equivalent, equations of motion in these frames should be the same. This means that Lagrange function transformed from one frame to the other should have the same form up to the irrelevant full time derivative. For instance, Lagrangian of a free particle $\mathcal{L} = m\mathbf{v}^2/2$ can be transformed as follows:

$$\mathcal{L} = \frac{m\mathbf{v}^2}{2} = \frac{m(\mathbf{v}' + \mathbf{V})^2}{2} = \frac{m\mathbf{v}'^2}{2} + m\mathbf{v}' \cdot \mathbf{V} + \frac{m\mathbf{V}^2}{2}$$
$$= \mathcal{L}' + \frac{d}{dt}\left(m\mathbf{r}' \cdot \mathbf{V} + \frac{m\mathbf{V}^2}{2}t\right), \tag{2.3.2}$$

where the Lagrangian in primed frame $\mathcal{L}' = m\mathbf{v}'^2/2$ has the same functional form and the irrelevant full derivative can be discarded.

If one requires that all equations of (non-relativistic) physics are covariant (i.e., have the same form) with respect to Galilean transformations, the principle of Galilean invariance becomes non-trivial and productive. For instance, a charged particle in a magnetic field is circling. Galilean transformation into another frame with $\mathbf{V}\perp\mathbf{B}$ makes non-closed orbits in the form of cycloids. What should be the corresponding equation of motion? We know that cycloidal orbits emerge in the case of electric field $\mathbf{E}\perp\mathbf{B}$. Thus, to keep equations covariant, one has to require that electromagnetic field is getting changed by Galilean transformations, so that in the moving frame, there is an electric field making trajectories cycloidal.

2.3.2 *Time invariance and energy conservation*

If the problem has no explicit time dependence (i.e., is invariant with respect to time shift) and the forces are potential, the total energy of the system $E = E_k + U$ is conserved, as we have seen in Newtonian mechanics above. Within Lagrangian formalism, one can define

$$E \equiv \sum_i \dot{q}_i \frac{\partial \mathcal{L}}{\partial \dot{q}_i} - \mathcal{L}. \tag{2.3.3}$$

With the help of Lagrange equations, it can be shown that in the case of $\mathcal{L} = \mathcal{L}(q, \dot{q})$, one obtains $\dot{E} = 0$ from Lagrange equations. One proceeds as follows.

$$\dot{E} = \sum_i \ddot{q}_i \frac{\partial \mathcal{L}}{\partial \dot{q}_i} + \sum_i \dot{q}_i \frac{d}{dt}\frac{\partial \mathcal{L}}{\partial \dot{q}_i} - \sum_i \frac{\partial \mathcal{L}}{\partial \dot{q}_i}\ddot{q}_i - \sum_i \frac{\partial \mathcal{L}}{\partial q_i}\dot{q}_i$$
$$= \sum_i \dot{q}_i \left(\frac{d}{dt}\frac{\partial \mathcal{L}}{\partial \dot{q}_i} - \frac{\partial \mathcal{L}}{\partial q_i}\right) = 0. \tag{2.3.4}$$

Since Lagrangian is a bilinear function of $\dot{q}_i$, the above-defined E is indeed the total energy:

$$\sum_i \dot{q}_i \frac{\partial \mathcal{L}}{\partial \dot{q}_i} - \mathcal{L} = 2E_k - (E_k - U) = E_k + U = E. \qquad (2.3.5)$$

2.3.3 *Translational invariance and momentum conservation*

Lagrangian of an isolated system is invariant under spatial translations

$$\mathbf{r}_i \Rightarrow \mathbf{r}_i + \boldsymbol{\varepsilon}. \qquad (2.3.6)$$

This actually means that potential energy depends only on the differences $\mathbf{r}_i - \mathbf{r}_j$. That is, there is only potential energy due to interaction between the parts of the system but no external potential energy. Considering $\boldsymbol{\varepsilon}$ as infinitesimal with $\delta\mathcal{L} = 0$ as the result of translation, one obtains

$$0 = \delta\mathcal{L} = \boldsymbol{\varepsilon} \cdot \sum_i \frac{\partial \mathcal{L}}{\partial \mathbf{r}_i} \quad \Rightarrow \quad \sum_i \frac{\partial \mathcal{L}}{\partial \mathbf{r}_i} = 0, \qquad (2.3.7)$$

as the direction of $\boldsymbol{\varepsilon}$ is arbitrary. Further, using $\partial\mathcal{L}/\partial\mathbf{r}_i = -\partial U/\partial\mathbf{r}_i = \mathbf{f}_i$, one obtains Newton's third law $\sum_i \mathbf{f}_i = 0$. On the other hand, from Lagrange equations follows

$$0 = \sum_i \left(\frac{d}{dt} \frac{\partial \mathcal{L}}{\partial \mathbf{v}_i} - \frac{\partial \mathcal{L}}{\partial \mathbf{r}_i} \right) = \frac{d}{dt} \sum_i \frac{\partial \mathcal{L}}{\partial \mathbf{v}_i} = \frac{d}{dt} \sum_i \mathbf{p}_i = \frac{d}{dt} \mathbf{P}, \qquad (2.3.8)$$

thus the total momentum is conserved: $\mathbf{P} = \text{const}$.

2.3.4 *Cyclic coordinates and generalized momenta*

The results of the preceding section can be generalized for the case when the Lagrangian is independent of one or several generalized coordinates q_i. In a similar way, it then follows that

$$\frac{\partial \mathcal{L}}{\partial q_i} = 0 \quad \Rightarrow \quad \frac{\partial \mathcal{L}}{\partial \dot{q}_i} = p_i = \text{const.}, \qquad (2.3.9)$$

where p_i are *generalized momenta*. In particular, when either a polar or spherical coordinate system is used and Lagrangian does not depend on the angle φ, as for the pendulum of Eq. (2.1.9) without gravity, then the conserved generalized momentum corresponding to φ is angular momentum:

$$\frac{\partial \mathcal{L}}{\partial \dot{\varphi}} = p_\varphi = l_\varphi. \qquad (2.3.10)$$

For a particle on a sphere with $U = U(\theta)$, considered in Newtonian mechanics, $l_\varphi = l_z$. For a charged particle in electromagnetic field, Lagrangian is given by Eq. (2.2.3), and generalized momentum becomes

$$\mathbf{p} = m\mathbf{v} + q\mathbf{A}. \tag{2.3.11}$$

2.3.5 *Rotational invariance and angular momentum conservation*

In the case of rotational invariance, infinitesimal rotation by an angle $\delta\boldsymbol{\varphi}$,

$$\delta\mathbf{r}_i = \delta\boldsymbol{\varphi} \times \mathbf{r}_i, \quad \delta\mathbf{v}_i = \delta\boldsymbol{\varphi} \times \mathbf{v}_i \tag{2.3.12}$$

leaves the Lagrangian unchanged. Using Lagrange equations and the definition of the generalized momentum, one proceeds as

$$\begin{aligned} 0 = \delta\mathcal{L} &= \sum_i \left(\frac{\partial\mathcal{L}}{\partial\mathbf{r}_i} \cdot \delta\mathbf{r}_i + \frac{\partial\mathcal{L}}{\partial\mathbf{v}_i} \cdot \delta\mathbf{v}_i \right) = \sum_i \left(\frac{d}{dt}\frac{\partial\mathcal{L}}{\partial\mathbf{v}_i} \cdot \delta\mathbf{r}_i + \frac{\partial\mathcal{L}}{\partial\mathbf{v}_i} \cdot \delta\mathbf{v}_i \right) \\ &= \sum_i (\dot{\mathbf{p}}_i \cdot \delta\mathbf{r}_i + \mathbf{p}_i \cdot \delta\mathbf{v}_i) = \sum_i (\dot{\mathbf{p}}_i \cdot [\delta\boldsymbol{\varphi} \times \mathbf{r}_i] + \mathbf{p}_i \cdot [\delta\boldsymbol{\varphi} \times \mathbf{v}_i]) \\ &= \delta\boldsymbol{\varphi} \cdot \sum_i (\mathbf{r}_i \times \dot{\mathbf{p}}_i + \mathbf{v}_i \times \mathbf{p}_i) = \delta\boldsymbol{\varphi} \cdot \frac{d}{dt} \sum_i \mathbf{r}_i \times \mathbf{p}_i = \delta\boldsymbol{\varphi} \cdot \dot{\mathbf{L}}. \end{aligned} \tag{2.3.13}$$

As the direction of $\delta\boldsymbol{\varphi}$ is arbitrary, one concludes that the total angular momentum of the system is conserved: $\mathbf{L} = \sum_i [\mathbf{r}_i \times \mathbf{p}_i] = \text{const}$.

2.3.6 *Emma Noether's Theorem*

Integrals of motion resulting from invariance of the Lagrange function with respect to different transformations are particular cases of a general result known as Emma Noether's theorem. Consider a general transformation

$$q(t) \Rightarrow q'(t) = Q(s, t), \tag{2.3.14}$$

where s is a continuous transformation parameter and $Q(0, t) = q(t)$. In the following, we consider s as infinitesimal and say that if

$$\frac{d}{ds}\mathcal{L}\left(Q(s,t), \dot{Q}(s,t), t\right) = 0, \tag{2.3.15}$$

the transformation applied is a continuous symmetry of the Lagrangian. The integral of motion corresponding to this symmetry or invariance can be found by a direct calculation using Lagrange equations:

$$0 = \left.\frac{d\mathcal{L}}{ds}\right|_{s=0} = \sum_i \left(\frac{\partial \mathcal{L}}{\partial q_i}\frac{\partial Q_i}{\partial s} + \frac{\partial \mathcal{L}}{\partial \dot{q}_i}\frac{\partial \dot{Q}_i}{\partial s}\right)\Bigg|_{s=0}$$

$$= \sum_i \left(\frac{d}{dt}\frac{\partial \mathcal{L}}{\partial \dot{q}_i}\frac{\partial Q_i}{\partial s} + \frac{\partial \mathcal{L}}{\partial \dot{q}_i}\frac{\partial \dot{Q}_i}{\partial s}\right)\Bigg|_{s=0} = \frac{d}{dt}\sum_i \left.\frac{\partial \mathcal{L}}{\partial \dot{q}_i}\frac{\partial Q_i}{\partial s}\right|_{s=0}.$$

Thus, the integral of motion is

$$\sum_i \left.\frac{\partial \mathcal{L}}{\partial \dot{q}_i}\frac{\partial Q_i}{\partial s}\right|_{s=0} = \text{const.} \tag{2.3.16}$$

Components of vector integrals of motion considered above can be obtained from this formula. For instance, if the Lagrangian is invariant with respect to $r_x \Rightarrow r'_x = r_x + s$, then the corresponding integral of motion is $\partial\mathcal{L}/\partial v_x = p_x = \text{const}$. With all its conceptual significance, Emma Noether's theorem provides no way of regularly obtaining integrals of motion additional to those we already know.

2.4 Problems on Lagrangian mechanics

2.4.1 *Bead on a rotating ring*

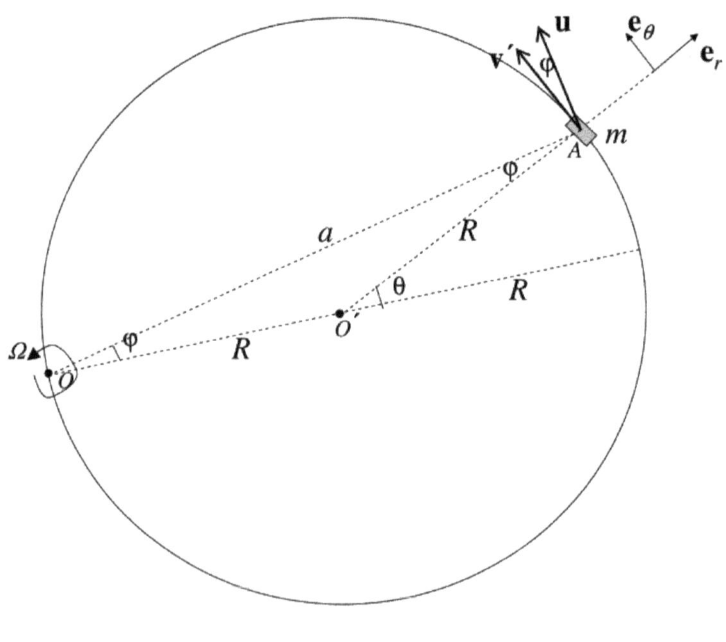

A ring of radius R is rotating in its plane with the constant angular velocity Ω around a point O. A bead of mass m is sliding along the ring without friction. Describing the position of the bead on the ring with the angle θ:
(a) Construct Lagrange function and obtain equations of motion.
(b) Identify effective kinetic, potential, and total energies.

Solution:
(a) In this problem, the potential energy is absent, thus the Lagrange function has the form

$$\mathcal{L} = \frac{m\mathbf{v}^2}{2}, \tag{2.4.1}$$

where $\mathbf{v}$ is the bead's velocity that consists of two contributions, sliding of the bead and rotating of the ring, respectively:

$$\mathbf{v} = \mathbf{u} + \mathbf{v}'. \tag{2.4.2}$$

Thus,

$$\mathcal{L} = \frac{m}{2}\left(u^2 + 2\mathbf{u}\cdot\mathbf{v}' + v'^2\right). \tag{2.4.3}$$

Here,

$$v' = R\dot{\theta} \tag{2.4.4}$$

and, from the triangles,

$$u = a\Omega = 2R\cos\varphi\Omega = 2R\cos\frac{\theta}{2}\Omega. \tag{2.4.5}$$

The angle between $\mathbf{v}'$ and $\mathbf{u}$ is $\varphi = \theta/2$, so that the Lagrange function becomes

$$\begin{aligned}
\mathcal{L} &= \frac{m}{2}\left(u^2 + 2uv'\cos\frac{\theta}{2} + v'^2\right) \\
&= \frac{mR^2}{2}\left(4\cos^2\frac{\theta}{2}\Omega^2 + 4\cos^2\frac{\theta}{2}\Omega\dot{\theta} + \dot{\theta}^2\right) \\
&= mR^2\left[\Omega^2\left(1+\cos\theta\right) + \Omega\dot{\theta}\left(1+\cos\theta\right) + \frac{1}{2}\dot{\theta}^2\right] \\
&= mR^2\left[\Omega^2\left(1+\cos\theta\right) + \frac{1}{2}\dot{\theta}^2\right].
\end{aligned} \tag{2.4.6}$$

The cross-term in the above expression has been dropped since it is a full time derivative:

$$\dot{\theta}\,(1+\cos\theta) = \frac{d}{dt}\,(\theta+\sin\theta), \tag{2.4.7}$$

which does not make a contribution into the Lagrange equation that can be checked directly. The Lagrange equation

$$\frac{d}{dt}\frac{\partial\mathcal{L}}{\partial\dot{\theta}} - \frac{\partial\mathcal{L}}{\partial\theta} = 0 \tag{2.4.8}$$

has the form

$$\ddot{\theta} + \Omega^2\sin\theta = 0, \tag{2.4.9}$$

the equation of motion for the pendulum.

(b) Already from the final expression for the Lagrangian, it is clear that the problem is equivalent to that of a pendulum and the effective kinetic and potential energies are given by

$$E_{k,\mathrm{eff}} = \frac{1}{2}mR^2\dot{\theta}^2, \quad U_{\mathrm{eff}} = -mR^2\Omega^2\,(1+\cos\theta). \tag{2.4.10}$$

The total effective energy

$$E_{\mathrm{eff}} = \frac{1}{2}mR^2\dot{\theta}^2 - mR^2\Omega^2\,(1+\cos\theta) \tag{2.4.11}$$

is conserved. Note that the true total energy is just $\mathcal{L}$ and it is not conserved.

2.4.2 *Two masses on circles with a spring*

Mass m_1 can slide without friction on a ring of radius R_1 in the plane $z=0$ with its center at $(x,y)=(0,0)$. Another mass m_2 is restricted to the circle of radius R_2 with its center at $(x,y)=(0,a)$. The two masses are connected by a spring so that their interaction energy is $U=\frac{1}{2}kd^2$, where d is the distance between the particles. Write down the Lagrangian and equations of motion. What integral of motion arises in the case $a=0$? For $m_1=m_2=m$, $R_1=R_2=R$, and $a\le 2R$, find the frequencies of small oscillations.

Solution: Use polar coordinates for each mass:

$$x_1 = R_1\cos\varphi_1, \quad y_1 = R_1\sin\varphi_1 \tag{2.4.12}$$

and

$$x_2 = R_2 \cos \varphi_2, \quad y_2 = a + R_2 \sin \varphi_2. \tag{2.4.13}$$

The distance d between the particles is given by

$$d^2 = (x_1 - x_2)^2 + (y_1 - y_2)^2 = (R_1 \cos \varphi_1 - R_2 \cos \varphi_2)^2 + (R_1 \sin \varphi_1 - a - R_2 \sin \varphi_2)^2 \tag{2.4.14}$$

and simplifies to

$$d^2 = R_1^2 + R_2^2 - 2R_1R_2 \cos(\varphi_1 - \varphi_2) + a^2 - 2a(R_1 \sin \varphi_1 - R_2 \sin \varphi_2). \tag{2.4.15}$$

Kinetic energy of the system is given by

$$E_k = \frac{1}{2} m_1 R_1^2 \dot{\varphi}_1^2 + \frac{1}{2} m_2 R_2^2 \dot{\varphi}_2^2, \tag{2.4.16}$$

so that the Lagrangian reads

$$\mathcal{L} = E_k - U = \frac{1}{2} m_1 R_1^2 \dot{\varphi}_1^2 + \frac{1}{2} m_2 R_2^2 \dot{\varphi}_2^2 + k \left[R_1 R_2 \cos(\varphi_1 - \varphi_2) + a(R_1 \sin \varphi_1 - R_2 \sin \varphi_2)\right]. \tag{2.4.17}$$

In the case $a = 0$, one can change to the sum and difference angles as dynamic variables:

$$\varphi_+ = \varphi_1 + \varphi_2, \quad \varphi_- = \varphi_1 - \varphi_2. \tag{2.4.18}$$

The Lagrangian depends only on φ_-, thus φ_+ is a cyclic variable and the corresponding generalized momentum is conserved. To find the latter, rewrite the kinetic energy using

$$\varphi_1 = \frac{1}{2}(\varphi_+ + \varphi_-), \quad \varphi_2 = \frac{1}{2}(\varphi_+ - \varphi_-). \tag{2.4.19}$$

One obtains

$$\begin{aligned} \mathcal{L} &= \frac{1}{8} m_1 R_1^2 (\dot{\varphi}_+ + \dot{\varphi}_-)^2 + \frac{1}{8} m_2 R_2^2 (\dot{\varphi}_+ - \dot{\varphi}_-)^2 + k R_1 R_2 \cos \varphi_- \\ &= \frac{1}{8}\left(m_1 R_1^2 + m_2 R_2^2\right)\left(\dot{\varphi}_+^2 + \dot{\varphi}_-^2\right) \\ &\quad + \frac{1}{4}\left(m_1 R_1^2 - m_2 R_2^2\right) \dot{\varphi}_+ \dot{\varphi}_- + k R_1 R_2 \cos \varphi_-. \end{aligned} \tag{2.4.20}$$

The integral of motion is

$$\begin{aligned} p_+ &= \frac{\partial \mathcal{L}}{\partial \dot{\varphi}_+} = \frac{1}{4}\left(m_1 R_1^2 + m_2 R_2^2\right)\dot{\varphi}_+ + \frac{1}{4}\left(m_1 R_1^2 - m_2 R_2^2\right)\dot{\varphi}_- \\ &= \frac{1}{4}\left(m_1 R_1^2 + m_2 R_2^2\right)(\dot{\varphi}_1 + \dot{\varphi}_2) + \frac{1}{4}\left(m_1 R_1^2 - m_2 R_2^2\right)(\dot{\varphi}_1 - \dot{\varphi}_2) \\ &= \frac{1}{2} m_1 R_1^2 \dot{\varphi}_1 + \frac{1}{2} m_2 R_2^2 \dot{\varphi}_2. \end{aligned} \tag{2.4.21}$$

This is nothing else than half the *angular momentum* of the system. This is an expected result because the external torque *with respect to the common center* acting on this system is zero. For $a \neq 0$, there is no common center and the angular momentum is not conserved.

The Lagrange equations have the form

$$\begin{aligned} 0 &= \frac{d}{dt}\frac{\partial \mathcal{L}}{\partial \dot{\varphi}_1} - \frac{\partial \mathcal{L}}{\partial \varphi_1} = m_1 R_1^2 \ddot{\varphi}_1 + k R_1 R_2 \sin(\varphi_1 - \varphi_2) - a R_1 \cos\varphi_1, \\ 0 &= \frac{d}{dt}\frac{\partial \mathcal{L}}{\partial \dot{\varphi}_2} - \frac{\partial \mathcal{L}}{\partial \varphi_2} = m_2 R_2^2 \ddot{\varphi}_2 + k R_1 R_2 \sin(\varphi_2 - \varphi_1) - a R_2 \cos\varphi_2. \end{aligned} \tag{2.4.22}$$

Now, consider small oscillations around the point of the minimal potential energy, i.e., the distance between the masses. In the case $R_1 = R_2 = R$ and $a \leq 2R$, the minimum of the distance corresponds to the intersection of the two circles at

$$\varphi_1 = \varphi_0 = \arcsin\frac{a}{2R}, \quad \varphi_1 = -\varphi_0, \tag{2.4.23}$$

as well as at $\varphi_1 = \pi - \varphi_0$ and $\varphi_1 = \pi + \varphi_0$. We will consider the former case using

$$\varphi_1 = \varphi_0 + \delta\varphi_1, \quad \varphi_2 = -\varphi_0 + \delta\varphi_2 \tag{2.4.24}$$

and expand the Lagrangian

$$\mathcal{L} = \frac{1}{2} m R^2 \left(\dot{\varphi}_1^2 + \dot{\varphi}_2^2\right) + kR\left[R\cos(\varphi_1 - \varphi_2) + a\left(\sin\varphi_1 - \sin\varphi_2\right)\right], \tag{2.4.25}$$

i.e.,

$$\begin{aligned} \mathcal{L} = \frac{1}{2} m R^2 \left(\delta\dot{\varphi}_1^2 + \delta\dot{\varphi}_2^2\right) + kR\left[R\cos(2\varphi_0 + \delta\varphi_1 - \delta\varphi_2)\right. \\ \left. + a\left(\sin(\varphi_0 + \delta\varphi_1) - \sin(-\varphi_0 + \delta\varphi_2)\right)\right] \end{aligned} \tag{2.4.26}$$

in small $\delta\varphi_1$ and $\delta\varphi_2$ up to second order. First, the constant in the potential energy has to be discarded. Second, one has to check that at first order, the result disappears, as it should be at the energy minimum. One has

$$
\begin{aligned}
& R\cos(2\varphi_0+\delta\varphi_1-\delta\varphi_2)+a(\sin(\varphi_0+\delta\varphi_1)+\sin(\varphi_0-\delta\varphi_2)) \\
& \quad\Rightarrow -R\sin(2\varphi_0)(\delta\varphi_1-\delta\varphi_2)+a\cos(\varphi_0)(\delta\varphi_1-\delta\varphi_2) \\
& \quad=\cos(\varphi_0)(\delta\varphi_1-\delta\varphi_2)(-2R\sin\varphi_0+a)=0.
\end{aligned}
\tag{2.4.27}
$$

Now, calculate quadratic terms:

$$
\begin{aligned}
& R\cos(2\varphi_0+\delta\varphi_1-\delta\varphi_2)+a(\sin(\varphi_0+\delta\varphi_1)+\sin(\varphi_0-\delta\varphi_2)) \\
& \quad\Rightarrow -\frac{1}{2}R\cos(2\varphi_0)(\delta\varphi_1-\delta\varphi_2)^2-\frac{1}{2}a\sin(\varphi_0)\left(\delta\varphi_1^2+\delta\varphi_2^2\right) \\
& \quad=-\frac{1}{2}R\left(1-2\sin^2(\varphi_0)\right)(\delta\varphi_1-\delta\varphi_2)^2-\frac{1}{2}a\sin(\varphi_0)\left(\delta\varphi_1^2+\delta\varphi_2^2\right) \\
& \quad=-\frac{1}{2}R\left(1-2\left(\frac{a}{2R}\right)^2\right)(\delta\varphi_1-\delta\varphi_2)^2-\frac{1}{2}a\frac{a}{2R}\left(\delta\varphi_1^2+\delta\varphi_2^2\right) \\
& \quad=-\frac{1}{2}R(\delta\varphi_1-\delta\varphi_2)^2+\frac{a^2}{4R}\left[(\delta\varphi_1-\delta\varphi_2)^2-\delta\varphi_1^2-\delta\varphi_2^2\right] \\
& \quad=-\frac{1}{2}R(\delta\varphi_1-\delta\varphi_2)^2-\frac{a^2}{2R}\delta\varphi_1\delta\varphi_2 \\
& \quad=-\frac{1}{2}R\left(\delta\varphi_1^2+\delta\varphi_2^2\right)+\left(R-\frac{a^2}{2R}\right)\delta\varphi_1\delta\varphi_2.
\end{aligned}
\tag{2.4.28}
$$

Now, to second order in small deviations, one has

$$
\mathcal{L}=\frac{1}{2}mR^2\left(\delta\dot{\varphi}_1^2+\delta\dot{\varphi}_2^2\right)+kR\left[-\frac{1}{2}R\left(\delta\varphi_1^2+\delta\varphi_2^2\right)+\left(R-\frac{a^2}{2R}\right)\delta\varphi_1\delta\varphi_2\right].
\tag{2.4.29}
$$

In matrix form, this reads

$$
\mathcal{L}=\frac{1}{2}\left(\delta\dot{\boldsymbol{\varphi}}^{\boldsymbol{T}}\cdot\mathbb{M}\cdot\delta\dot{\boldsymbol{\varphi}}-\delta\boldsymbol{\varphi}^{\boldsymbol{T}}\cdot\mathbb{K}\cdot\delta\boldsymbol{\varphi}\right),
\tag{2.4.30}
$$

where

$$
\mathbb{M}=mR^2\begin{pmatrix}1 & 0\\ 0 & 1\end{pmatrix},\quad \mathbb{K}=kR^2\begin{pmatrix}1 & -1+\frac{a^2}{2R^2}\\ -1+\frac{a^2}{2R^2} & 1\end{pmatrix}.
\tag{2.4.31}
$$

The eigenvalue equation has the form

$$
\left|-\omega^2\mathbb{M}+\mathbb{K}\right|=0,
\tag{2.4.32}
$$

i.e.,

$$\begin{vmatrix} -\omega^2+\omega_0^2 & \left(-1+\frac{a^2}{2R^2}\right)\omega_0^2 \\ \left(-1+\frac{a^2}{2R^2}\right)\omega_0^2 & -\omega^2+\omega_0^2 \end{vmatrix} = 0, \tag{2.4.33}$$

where $\omega_0^2 = k/m$. Further, one obtains

$$\begin{aligned} 0 &= \left(\omega^2-\omega_0^2\right)^2 - \left(1-\frac{a^2}{2R^2}\right)^2 \omega_0^4 \\ &= \left[\omega^2-\omega_0^2+\left(1-\frac{a^2}{2R^2}\right)\omega_0^2\right]\left[\omega^2-\omega_0^2-\left(1-\frac{a^2}{2R^2}\right)\omega_0^2\right] \\ &= \left(\omega^2-\frac{a^2}{2R^2}\omega_0^2\right)\left(\omega^2-\left(2-\frac{a^2}{2R^2}\right)\omega_0^2\right). \end{aligned} \tag{2.4.34}$$

Thus, the two oscillation modes are as follows:

$$\omega_1 = \frac{a}{\sqrt{2}R}\omega_0, \quad \omega_2 = \sqrt{2}\sqrt{1-\left(\frac{a}{2R}\right)^2}\,\omega_0. \tag{2.4.35}$$

In the limit $a \to 0$, the first mode softens, which is related to the emergence of the integral of motion, the angular momentum. It is obvious that the first mode is the in-phase mode. The second mode softens in the limit $a \to 2R$, where the intersection of the two circles disappears. In the region $a > 2R$, one has to perform a separate calculation that is simpler than above since $\varphi_0 = \pi/2$. There are also two oscillation modes, one of which softens when a approaches $2R$ from above.

2.4.3 *Mass on a plane connected to another mass hanging through a hole*

A particle of mass m_1 sits on a smooth horizontal table and is connected by a light rope of length a to another particle of mass m_2 through a hole, through which the rope is threaded. The second particle hangs straight beneath the hole. Write down the Lagrangian, identify cyclic variables and conserved quantities, and set up equations of motion.

Solution: Use a polar coordinate system for the first particle and denote its distance from the hole as r. The kinetic energy of the system is

$$E_k = \frac{m_1}{2}\left(\dot{r}^2 + r^2\dot{\varphi}^2\right) + \frac{m_2}{2}\dot{r}^2, \tag{2.4.36}$$

while the potential energy is given by

$$U = m_2 gr + \text{const}. \tag{2.4.37}$$

The rope length a is irrelevant. The Lagrangian has the form

$$\mathcal{L} = E_k - U = \frac{m_1}{2}\left(\dot{r}^2 + r^2\dot{\varphi}^2\right) + \frac{m_2}{2}\dot{r}^2 - m_2 gr. \tag{2.4.38}$$

The cyclic variable is φ. The corresponding generalized momentum is angular momentum,

$$p_\varphi = \frac{\partial \mathcal{L}}{\partial \dot{\varphi}} = m_1 r^2 \dot{\varphi} = l = \text{const}. \tag{2.4.39}$$

The Lagrange equation for r has the form

$$0 = \frac{d}{dt}\frac{\partial \mathcal{L}}{\partial \dot{r}} - \frac{\partial \mathcal{L}}{\partial r} = (m_1 + m_2)\,\ddot{r} - m_1 r\dot{\varphi}^2 + m_2 g. \tag{2.4.40}$$

Eliminating $\dot{\varphi}$, one obtains a closed differential equation for r

$$(m_1 + m_2)\,\ddot{r} - \frac{l^2}{m_1 r^3} + m_2 g = 0. \tag{2.4.41}$$

This equation can be integrated by multiplying by the integrating factor $\dot{r}$ and obtaining the energy integral of motion:

$$(m_1 + m_2)\,\ddot{r}\dot{r} - \frac{l^2}{m_1 r^3}\dot{r} + m_2 g\dot{r} = \frac{d}{dt}\left(\frac{m_1 + m_2}{2}\dot{r}^2 + \frac{l^2}{2m_1 r^2} + m_2 gr\right) = 0, \tag{2.4.42}$$

thus

$$\frac{m_1 + m_2}{2}\dot{r}^2 + \frac{l^2}{2m_1 r^2} + m_2 gr = E = \text{const}. \tag{2.4.43}$$

For m_1, this is a motion in a central field with $U \propto r$. The orbits are non-closed. As usual, non-zero angular momentum prevents m_1 from falling onto the hole.

2.4.4 *Pendulum on a wheel*

The pivot of a simple pendulum is attached to a disc of radius R, which rotates in the plane of the pendulum with angular velocity ω (see the following diagram). Write down the Lagrangian and derive the equations of motion for dynamical variable θ.

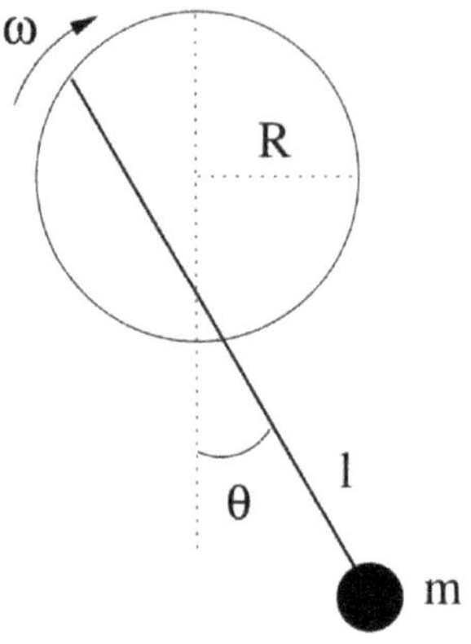

Solution: In terms of θ, the coordinates of the mass m read

$$\begin{aligned} x &= l\sin\theta + R\sin(\omega t), \\ y &= -\, l\cos\theta + R\cos(\omega t) \end{aligned} \tag{2.4.44}$$

(at $t = 0$ the pendulum support is at the top of the circle). Differentiating this, one obtains

$$\begin{aligned} \dot{x} &= l\cos\theta\dot{\theta} + R\omega\cos(\omega t), \\ \dot{y} &= l\sin\theta\dot{\theta} - R\omega\sin(\omega t). \end{aligned} \tag{2.4.45}$$

Now, $E_k = m\left(\dot{x}^2 + \dot{y}^2\right)/2$ and $U = mgy$, so that the Lagrange function becomes

$$\mathcal{L} = \frac{m}{2}\left[l^2\dot{\theta}^2 + 2lR\omega\left(\cos\theta\cos(\omega t) - \sin\theta\sin(\omega t)\right)\dot{\theta}\right] + mgl\cos\theta \tag{2.4.46}$$

or

$$\mathcal{L} = \frac{m}{2}\left[l^2\dot{\theta}^2 + 2lR\omega\cos(\theta + \omega t)\,\dot{\theta}\right] + mgl\cos\theta \tag{2.4.47}$$

up to irrelevant terms. The Lagrange equation has the form

$$0 = \frac{d}{dt}\frac{\partial\mathcal{L}}{\partial\dot{\theta}} - \frac{\partial\mathcal{L}}{\partial\theta} = ml^2\ddot{\theta} - mlR\omega^2\sin(\theta + \omega t) + mgl\sin\theta, \tag{2.4.48}$$

and finally,

$$\ddot{\theta} + \frac{g}{l}\sin\theta = \frac{R}{l}\omega^2 \sin(\theta + \omega t). \tag{2.4.49}$$

This is an equation of motion for a pendulum with an external force depending on the pendulum's angle.

2.4.5 *Double pendulum*

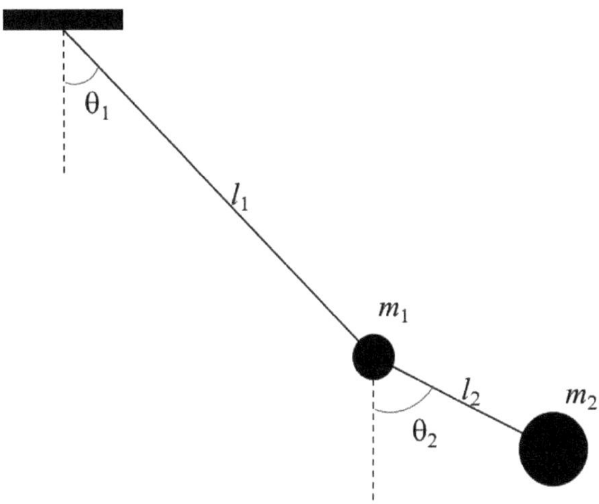

Consider a double pendulum that consists of pendulum 1 attached in a standard way and pendulum 2 attached to the end of pendulum 1. For simplicity, consider the case $m_1 = m_2 = m$ and $l_1 = l_2 = l$.
(a) Find the Lagrange function.
(b) Find the Lagrange equations.
(c) Linearize the Lagrange equations near the equilibrium.
(d) Resolve the resulting system of equations for $\ddot{\theta}_1$ and $\ddot{\theta}_2$.

Solution:
(a) Describing the pendula by the angles θ_1 and θ_2, one writes

$$\begin{aligned} x_1 &= l_1 \sin\theta_1, \quad x_2 = l\sin\theta_1 + l\sin\theta_2, \\ y_1 &= -l\cos\theta_1, \quad y_2 = -l\cos\theta_1 - l\cos\theta_2. \end{aligned} \tag{2.4.50}$$

Potential energy reads

$$U = mg\,(y_1 + y_2) = -mgl\,(2\cos\theta_2 + \cos\theta_1), \tag{2.4.51}$$

and the kinetic energy is given by

$$\begin{aligned} E_k &= \frac{m}{2}\left[\dot{x}_1^2 + \dot{y}_1^2 + \dot{x}_2^2 + \dot{y}_2^2\right] \\ &= \frac{ml^2}{2}\left[\dot{\theta}_1^2 + \left(\cos\theta_1\dot{\theta}_1 + \cos\theta_2\dot{\theta}_2\right)^2 + \left(\sin\theta_1\dot{\theta}_1 + \sin\theta_2\dot{\theta}_2\right)^2\right] \\ &= \frac{ml^2}{2}\left[2\dot{\theta}_1^2 + \dot{\theta}_2^2 + 2\left(\cos\theta_1\cos\theta_2 + \sin\theta_1\sin\theta_2\right)\dot{\theta}_1\dot{\theta}_2\right] \\ &= \frac{ml^2}{2}\left[2\dot{\theta}_1^2 + \dot{\theta}_2^2 + 2\cos\left(\theta_1 - \theta_2\right)\dot{\theta}_1\dot{\theta}_2\right]. \end{aligned} \tag{2.4.52}$$

The Lagrange function reads

$$\mathcal{L} = E_k - U = \frac{ml^2}{2}\left[2\dot{\theta}_1^2 + \dot{\theta}_2^2 + 2\cos\left(\theta_1 - \theta_2\right)\dot{\theta}_1\dot{\theta}_2\right] + mgl\left(2\cos\theta_2 + \cos\theta_1\right). \tag{2.4.53}$$

(b) The Lagrange equations

$$\frac{d}{dt}\frac{\partial\mathcal{L}}{\partial\dot{\theta}_{1,2}} - \frac{\partial\mathcal{L}}{\partial\theta_{1,2}} = 0 \tag{2.4.54}$$

are given by

$$\begin{aligned} 2l\ddot{\theta}_1 + l\frac{d}{dt}\left[\cos\left(\theta_1 - \theta_2\right)\dot{\theta}_2\right] + g\sin\theta_1 &= 0, \\ l\ddot{\theta}_2 + l\frac{d}{dt}\left[\cos\left(\theta_1 - \theta_2\right)\dot{\theta}_1\right] + 2g\sin\theta_2 &= 0, \end{aligned} \tag{2.4.55}$$

and finally,

$$\begin{aligned} 2\ddot{\theta}_1 + \cos\left(\theta_1 - \theta_2\right)\ddot{\theta}_2 - \sin\left(\theta_1 - \theta_2\right)\left(\dot{\theta}_1 - \dot{\theta}_2\right)\dot{\theta}_2 + \omega_0^2\sin\theta_1 &= 0, \\ \ddot{\theta}_2 + \cos\left(\theta_1 - \theta_2\right)\ddot{\theta}_1 - \sin\left(\theta_1 - \theta_2\right)\left(\dot{\theta}_1 - \dot{\theta}_2\right)\dot{\theta}_1 + 2\omega_0^2\sin\theta_2 &= 0, \end{aligned} \tag{2.4.56}$$

where $\omega_0 \equiv \sqrt{g/l}$. These equations are rather complicated and it would be difficult to obtain them within the Newtonian approach. This problem is a classical example of the efficiency of the Lagrangian mechanics.

(c) Near equilibrium ($|\theta_{1,2}| \ll 1$, $|p_{1,2}| \ll 1$), linearization of these equations results in

$$\begin{aligned} 2\ddot{\theta}_1 + \ddot{\theta}_2 + \omega_0^2\theta_1 &= 0, \\ \ddot{\theta}_2 + \ddot{\theta}_1 + 2\omega_0^2\theta_2 &= 0. \end{aligned} \tag{2.4.57}$$

(d) Combining these equations, one obtains

$$\begin{aligned} \ddot{\theta}_1 + \omega_0^2 (\theta_1 - 2\theta_2) &= 0, \\ \ddot{\theta}_2 + \omega_0^2 (-\theta_1 + 4\theta_2) &= 0. \end{aligned} \tag{2.4.58}$$

2.4.6 *Two connected masses on a double incline*

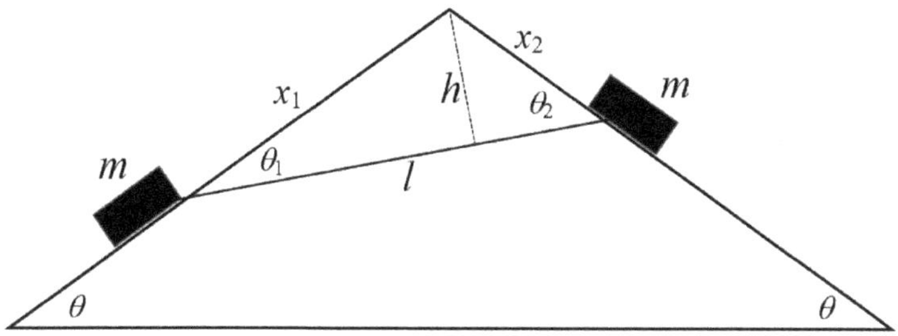

Two masses m on different sides of a double incline with the angle θ are connected by a light rod of length l. Find the Lagrangian, equations of motion and the frequency of small oscillations around the equilibrium configuration. *Tip*: As the dynamic variable, choose the angle θ_1.

Solution: With respect to the top of the incline, potential energy has the form

$$U = -mg\sin\theta\,(x_1 + x_2). \tag{2.4.59}$$

Let us express x_1 and x_2 via the angles θ_1 and θ_2. From the cosine theorem, one obtains

$$l^2 = x_1^2 + x_2^2 - 2x_1x_2\cos(\pi - 2\theta) = x_1^2 + x_2^2 + 2x_1x_2\cos(2\theta). \tag{2.4.60}$$

Expressing the triangle height h in both ways yields another equation

$$x_1\sin\theta_1 = x_2\sin\theta_2. \tag{2.4.61}$$

From here, one finds

$$x_1 = \frac{l\sin\theta_2}{\sqrt{\sin^2\theta_1 + \sin^2\theta_2 + 2\sin\theta_1\sin\theta_2\cos(2\theta)}} \tag{2.4.62}$$

and a similar result for x_2. Taking into account the sum rule

$$\theta_1 + \theta_2 = 2\theta, \tag{2.4.63}$$

one can prove that

$$\sin^2\theta_1 + \sin^2\theta_2 + 2\sin\theta_1\sin\theta_2\cos(2\theta) = \sin^2(2\theta). \tag{2.4.64}$$

After this, the final result becomes

$$x_1 = l\frac{\sin\theta_2}{\sin(2\theta)}, \quad x_2 = l\frac{\sin\theta_1}{\sin(2\theta)}. \tag{2.4.65}$$

Now, the potential energy takes the form

$$U = -mgl\sin\theta\frac{\sin\theta_1 + \sin\theta_2}{\sin(2\theta)}. \tag{2.4.66}$$

Eliminating θ_2, after simplifications, one finally obtains

$$U = -mgl\tan\theta\cos(\theta - \theta_1). \tag{2.4.67}$$

The angle θ_1 changes within the interval $(0, 2\theta)$, and at the boundaries of this interval, one has $U = -mgl\sin\theta$, as it should be. The minimum of U is at $\theta_1 = \theta$, as could be expected on symmetry grounds.

Kinetic energy of the system can be obtained from Eq. (2.4.65):

$$E_k = \frac{m}{2}\left(\dot{x}_1^2 + \dot{x}_2^2\right) = \frac{ml^2}{2}\frac{\left[\cos^2(2\theta - \theta_1) + \cos^2\theta_1\right]\dot{\theta}_1^2}{\sin^2(2\theta)}. \tag{2.4.68}$$

After simplification, one obtains

$$E_k = \frac{ml^2}{2}\frac{[1 + \cos(2\theta)\cos(2\theta - 2\theta_1)]\,\dot{\theta}_1^2}{\sin^2(2\theta)}. \tag{2.4.69}$$

Thus, the Lagrangian has the form

$$\mathcal{L} = E_k - U = \frac{ml^2}{2}\frac{[1 + \cos(2\theta)\cos(2\theta - 2\theta_1)]\,\dot{\theta}_1^2}{\sin^2(2\theta)} + mgl\tan\theta\cos(\theta - \theta_1). \tag{2.4.70}$$

The Lagrange equation

$$\frac{d}{dt}\frac{\partial\mathcal{L}}{\partial\dot{\theta}_1} - \frac{\partial\mathcal{L}}{\partial\theta_1} = 0 \tag{2.4.71}$$

becomes

$$\frac{ml^2}{\sin^2(2\theta)}\left\{[1 + \cos(2\theta)\cos(2\theta - 2\theta_1)]\,\ddot{\theta}_1 + 2\cos(2\theta)\sin(2\theta - 2\theta_1)\,\dot{\theta}_1^2\right\}$$
$$- mgl\tan\theta\sin(\theta - \theta_1) = 0 \tag{2.4.72}$$

or

$$[1 + \cos(2\theta)\cos(2\theta - 2\theta_1)]\,\ddot{\theta}_1 + 2\cos(2\theta)\sin(2\theta - 2\theta_1)\,\dot{\theta}_1^2 \\ - \frac{g}{l}\tan\theta\sin(\theta - \theta_1)\sin^2(2\theta) = 0. \tag{2.4.73}$$

Near equilibrium, one can expand the equation of motion in small $\delta\theta_1$, setting $\theta_1 = \theta + \delta\theta_1$. Keeping only linear terms, one obtains

$$[1 + \cos(2\theta)]\,\delta\ddot{\theta}_1 + \frac{g}{l}\tan\theta\sin^2(2\theta)\,\delta\theta_1 = 0 \tag{2.4.74}$$

or

$$\delta\ddot{\theta}_1 + 2\tan\theta\sin^2\theta\frac{g}{l}\delta\theta_1 = 0. \tag{2.4.75}$$

Thus, the oscillation frequency is given by

$$\omega_0 = \sqrt{2\tan\theta\sin^2\theta\frac{g}{l}}. \tag{2.4.76}$$

For $\theta \ll 1$, the frequency becomes small, $\omega_0 \propto \theta^{3/2}$.

Chapter 3

Oscillations

Basic harmonic oscillator with damping and an external force was considered at the beginning of this book (see Sec. 1.1.3 of chapter 1). Here, oscillating systems with one degree of freedom will be reviewed in more depth, including harmonic oscillator with a periodic force, parametric resonance, and motion in a rapidly oscillating field. Then small oscillations in general systems with many degrees of freedom will be considered.

3.1 Forced harmonic oscillator

Dynamics of a harmonic oscillator with an arbitrary force is defined by Eq. (1.1.46). Let us consider the important case of a sinusoidal force

$$f(t) = f_0 \sin \omega t, \tag{3.1.1}$$

applied starting at $t = 0$. To compute $x_{\text{forced}}(t)$, it is convenient to convert everything into the exponential form using $\sin \omega t = \left(e^{i\omega t} - e^{-i\omega t}\right)/(2i)$, after which integration simplifies and one obtains

$$x_{\text{forced}}(t) = \frac{f_0}{4\tilde{\omega}_0} \left[-\frac{e^{i\omega t} - e^{(-\Gamma + i\tilde{\omega}_0)t}}{\Gamma - i\tilde{\omega}_0 + i\omega} + \frac{e^{i\omega t} - e^{(-\Gamma - i\tilde{\omega}_0)t}}{\Gamma + i\tilde{\omega}_0 + i\omega} \right] + \text{c.c.}, \tag{3.1.2}$$

where c.c. is complex conjugate. The first term in this expression is the so-called resonant term in which the denominator becomes small for ω close to ω_0. The other term is the non-resonant term, which differs from the first one by replacement, $\tilde{\omega}_0 \Rightarrow -\tilde{\omega}_0$. It is sufficient to compute one of these terms, and then the other one can be easily obtained from the first one.

The explicitly real form of $x_{\text{forced}}(t)$ is

$$x_{\text{forced}}(t) = -\frac{f_0}{2\tilde{\omega}_0} \times \frac{\Gamma\left[\cos(\omega t) - \cos(\tilde{\omega}_0 t)\, e^{-\Gamma t}\right] - (\tilde{\omega}_0 - \omega)\left[\sin(\omega t) - \sin(\tilde{\omega}_0 t)\, e^{-\Gamma t}\right]}{(\omega - \tilde{\omega}_0)^2 + \Gamma^2} + \cdots, \tag{3.1.3}$$

where the dots stand for the non-resonant terms. At the times longer than the relaxation time of the oscillator $\tau \equiv 1/\Gamma$, the terms in the above formula which are oscillating at the oscillator's own frequency $\tilde{\omega}_0$ die out and only the forced terms oscillating at the frequency ω remain:

$$x_{\text{forced}}(t) = \frac{f_0}{2\tilde{\omega}_0}\left[-\frac{\Gamma\cos(\omega t) + (\omega - \tilde{\omega}_0)\sin(\omega t)}{(\omega - \tilde{\omega}_0)^2 + \Gamma^2} + \frac{\Gamma\cos(\omega t) + (\omega + \tilde{\omega}_0)\sin(\omega t)}{(\omega + \tilde{\omega}_0)^2 + \Gamma^2}\right]. \tag{3.1.4}$$

Here, the terms with $\cos(\omega t)$ are shifted by quarter of the period with respect to the harmonic force. Exactly at resonance, $\omega = \tilde{\omega}_0$, only this term in the resonant part of the expression becomes dominant and reaches its maximum. Near the resonance, $|\omega - \tilde{\omega}_0| \sim \Gamma \ll \omega_0$, the resonant term is large, and the much smaller non-resonant term can be neglected. Still keeping the non-resonant term, one can add the two terms in the above equation to obtain

$$x_{\text{forced}}(t) = f_0 \frac{-2\omega\Gamma\cos(\omega t) + (-\omega^2 + \tilde{\omega}_0^2 + \Gamma^2)\sin(\omega t)}{[(\omega - \tilde{\omega}_0)^2 + \Gamma^2][(\omega + \tilde{\omega}_0)^2 + \Gamma^2]}. \tag{3.1.5}$$

Using $\tilde{\omega}_0^2 \equiv \omega_0^2 - \Gamma^2$, one can further simplify this as

$$x_{\text{forced}}(t) = -f_0 \frac{2\Gamma\omega\cos\omega t + (\omega^2 - \omega_0^2)\sin\omega t}{(\omega^2 - \omega_0^2)^2 + 4\Gamma^2\omega^2}. \tag{3.1.6}$$

This solution contains both resonant and non-resonant terms. Near the resonance, one can write

$$\omega^2 - \omega_0^2 = (\omega - \omega_0)(\omega + \omega_0) \cong 2\omega_0(\omega - \omega_0), \tag{3.1.7}$$

and thus,

$$x(t) \cong -\frac{f_0}{2\omega_0}\frac{\Gamma\cos\omega t + (\omega - \omega_0)\sin\omega t}{(\omega - \omega_0)^2 + \Gamma^2}, \tag{3.1.8}$$

which practically coincides with the resonant term in Eq. (3.1.4) as $\Gamma \ll \omega_0$ and $\tilde{\omega}_0 \cong \omega_0$.

The power absorbed by the oscillator can be calculated via the work of the external force done on the oscillator:

$$P_{\text{abs}} = \frac{1}{T}\int_0^T dt\frac{\delta A}{dt} = \frac{1}{T}\int_0^T dtF(t)\dot{x}(t), \tag{3.1.9}$$

where $T \equiv 2\pi/\omega$. In the stationary state near the resonance, $x(t)$ is given by the first term of Eq. (3.1.4) and $F(t) = mf_0 \sin\omega t$. One obtains

$$P_{\text{abs}} = \frac{1}{T}\int_0^T dt\,\frac{mf_0^2}{2\omega_0}\frac{\Gamma}{(\omega-\omega_0)^2+\Gamma^2}\omega\sin^2\omega t \cong \frac{mf_0^2}{4}\frac{\Gamma}{(\omega-\omega_0)^2+\Gamma^2}. \tag{3.1.10}$$

Here, approximation $\omega \cong \tilde{\omega}_0 \cong \omega_0$ was used, which is mandatory since the whole approach ignoring the non-resonant term is valid near the resonance only under the condition that the resonance is narrow, $\Gamma \ll \omega_0$.

The stationary (settled) state of the oscillator under the influence of a harmonic force can be obtained in a much easier way by just searching for the solution in the form

$$x(t) = A\sin\omega t + B\cos\omega t. \tag{3.1.11}$$

Inserting it into Eq. (1.1.25), using Eq. (3.1.1), and then equating the coefficients in front of $\sin\omega t$ and $\cos\omega t$, one obtains the system of linear equations

$$\begin{aligned} -A\omega^2 - 2\Gamma B\omega + A\omega_0^2 &= f_0, \\ -B\omega^2 + 2\Gamma A\omega + B\omega_0^2 &= 0. \end{aligned} \tag{3.1.12}$$

The solution is

$$B = -f_0\frac{2\Gamma\omega}{(\omega^2-\omega_0^2)^2+4\Gamma^2\omega^2}, \qquad A = -f_0\frac{\omega^2-\omega_0^2}{(\omega^2-\omega_0^2)^2+4\Gamma^2\omega^2}, \tag{3.1.13}$$

which results in Eq. (3.1.6).

3.2 Parametric resonance

Consider a harmonic oscillator with time-dependent frequency

$$\ddot{x} + \omega_0^2(t)x = 0. \tag{3.2.1}$$

Let the latter be

$$\omega_0(t) = \omega_0[1 + \alpha \cos \omega t], \qquad \alpha \ll 1. \tag{3.2.2}$$

As we shall see, if ω is close to $2\omega_0$, the amplitude of oscillations will exponentially increase with time. This phenomenon is called *parametric resonance* since a parameter of the problem, in this case, the oscillator's frequency, is changing. It is convenient to set

$$\omega = 2\omega_0 + \epsilon, \tag{3.2.3}$$

with a small resonance detuning ϵ. Since $\alpha \ll 1$, one can simplify the equation of motion to

$$\ddot{x} + \omega_0^2 (1 + 2\alpha \cos \omega t) x = 0, \tag{3.2.4}$$

dropping the term $\sim \alpha^2$. The solution of this equation can be searched for in the form

$$x(t) = a(t) \cos\left(\frac{\omega}{2}t\right) + b(t) \sin\left(\frac{\omega}{2}t\right). \tag{3.2.5}$$

The functions $a(t)$ and $b(t)$ can be expected to change slowly with time for small α and ϵ, thus their second derivatives can be dropped. The derivatives of x are given by

$$\dot{x} = \dot{a} \cos\left(\frac{\omega}{2}t\right) + \dot{b} \sin\left(\frac{\omega}{2}t\right) - a\frac{\omega}{2} \sin\left(\frac{\omega}{2}t\right) + b\frac{\omega}{2} \cos\left(\frac{\omega}{2}t\right) \tag{3.2.6}$$

and

$$\begin{aligned} \ddot{x} \cong & -2\dot{a}\frac{\omega}{2} \sin\left(\frac{\omega}{2}t\right) + 2\dot{b}\frac{\omega}{2} \cos\left(\frac{\omega}{2}t\right) - a\left(\frac{\omega}{2}\right)^2 \cos\left(\frac{\omega}{2}t\right) \\ & - b\left(\frac{\omega}{2}\right)^2 \sin\left(\frac{\omega}{2}t\right). \end{aligned} \tag{3.2.7}$$

Writing

$$\left(\frac{\omega}{2}\right)^2 = \left(\omega_0 + \frac{\epsilon}{2}\right)^2 \cong \omega_0^2 + \epsilon\omega_0 \tag{3.2.8}$$

(dropping terms of order ϵ^2) and replacing $\omega/2 \Rightarrow \omega_0$ in the terms with derivatives of the slow variables, one obtains

$$\begin{aligned} \ddot{x} \cong & -\omega_0^2 x - a\omega_0\epsilon \cos\left(\frac{\omega}{2}t\right) - b\omega_0\epsilon \sin\left(\frac{\omega}{2}t\right) - 2\dot{a}\omega_0 \sin\left(\frac{\omega}{2}t\right) \\ & + 2\dot{b}\omega_0 \cos\left(\frac{\omega}{2}t\right). \end{aligned}$$

Products of trigonometric functions in Eq. (3.2.4) should be reduced to single trigonometric functions with combined arguments:

$$2\cos(\omega t)\cos\left(\frac{\omega}{2}t\right) = \cos\left(\frac{\omega}{2}t\right) + \cos\left(\frac{3\omega}{2}t\right),$$
$$2\cos(\omega t)\sin\left(\frac{\omega}{2}t\right) = -\sin\left(\frac{\omega}{2}t\right) + \sin\left(\frac{3\omega}{2}t\right). \tag{3.2.9}$$

Here, functions with triple arguments should be dropped as it can be shown that these terms are of a higher order in small α. After this, Eq. (3.2.4) takes the form

$$\left(-a\epsilon + 2\dot{b} + \alpha\omega_0 a\right)\omega_0\cos\left(\frac{\omega}{2}t\right) + \left(-b\epsilon - 2\dot{a} - \alpha\omega_0 b\right)\omega_0\sin\left(\frac{\omega}{2}t\right) = 0. \tag{3.2.10}$$

This equation is satisfied if and only if the coefficients in front of the sin and cos functions are zero:

$$2\dot{a} + (\alpha\omega_0 + \epsilon)\,b = 0,$$
$$(\alpha\omega_0 - \epsilon)\,a + 2\dot{b} = 0. \tag{3.2.11}$$

The solution of this system of equations can be searched for in the form

$$a(t) = a_0 e^{\lambda t}, \quad b(t) = b_0 e^{\lambda t}. \tag{3.2.12}$$

The increment of the parametric resonance λ satisfies the characteristic equation

$$\begin{vmatrix} 2\lambda & \alpha\omega_0 + \epsilon \\ \alpha\omega_0 - \epsilon & 2\lambda \end{vmatrix} = (2\lambda)^2 - (\alpha\omega_0)^2 + \epsilon^2 = 0. \tag{3.2.13}$$

Thus,

$$\lambda = \pm\frac{1}{2}\sqrt{(\alpha\omega_0)^2 - \epsilon^2}. \tag{3.2.14}$$

In the frequency range near the resonance

$$|\epsilon| \equiv |2\omega_0 - \omega| < \alpha\omega_0, \tag{3.2.15}$$

see Eq. (3.2.3), λ takes positive and negative real values. The root $\lambda > 0$ corresponds to the parametric instability that manifests itself in the exponential increase of the amplitude of oscillations. Note that inside the window of parametric instability, the oscillator locks into the frequency $\omega/2$, which differs from its own frequency ω_0. Outside the instability window, λ is imaginary:

$$\lambda = \mathrm{i}\Omega, \quad \Omega = \frac{1}{2}\sqrt{\epsilon^2 - (\alpha\omega_0)^2}. \tag{3.2.16}$$

In this case, the solution $x(t)$ contains the terms such as

$$\sin\left[\left(\omega_0 + \frac{\epsilon}{2} - \Omega\right)t\right] \quad \text{and} \quad \cos\left[\left(\omega_0 + \frac{\epsilon}{2} - \Omega\right)t\right], \tag{3.2.17}$$

i.e., the oscillation frequency is neither ω_0 nor $\omega/2$. One can see that in the limit $\alpha \to 0$, the oscillator oscillates at its own frequency ω_0.

If the oscillator is damped with damping constant Γ, then the straightforward generalization of the consideration above leads to

$$a(t) = a_0 e^{(\lambda-\Gamma)t}, \quad b(t) = b_0 e^{(\lambda-\Gamma)t}, \tag{3.2.18}$$

where λ is the same as above. Then the condition of the parametric instability has the form

$$\lambda^* \equiv \lambda_+ - \Gamma = \frac{1}{2}\sqrt{(\alpha\omega_0)^2 - \epsilon^2} - \Gamma > 0. \tag{3.2.19}$$

The parametric-resonance window now becomes

$$\left|\omega_0 - \frac{\omega}{2}\right| < \sqrt{\left(\frac{1}{2}\alpha\omega_0\right)^2 - \Gamma^2}, \tag{3.2.20}$$

i.e., narrower than the undamped result of Eq. (3.2.15). The resonance window disappears at

$$\alpha\omega_0 = 2\Gamma, \tag{3.2.21}$$

which defines the threshold of the parametric resonance on the amplitude of the pumping α. This condition provides a method of measuring the damping constant Γ.

A physical example of parametric resonance is a pendulum with its length l oscillating in time. A person sitting on a swing and shifting his/her center of mass up and down with the double frequency is doing exactly this. Near the vertical position ($\varphi = 0$), l is being decreased by Δl, while near turning points ($\varphi = \pm\varphi_{\max}$) l is being increased by Δl. Working against gravity, the person is injecting the energy $\Delta E = 2mg\Delta l\,(1 - \cos\varphi_{\max})$ during the period of motion. The greater the amplitude $\varphi_{\max}$, the higher is the energy input. This is why the amplitude increases exponentially.

In spite of the exponential increase of the amplitude, overall the parametric resonance is a weaker phenomenon than a regular resonance caused by an applied sinusoidal force. With comparable forcings, it takes a longer time to develop a given amplitude under conditions of parametric resonance than regular resonance.

3.3 Harmonic oscillator via complex variables

The energy of the one-dimensional forced harmonic oscillator expressed via its momentum and coordinate (i.e., its Hamiltonian function, see Chapter V) has the form

$$\mathcal{H} = \frac{p^2}{2m} + \frac{kx^2}{2} - Fx = \frac{p^2}{2m} + \frac{m\omega_0^2 x^2}{2} - Fx, \tag{3.3.1}$$

where $\omega_0 = \sqrt{k/m}$ is the resonance frequency and F is the force. The Hamiltonian equations of motion for the oscillator are

$$\dot{p} = -\frac{\partial \mathcal{H}}{\partial x} = -kx + F, \quad \dot{x} = \frac{\partial \mathcal{H}}{\partial p} = \frac{p}{m}. \tag{3.3.2}$$

At this stage, it is convenient to include the damping Γ in the first of these equations, which becomes $\dot{p} = -kx + F - 2m\Gamma\dot{x}$. Differentiating the second equation over time and using the first equation, one obtains

$$\ddot{x} = \frac{\dot{p}}{m} = -\frac{kx}{m} + \frac{F}{m} - 2\Gamma\dot{x} = -\omega_0^2 x - 2\Gamma\dot{x} + f, \tag{3.3.3}$$

which reproduces the Newtonian equation of motion for the damped harmonic oscillator (1.1.25).

Let us introduce the complex variable

$$a \equiv \frac{1}{2}\left(\frac{x}{x_0} + i\frac{p}{p_0}\right), \quad x_0 \equiv \sqrt{\frac{\hbar}{2m\omega_0}}, \quad p_0 \equiv \sqrt{\frac{\hbar m\omega_0}{2}}, \tag{3.3.4}$$

where the Planck constant $\hbar$ having the unit of J · s was included to make a dimensionless. The coordinate and momentum can be expressed via a as

$$x = (a + a^*)\, x_0, \quad p = -i\,(a - a^*)\, p_0 = 2\mathrm{Im} a p_0, \tag{3.3.5}$$

whereas the energy in terms of a has the form

$$\mathcal{H} = \hbar\omega_0 a a^* - f\sqrt{\frac{\hbar m}{2\omega_0}}\,(a + a^*), \tag{3.3.6}$$

where a^* is complex conjugate of a. The time derivative of a is given by

$$\dot{a} = \frac{1}{2}\left(\frac{\dot{x}}{x_0} + i\frac{\dot{p}}{p_0}\right) = -i\omega_0 a - \Gamma\,(a - a^*) + if\sqrt{\frac{m}{2\hbar\omega_0}}, \tag{3.3.7}$$

where we used $\dot{p} = -kx + F - 2\Gamma p$. One can see that in the absence of damping, this equation of motion can be obtained as $i\hbar\dot{a} = \partial\mathcal{H}/\partial a^*$. For small damping, $\Gamma \ll \omega_0$, one can keep only the resonant term $-\Gamma a$ and drop the non-resonant term Γa^*. The resulting equation for the complex variable a reads

$$\dot{a} + (i\omega_0 + \Gamma)\, a = if\sqrt{\frac{m}{2\hbar\omega_0}}. \tag{3.3.8}$$

Beyond the resonance approximation, one should keep the term Γa^* in Eq. (3.3.7) and add the equation of motion for a^*, which is the complex conjugate of Eq. (3.3.7).

The advantage of writing the equation of motion for the harmonic oscillator in terms of the complex variable a is that in the resonance approximation, one has a first-order differential equation instead of the second-order one. The forced solution can be immediately written as

$$a_{\text{forced}}(t) = i\sqrt{\frac{m}{2\hbar\omega_0}} \int_0^t dt' e^{(i\omega_0+\Gamma)(t-t')} f(t'), \tag{3.3.9}$$

which yields $x(t)$ with the help of Eq. (3.3.5) in the form

$$x_{\text{forced}}(t) = 2\text{Re}ax_0 = \text{Re}\left[\frac{i}{\omega_0}\int_0^t dt' e^{(i\omega_0+\Gamma)(t-t')} f(t')\right]. \tag{3.3.10}$$

Here, as it should be, the Planck constant is canceled.

3.4 Parametric resonance via complex variables

Using complex variables for the harmonic oscillator allows a more elegant and natural description of the parametric resonance than the straightforward method of Sec. 3.2.[1] Consider a harmonic oscillator with a time-dependent frequency

$$\omega_0^2(t) = \omega_0^2\left[1 + 2\mu(t)\right], \tag{3.4.1}$$

with $\mu(t) \ll 1$. The additional term in the energy, $\delta\mathcal{H} = m\omega_0^2 x^2 \mu(t)$, with the help of Eq. (3.3.5), transforms into

$$\delta\mathcal{H} = \frac{1}{2}\hbar\omega_0\mu(t)\,(a + a^*)^2. \tag{3.4.2}$$

This generates the additional term in the equation of motion for a which can be obtained in the easiest way from the generic equation $i\hbar\dot{a} = \partial\mathcal{H}/\partial a^*$.

[1] V. L. Safonov, M. E. McConney, and M. R. Page, *Physical Review E*, **100**, 040201(R) (2019).

Instead of Eq. (3.3.8), one obtains

$$\dot{a} + (i\omega_0 + \Gamma)\, a = -i\omega_0 \mu(t)(a + a^*). \tag{3.4.3}$$

As we shall see, parametric resonance is excited by the term with a^* in the rhs of this equation. The term with a is "innocent" and can be eliminated by the variable transformation $a \Rightarrow b$, where

$$a(t) = b(t) \exp\left[-i\omega_0 \int_0^t dt' \mu(t')\right]. \tag{3.4.4}$$

One obtains the equation

$$\dot{b} + (i\omega_0 + \Gamma)\, b = -i\omega_0 \tilde{\mu}(t) b^*, \tag{3.4.5}$$

with

$$\tilde{\mu}(t) \equiv \mu(t) \exp\left[2i\omega_0 \int_0^t dt' \mu(t')\right]. \tag{3.4.6}$$

If $\mu(t)$ is a harmonic function, e.g., $\mu(t) = \alpha \cos \omega t$, then $\tilde{\mu}(t)$ contains multiples of ω because of the exponential term in the above formula. One can expand $\tilde{\mu}(t)$ in a Fourier series

$$\tilde{\mu}(t) = \sum_{n=-\infty}^{\infty} \tilde{\mu}_n e^{in\omega t}. \tag{3.4.7}$$

We will see that parametric resonance can occur if one of the harmonics has a frequency close to the double oscillator frequency, $n\omega \cong 2\omega_0$. Keeping only this harmonic, one can make another variable transformation

$$b(t) = c(t) e^{-in\omega t/2}, \tag{3.4.8}$$

which results in the equation

$$\dot{c} + \left[i\left(\omega_0 - \frac{n\omega}{2}\right) + \Gamma\right] c = -i\omega_0 \tilde{\mu}_n c^*. \tag{3.4.9}$$

Here, c is a slow variable capturing the gradual excitation of the parametric resonance, whereas other terms of the Fourier series in Eq. (3.4.7), discarded here, are rapidly oscillating and average to zero. The resonance at the double oscillator frequency ω_0 arises because of the term b^* in Eq. (3.4.5), which has the phase changing in the opposite direction with respect to that of b. To solve Eq. (3.4.9), one needs to add the equation of motion for

c^* as both equations are coupled. Searching for the solution in the form $c(t) \propto e^{\lambda t}$, one obtains the equation for the increment λ:

$$\begin{vmatrix} i\left(\omega_0 - \frac{n\omega}{2}\right) + \Gamma + \lambda & i\omega_0\tilde{\mu}_n \\ -i\omega_0\tilde{\mu}_n^* & -i\left(\omega_0 - \frac{n\omega}{2}\right) + \Gamma + \lambda \end{vmatrix} = 0. \tag{3.4.10}$$

The largest of the two solutions reads

$$\lambda = -\Gamma + \sqrt{\omega_0^2 \left|\tilde{\mu}_n\right|^2 - \left(\omega_0 - \frac{n\omega}{2}\right)^2}. \tag{3.4.11}$$

The condition for the parametric resonance is $\lambda > 0$. The corresponding resonance window is defined by

$$\left|\omega_0 - \frac{n\omega}{2}\right| < \sqrt{\omega_0^2 \left|\tilde{\mu}_n\right|^2 - \Gamma^2}. \tag{3.4.12}$$

Exactly at resonance, the parametric instability requires $\omega_0 \left|\tilde{\mu}_n\right| > \Gamma$.

For $\mu(t) = \alpha \cos\omega t$, in Eq. (3.4.6), one has $\int_0^t dt'\mu(t') = (\alpha/\omega)\sin\omega t$, and the coefficients of the Fourier series in Eq. (3.4.7) can be computed by expanding the exponential in Eq. (3.4.6) in powers of $\mu(t) \ll 1$. This yields

$$\tilde{\mu}_n \cong \frac{\alpha}{2\,(n-1)!}\left(\frac{\alpha\omega_0}{\omega}\right)^{n-1}. \tag{3.4.13}$$

For the main resonance, one obtains $\tilde{\mu}_1 = \alpha/2$, which amounts to discarding the exponential in $\tilde{\mu}(t)$ and reproduces the results for the parametric resonance obtained by the straightforward method, Eq. (3.2.20).

3.5 Motion in a fast oscillating field

If a fast oscillating force is applied to a system, the resulting motion is a small amplitude fast oscillating motion as a response to the fast oscillating force plus a regular motion. The former can be integrated out, leading to the averaged regular motion with an additional potential energy due to the oscillating force. As the basic model, consider the one-dimensional motion described by the Newtonian equation

$$m\ddot{x} = -\frac{\partial U}{\partial x} + f(x,t), \quad f(x,t) = f_0(x)\cos(\omega t + \varphi_0). \tag{3.5.1}$$

In accordance with the comment above, one can search for the solution in the form

$$x(t) = X(t) + \xi(t), \tag{3.5.2}$$

where, ξ is the fast-oscillating response to f, while X describes the averaged motion having a regular time dependence plus smaller and thus irrelevant fast oscillating terms. Due to large ω (in comparison to the frequency of the regular motion), the amplitude of ξ is small, which will be confirmed by the calculation. Thus, one can consider ξ perturbatively and expand the equation up to the first order in ξ:

$$m\ddot{X} + m\ddot{\xi} = -\frac{\partial U(X)}{\partial X} - \xi\frac{\partial^2 U(X)}{\partial X^2} + f(X,t) + \xi\frac{\partial f(X,t)}{\partial X}. \tag{3.5.3}$$

Splitting the dominant fast oscillating terms, one obtains the equation that is, in fact, the definition of ξ:

$$m\ddot{\xi} = f(X,t). \tag{3.5.4}$$

Using Eq. (3.5.1), this equation can be immediately integrated to yield

$$\xi = -\frac{f(X,t)}{m\omega^2}. \tag{3.5.5}$$

Thus, ξ is, indeed, small. The remaining terms in Eq. (3.5.3) are

$$m\ddot{X} = -\frac{\partial U(X)}{\partial X} - \xi\frac{\partial^2 U(X)}{\partial X^2} + \xi\frac{\partial f(X,t)}{\partial X}. \tag{3.5.6}$$

Here, the second term in the rhs gives rise to a small fast-oscillating contribution to X that has been discarded in Eq. (3.5.4). We are interested, however, in the averaged value $\langle X\rangle$ that will still be denoted as X for the sake of brevity. Averaging over fast oscillations eliminates the second term in the rhs of the equation above. However, the third term yields a non-vanishing contribution, so that

$$m\ddot{X} = -\frac{\partial U(X)}{\partial X} + \left\langle \xi\frac{\partial f(X,t)}{\partial X}\right\rangle_t = -\frac{\partial U(X)}{\partial X} - \left\langle \frac{f(X,t)}{m\omega^2}\frac{\partial f(X,t)}{\partial X}\right\rangle_t. \tag{3.5.7}$$

Finally, the averaged equation of motion can be represented in the form

$$m\ddot{X} = -\frac{\partial U_{\text{eff}}}{\partial X}, \tag{3.5.8}$$

where the effective potential energy reads

$$U_{\text{eff}}(X) = U(X) + \frac{\langle f^2(X,t)\rangle_t}{2m\omega^2} = U(X) + \frac{f_0^2(X)}{4m\omega^2}. \tag{3.5.9}$$

To elucidate the physical meaning of this result, one can rewrite it in terms of ξ using Eqs. (3.5.4) and (3.5.5):

$$U_{\text{eff}}(X) = U(X) - \frac{m}{2}\langle \xi\ddot{\xi}\rangle_t = U(X) + \frac{m}{2}\langle \dot{\xi}^2\rangle_t. \tag{3.5.10}$$

Thus, the additional effective potential energy is the kinetic energy of fast oscillation that depends on X.

As an example, consider a pendulum with the support point oscillating in the vertical direction with amplitude a and a high frequency γ. With

$$x = l\sin\varphi, \quad y = a\sin(\gamma t) - l\cos\varphi, \tag{3.5.11}$$

and $U = mgy$, the Lagrange function becomes

$$\mathcal{L} = \frac{m}{2}\left(\dot{x}^2 + \dot{y}^2\right) - U = \frac{m}{2}\left(l^2\dot{\varphi}^2 - 2a\gamma l\dot{\varphi}\cos(\gamma t)\sin\varphi\right) + mgl\cos\varphi, \tag{3.5.12}$$

where irrelevant terms have been dropped. Lagrange equation has the form

$$\frac{d}{dt}\frac{\partial\mathcal{L}}{\partial\dot{\varphi}} - \frac{\partial\mathcal{L}}{\partial\varphi} = ml^2\ddot{\varphi} + mal\gamma^2\sin(\gamma t)\sin\varphi + mgl\sin\varphi = 0. \tag{3.5.13}$$

This equation can be put into the Newtonian form

$$ml^2\ddot{\varphi} = -\frac{\partial}{\partial\varphi}(-mgl\cos\varphi) + f(\varphi, t), \quad f(\varphi, t) \equiv -mal\gamma^2\sin(\gamma t)\sin\varphi, \tag{3.5.14}$$

similar to Eq. (3.5.1) with $m \Rightarrow ml^2$ and $x \Rightarrow \varphi$. Thus, the effective potential energy following from Eq. (3.5.9) becomes

$$U_{\text{eff}}(\varphi) = -mgl\cos\varphi + \frac{\left(mal\gamma^2\sin\varphi\right)^2}{4ml^2\gamma^2} = mgl\left(-\cos\varphi + \frac{a^2\gamma^2}{4gl}\sin^2\varphi\right). \tag{3.5.15}$$

Additional potential energy can stabilize the vertical up position of the pendulum, $\varphi = \pi$. Setting $\varphi = \pi + \varphi'$ with $|\varphi'| \ll 1$ and expanding in φ', one obtains

$$U_{\text{eff}}(\varphi') = mgl\left(\cos\varphi' + \frac{a^2\gamma^2}{4gl}\sin^2\varphi'\right) \cong mgl\left(1 - \frac{\varphi'^2}{2} + \frac{a^2\gamma^2}{4gl}\varphi'^2\right). \tag{3.5.16}$$

One can see that this state is stable if

$$a^2\gamma^2 > 2gl. \tag{3.5.17}$$

3.6 Small oscillations: Many degrees of freedom

Consider a dynamical system with N degrees of freedom near a minimum of the potential energy. The Lagrangian has a general form

$$\mathcal{L} = \frac{1}{2}\sum_{ij}(m_{ij}\dot{x}_i\dot{x}_j - k_{ij}x_i x_j), \tag{3.6.1}$$

where $x_i \equiv q_i - q_i(0)$ are deviations from the minimum, while the tensor of masses and tensor of spring coefficients are symmetric: $m_{ij} = m_{ji}$ and $k_{ij} = k_{ji}$. Note that because of the tensor masses, Newtonian mechanics is not applicable here. The Lagrangian can be written in the matrix form as

$$\mathcal{L} = \frac{1}{2}(\dot{\mathbf{X}}^T \cdot \mathbb{M} \cdot \dot{\mathbf{X}} - \mathbf{X}^T \cdot \mathbb{K} \cdot \mathbf{X}), \tag{3.6.2}$$

where $\mathbb{M}$ is the mass matrix, $\mathbb{K}$ is the stiffness matrix, $\mathbf{X} = (x_1, x_2, \ldots, x_N)$ (a column) and $\mathbf{X}^T$ its transposition (a row). The Lagrange equations read

$$\sum_j (m_{ij}\ddot{x}_j + k_{ij}x_j) = 0, \quad i = 1, \ldots, N \tag{3.6.3}$$

or, in the matrix form,

$$\mathbb{M} \cdot \ddot{\mathbf{X}} + \mathbb{K} \cdot \mathbf{X} = 0. \tag{3.6.4}$$

We search for the solution in the form

$$\mathbf{X} = \mathbf{a}\sin(\omega t) \tag{3.6.5}$$

(or the same with $\cos(\omega t)$) and obtain the system of linear algebraic equations

$$\left(-\omega^2\mathbb{M} + \mathbb{K}\right) \cdot \mathbf{a} = 0. \tag{3.6.6}$$

This system of equations has a non-trivial solution for $\mathbf{a}$ only if

$$\left|-\omega^2\mathbb{M} + \mathbb{K}\right| = 0, \tag{3.6.7}$$

which defines N eigenfrequencies ω_α^2, $\alpha = 1, \ldots, N$. One can prove that ω_α^2 are positive if the potential energy is a positively defined bilinear form, as it is the case for the energy minimum. Then the vectors a_α are real. We will assume the simplest case of ω_α^2 non-degenerate, $\omega_\alpha^2 \neq \omega_\beta^2$ for $\alpha \neq \beta$. Technically, Eq. (3.6.6) is a generalized eigenvalue problem:

$$\mathbb{K} \cdot \mathbf{a} = \lambda\mathbb{M} \cdot \mathbf{a}, \tag{3.6.8}$$

and $\mathbf{a}$ is the eigenvector of $\mathbb{K}$ with respect to $\mathbb{M}$. The algorithm for solving this problem is implemented in Wolfram Mathematica.

The problem we are solving resembles an eigenvalue problem. It is more complicated, however, as we have two matrices instead of one. Whereas in eigenvalue problems, eigenvectors corresponding to different eigenvalues are orthogonal, here a generalized orthogonality relation takes place. To obtain it, write Eq. (3.6.6) with two different eigenvalues ω_α^2 and ω_β^2:

$$\begin{aligned}\omega_\alpha^2 \mathbb{M} \cdot \mathbf{a}_\alpha &= \mathbb{K} \cdot \mathbf{a}_\alpha, \\ \omega_\beta^2 \mathbb{M} \cdot \mathbf{a}_\beta &= \mathbb{K} \cdot \mathbf{a}_\beta. \end{aligned} \tag{3.6.9}$$

Now, multiply the first equation by $\mathbf{a}_\beta^T$ from the left, multiply the second equation by $\mathbf{a}_\alpha^T$ from the left, and subtract them from each other:

$$\omega_\alpha^2 \mathbf{a}_\beta^T \cdot \mathbb{M} \cdot \mathbf{a}_\alpha - \omega_\beta^2 \mathbf{a}_\alpha^T \cdot \mathbb{M} \cdot \mathbf{a}_\beta = \mathbf{a}_\beta^T \cdot \mathbb{K} \cdot \mathbf{a}_\alpha - \mathbf{a}_\alpha^T \cdot \mathbb{K} \cdot \mathbf{a}_\beta. \tag{3.6.10}$$

Using the symmetry of matrices $\mathbb{M}$ and $\mathbb{K}$, one can easily show $\mathbf{a}_\beta^T \cdot \mathbb{M} \cdot \mathbf{a}_\alpha = \mathbf{a}_\alpha^T \cdot \mathbb{M} \cdot \mathbf{a}_\beta$ and $\mathbf{a}_\beta^T \cdot \mathbb{K} \cdot \mathbf{a}_\alpha = \mathbf{a}_\alpha^T \cdot \mathbb{K} \cdot \mathbf{a}_\beta$. Thus, the rhs vanishes and one obtains

$$\left(\omega_\alpha^2 - \omega_\beta^2\right) \mathbf{a}_\beta^T \cdot \mathbb{M} \cdot \mathbf{a}_\alpha = 0. \tag{3.6.11}$$

This means that for $\alpha \neq \beta$, one has $\mathbf{a}_\beta^T \cdot \mathbb{M} \cdot \mathbf{a}_\alpha = 0$. It is also convenient to require $\mathbf{a}_\alpha^T \cdot \mathbb{M} \cdot \mathbf{a}_\alpha = 1$. This gives the generalized orthogonality condition

$$\mathbf{a}_\alpha^T \cdot \mathbb{M} \cdot \mathbf{a}_\beta = \delta_{\alpha\beta}. \tag{3.6.12}$$

Now, from the second of Eqs. (3.6.9), one obtains

$$\mathbf{a}_\alpha^T \cdot \mathbb{K} \cdot \mathbf{a}_\beta = \omega_\beta^2 \mathbf{a}_\alpha^T \cdot \mathbb{M} \cdot \mathbf{a}_\beta = \omega_\beta^2 \delta_{\alpha\beta}. \tag{3.6.13}$$

One can compose the $N \times N$ matrix $\mathbb{A}$ of vectors $\mathbf{a}_\alpha$ by stacking all $\mathbf{a}_\alpha$ together. In terms of $\mathbb{A}$, Eq. (3.6.12) takes the form

$$\mathbb{A}^T \cdot \mathbb{M} \cdot \mathbb{A} = \mathbb{I}, \tag{3.6.14}$$

where $\mathbb{I}$ is the unit matrix, whereas Eq. (3.6.13) becomes

$$\mathbb{A}^T \cdot \mathbb{K} \cdot \mathbb{A} = \boldsymbol{\Omega}^2, \tag{3.6.15}$$

where $\boldsymbol{\Omega} = \operatorname{diag}\{\omega_\alpha\}$.

Let us now introduce the normal-coordinate vector $\boldsymbol{\zeta}$ defined by

$$\mathbf{X} = \mathbb{A} \cdot \boldsymbol{\zeta}, \quad \mathbf{X}^T = (\mathbb{A} \cdot \boldsymbol{\zeta})^T = \boldsymbol{\zeta}^T \cdot \mathbb{A}^T. \tag{3.6.16}$$

These equations can be resolved for $\boldsymbol{\zeta}$ as

$$\boldsymbol{\zeta} = \mathbb{A}^{-1} \cdot \mathbf{X}. \tag{3.6.17}$$

Note that, according to Eq. (3.6.14), $\mathbb{A}^{-1} \neq \mathbb{A}^T$and $\left(\mathbb{A}^{-1}\right)_{\alpha i} \neq a_{\alpha i}^{-1}$. Inserting Eq. (3.6.16) into the Lagrangian, Eq. (3.6.2), and using Eqs. (3.6.14) and (3.6.15), one obtains

$$\mathcal{L} = \frac{1}{2}(\dot{\boldsymbol{\zeta}}^T \cdot \mathbb{A}^T \cdot \mathbb{M} \cdot \mathbb{A} \cdot \dot{\boldsymbol{\zeta}} - \boldsymbol{\zeta}^T \cdot \mathbb{A}^T \cdot \mathbb{K} \cdot \mathbb{A} \cdot \boldsymbol{\zeta}) = \frac{1}{2}(\dot{\boldsymbol{\zeta}}^T \cdot \dot{\boldsymbol{\zeta}} - \boldsymbol{\zeta}^T \cdot \boldsymbol{\Omega}^2 \cdot \boldsymbol{\zeta}) \tag{3.6.18}$$

or

$$\mathcal{L} = \frac{1}{2}\sum_{\alpha=1}^{N}(\dot{\zeta}_\alpha^2 - \omega_\alpha^2\zeta_\alpha^2), \tag{3.6.19}$$

where ζ_α are components of the normal-coordinate vector $\boldsymbol{\zeta}$ or simply normal coordinates. One can see that the Lagrangian is separable in terms of the normal coordinates, so that every normal coordinate ζ_α oscillates with its own frequency ω_α:

$$\zeta_\alpha(t) = C_{s\alpha}\sin\omega_\alpha t + C_{c\alpha}\cos\omega_\alpha t. \tag{3.6.20}$$

From Eq. (3.6.16), one obtains

$$x_i(t) = \sum_{\alpha=1}^{N} a_{i\alpha}\zeta_\alpha(t), \tag{3.6.21}$$

which becomes

$$x_i(t) = \sum_{\alpha=1}^{N} a_{i\alpha}(C_{s\alpha}\sin\omega_\alpha t + C_{c\alpha}\cos\omega_\alpha t). \tag{3.6.22}$$

To obtain the coefficients from the initial conditions, one can set $t = 0$ in Eq. (3.6.20) and its derivative

$$\dot{\zeta}_\alpha(t) = C_{s\alpha}\omega_\alpha\cos\omega_\alpha t - C_{c\alpha}\omega_\alpha\sin\omega_\alpha t. \tag{3.6.23}$$

Using Eq. (3.6.17) in the component form

$$\zeta_\alpha = \sum_i \left(\mathbb{A}^{-1}\right)_{\alpha i} x_i \tag{3.6.24}$$

and its derivative, one obtains

$$C_{c\alpha} = \sum_i \left(\mathbb{A}^{-1}\right)_{\alpha i} x_i(0), \quad C_{s\alpha} = \frac{1}{\omega_\alpha}\sum_i \left(\mathbb{A}^{-1}\right)_{\alpha i} \dot{x}_i(0). \tag{3.6.25}$$

3.7 Problems on oscillations

3.7.1 *Two masses on a double incline connected by a spring*

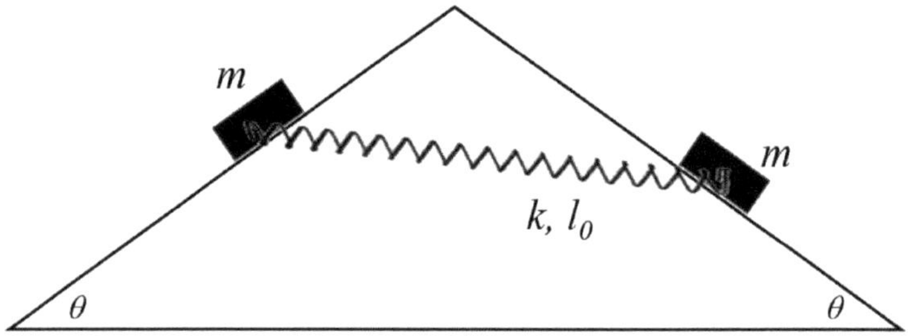

Two masses m on a double incline with the angle θ are connected by a spring of the equilibrium length l_0 and stiffness k. Find the oscillation frequencies. Consider the limit $k \to \infty$.

Solution: Let us take the top of the double incline for the origin O and measure the position of the masses on the incline from this point, as shown in the figure.

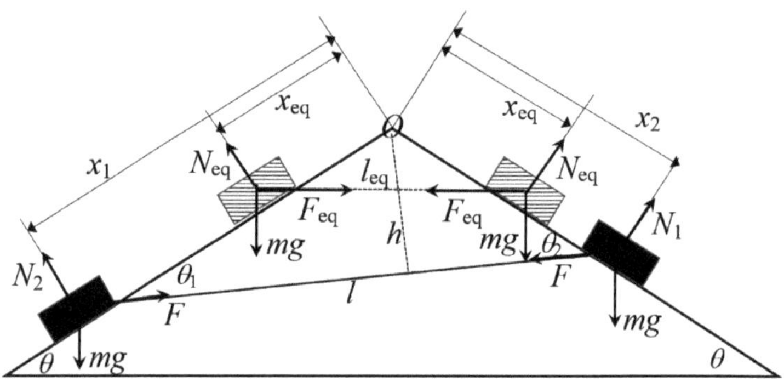

Newtonian solution below contains some involved geometry. The equilibrium position of the masses should be symmetric. From the balance of horizontal and vertical forces, one obtains

$$N_{\mathrm{eq}} \cos\theta = mg, \quad N_{\mathrm{eq}} \sin\theta = F_{\mathrm{eq}} = k(l_{\mathrm{eq}} - l_0). \tag{3.7.1}$$

From this, one finds

$$F_{\mathrm{eq}} = k(l_{\mathrm{eq}} - l_0) = mg\tan\theta. \tag{3.7.2}$$

Also, one can project the forces onto the direction of motion along the slopes and the direction perpendicular to it. In this case, the normal force

N becomes irrelevant and one obtains

$$mg\sin\theta = F_{\text{eq}}\cos\theta, \tag{3.7.3}$$

which leads to the same result. Equilibruim positions of the masses are characterized by

$$x_{\text{eq}} = l_{\text{eq}}/(2\cos\theta). \tag{3.7.4}$$

Equations of motion for the masses have the form (see the figure)

$$\begin{aligned} m\ddot{x}_1 &= mg\sin\theta - F\cos\theta_1, \\ m\ddot{x}_2 &= mg\sin\theta - F\cos\theta_2, \end{aligned} \tag{3.7.5}$$

where $F = k(l - l_0)$ and

$$l = \sqrt{x_1^2 + x_2^2 - 2x_1x_2\cos(\pi - 2\theta)} = \sqrt{x_1^2 + x_2^2 + 2x_1x_2\cos(2\theta)}. \tag{3.7.6}$$

The angles θ_1 and θ_2 satisfy the two equations:

$$\theta_1 + \theta_2 = 2\theta \tag{3.7.7}$$

and

$$x_1\sin\theta_1 = x_2\sin\theta_2, \tag{3.7.8}$$

which is the triangle height h expressed in two ways.

One can see that the problem is nonlinear, and even to obtain solvable equations of motion, one has first to solve the algebraic problem of finding θ_1 and θ_2 in terms of x_1 and x_2. Since we are interested in small oscillations around the equilibrium configuration, we expand everything up to the linear order in small deviations δx_1 and δx_2 defined by

$$x_1 = x_{\text{eq}} + \delta x_1, \quad x_2 = x_{\text{eq}} + \delta x_2. \tag{3.7.9}$$

Correspondingly,

$$\theta_1 = \theta + \delta\theta_1, \quad \theta_2 = \theta + \delta\theta_2. \tag{3.7.10}$$

Now,

$$\begin{aligned} l &\cong \sqrt{\begin{aligned} &x_{\text{eq}}^2 + 2x_{\text{eq}}\delta x_1 + x_{\text{eq}}^2 + 2x_{\text{eq}}\delta x_2 + 2x_{\text{eq}}^2\cos(2\theta) \\ &\quad + 2x_{\text{eq}}(\delta x_1 + \delta x_2)\cos(2\theta) \end{aligned}} \\ &= x_{\text{eq}}\sqrt{2[1 + \cos(2\theta)] + \frac{\delta x_1 + \delta x_2}{x_{\text{eq}}}2[1 + \cos(2\theta)]} \end{aligned}$$

$$= 2x_{\text{eq}} \cos\theta \sqrt{1 + \frac{\delta x_1 + \delta x_2}{x_{\text{eq}}}} = l_{\text{eq}} \sqrt{1 + \frac{\delta x_1 + \delta x_2}{x_{\text{eq}}}}$$

$$\cong l_{\text{eq}} + \cos\theta\,(\delta x_1 + \delta x_2) \equiv l_{\text{eq}} + \delta l,$$

where Eq. (3.7.4) was used. Linearized Eqs. (3.7.7) and (3.7.8) become

$$\begin{aligned} \delta\theta_1 + \delta\theta_2 &= 0, \\ (\delta x_1 - \delta x_2)\sin\theta &= -\, x_{\text{eq}} \cos\theta(\delta\theta_1 - \delta\theta_2). \end{aligned} \tag{3.7.11}$$

Solving these equations, one obtains

$$\delta\theta_1 = -\delta\theta_2 = -\frac{\delta x_1 - \delta x_2}{2x_{\text{eq}}} \tan\theta = -\frac{\delta x_1 - \delta x_2}{l_{\text{eq}}} \sin\theta. \tag{3.7.12}$$

After that, with Eqs. (3.7.5), one expands

$$\begin{aligned} F\cos\theta_1 &= (F_{\text{eq}} + k\delta l)\cos(\theta + \delta\theta_1) \\ &\cong F_{\text{eq}}\cos\theta + k\delta l\cos\theta - F_{\text{eq}}\sin\theta\delta\theta_1 \\ &= mg\sin\theta + k\cos^2\theta\,(\delta x_1 + \delta x_2) + mg\sin^2\theta\tan\theta\frac{\delta x_1 - \delta x_2}{l_{\text{eq}}} \\ &= mg\sin\theta + \left(k\cos^2\theta + \frac{mg\sin^2\theta\tan\theta}{l_{\text{eq}}}\right)\delta x_1 \\ &\quad + \left(k\cos^2\theta - \frac{mg\sin^2\theta\tan\theta}{l_{\text{eq}}}\right)\delta x_2, \end{aligned} \tag{3.7.13}$$

where Eq. (3.7.2) was used. Similarly, one obtains

$$\begin{aligned} F\cos\theta_2 \cong mg\sin\theta &+ \left(k\cos^2\theta - \frac{mg\sin^2\theta\tan\theta}{l_{\text{eq}}}\right)\delta x_1 \\ &+ \left(k\cos^2\theta + \frac{mg\sin^2\theta\tan\theta}{l_{\text{eq}}}\right)\delta x_2. \end{aligned} \tag{3.7.14}$$

Now, the equations of motion become

$$\begin{aligned} &m\delta\ddot{x}_1 + A_+\delta x_1 + A_-\delta x_2 = 0, \\ &m\delta\ddot{x}_2 + A_-\delta x_1 + A_+\delta x_2 = 0, \end{aligned} \tag{3.7.15}$$

where

$$A_\pm \equiv k\cos^2\theta \pm \frac{mg\sin^2\theta\tan\theta}{l_{\text{eq}}}. \tag{3.7.16}$$

The eingenfrequencies follow from the secular equation

$$\left(-m\omega^2 + A_+\right)^2 - A_-^2 = 0. \tag{3.7.17}$$

The two solutions are

$$\omega_\pm^2 = \frac{1}{m}(A_+ \pm A_-). \tag{3.7.18}$$

Explicitly, this becomes

$$\omega_+^2 = 2\cos^2\theta \frac{k}{m}, \quad \omega_-^2 = 2\sin^2\theta \tan\theta \frac{g}{l_{\text{eq}}}. \tag{3.7.19}$$

Thus, there are two decoupled modes. One mode is due to the stiffness of the spring and another is due to gravity. In the stiffness mode, the masses are moving anti-phase, while in the gravity mode, they are moving in-phase with $l = l_{\text{eq}} = \text{const}$. In the limit $k \to \infty$, the stiffness mode disappears and only the gravity mode remains. In this case, the system has only one degree of freedom.

Solution using Lagrangial mechanics below is more straightforward. Kinetic energy of the system is given by

$$E_k = \frac{m}{2}(\dot{x}_1^2 + \dot{x}_2^2). \tag{3.7.20}$$

The potential energy reads

$$U = -mg\sin\theta\,(x_1 + x_2) + \frac{1}{2}k\,(l - l_0)^2, \tag{3.7.21}$$

where

$$l = \sqrt{x_1^2 + x_2^2 - 2x_1x_2\cos(\pi - 2\theta)} = \sqrt{x_1^2 + x_2^2 + 2x_1x_2\cos(2\theta)}. \tag{3.7.22}$$

Thus, the Langrange function becomes

$$\mathcal{L} = \frac{m}{2}\left(\dot{x}_1^2 + \dot{x}_2^2\right) + mg\sin\theta\,(x_1 + x_2) - \frac{1}{2}k\,(l - l_0)^2. \tag{3.7.23}$$

The Lagrange equations have the form

$$0 = \frac{d}{dt}\frac{\partial\mathcal{L}}{\partial\dot{x}_i} - \frac{\partial\mathcal{L}}{\partial x_i} = m\ddot{x}_i - mg\sin\theta + k\left(1 - \frac{l_0}{l}\right)[x_i + x_{3-i}\cos(2\theta)], \tag{3.7.24}$$

where $i = 1, 2$.

Next, one has to expand the elastic term in small deviations from the equilibrium corresponding to the minimum of the potential energy. The latter can be searched for under the constraint $x_1 = x_2 = x$, so that

$$U = -2mg\sin\theta x + \frac{1}{2}k\left(x\sqrt{2\left[1+\cos(2\theta)\right]} - l_0\right)^2 \tag{3.7.25}$$

or

$$U = -2mg\sin\theta x + \frac{1}{2}k\left(2x\cos\theta - l_0\right)^2, \tag{3.7.26}$$

and the equilibrium value $x = x_{\text{eq}}$ is defined by

$$0 = \frac{\partial U}{\partial x} = -2mg\sin\theta + k\left(2x\cos\theta - l_0\right)2\cos\theta, \tag{3.7.27}$$

so that

$$x_{\text{eq}} = \frac{1}{2\cos\theta}\left(\frac{mg}{k}\tan\theta + l_0\right). \tag{3.7.28}$$

Now, one has to expand the elastic term up to linear terms in small deviations from the equilibrium:

$$x_1 = x_{\text{eq}} + \delta x_1, \quad x_2 = x_{\text{eq}} + \delta x_2. \tag{3.7.29}$$

One has

$$\begin{aligned} l &\cong \sqrt{\begin{aligned} &x_{\text{eq}}^2 + 2x_{\text{eq}}\delta x_1 + x_{\text{eq}}^2 + 2x_{\text{eq}}\delta x_2 + 2x_{\text{eq}}^2\cos(2\theta) \\ &\quad + 2x_{\text{eq}}\left(\delta x_1 + \delta x_2\right)\cos(2\theta) \end{aligned}} \\ &= x_{\text{eq}}\sqrt{2\left[1+\cos(2\theta)\right] + \frac{\delta x_1 + \delta x_2}{x_{\text{eq}}}2\left[1+\cos(2\theta)\right]} \\ &= 2x_{\text{eq}}\cos\theta\sqrt{1 + \frac{\delta x_1 + \delta x_2}{x_{\text{eq}}}} = l_{\text{eq}}\sqrt{1 + \frac{\delta x_1 + \delta x_2}{x_{\text{eq}}}} \\ &\cong l_{\text{eq}} + \cos\theta(\delta x_1 + \delta x_2), \end{aligned} \tag{3.7.30}$$

where $l_{\text{eq}} = 2x_{\text{eq}}\cos\theta$. Now,

$$1 - \frac{l_0}{l} \cong 1 - \frac{l_0}{l_{\text{eq}} + \cos\theta\left(\delta x_1 + \delta x_2\right)} \cong 1 - \frac{l_0}{l_{\text{eq}}}\left(1 - \frac{\cos\theta\left(\delta x_1 + \delta x_2\right)}{l_{\text{eq}}}\right), \tag{3.7.31}$$

and the Lagrange equations become

$$0 = m\delta\ddot{x}_i - mg\sin\theta + k\left[1 - \frac{l_0}{l_{\rm eq}}\left(1 - \frac{\cos\theta\,(\delta x_1 + \delta x_2)}{l_{\rm eq}}\right)\right]$$
$$\times\,[x_{\rm eq}(1 + \cos(2\theta)) + \delta x_i + \delta x_{3-i}\cos(2\theta)]. \tag{3.7.32}$$

The zero-order part of these equations,

$$-mg\sin\theta + k\left(1 - \frac{l_0}{l_{\rm eq}}\right)x_{\rm eq}\,(1 + \cos\,(2\theta)) = 0, \tag{3.7.33}$$

because of Eq. (3.7.28), so that at linear order, one has

$$0 = m\delta\ddot{x}_i + k\left(1 - \frac{l_0}{l_{\rm eq}}\right)(\delta x_i + \delta x_{3-i}\cos\,(2\theta))$$
$$+\,k\frac{l_0}{l_{\rm eq}}\frac{\cos\theta\,(\delta x_1 + \delta x_2)}{l_{\rm eq}}x_{\rm eq}\,(1 + \cos\,(2\theta)) \tag{3.7.34}$$

or

$$0 = m\delta\ddot{x}_i + k\left(1 - \frac{l_0}{l_{\rm eq}}\right)(\delta x_i + \delta x_{3-i}\cos\,(2\theta))$$
$$+\,k\frac{l_0}{l_{\rm eq}}\frac{\delta x_1 + \delta x_2}{2}(1 + \cos(2\theta)). \tag{3.7.35}$$

This becomes

$$m\delta\ddot{x}_1 + k\left(1 - \frac{l_0}{l_{\rm eq}}\frac{1 - \cos\,(2\theta)}{2}\right)\delta x_1$$
$$+\,k\left(\cos\,(2\theta) + \frac{l_0}{l_{\rm eq}}\frac{1 - \cos\,(2\theta)}{2}\right)\delta x_2 = 0,$$
$$m\delta\ddot{x}_2 + k\left(\cos\,(2\theta) + \frac{l_0}{l_{\rm eq}}\frac{1 - \cos\,(2\theta)}{2}\right)\delta x_1$$
$$+\,k\left(1 - \frac{l_0}{l_{\rm eq}}\frac{1 - \cos\,(2\theta)}{2}\right)\delta x_2 = 0. \tag{3.7.36}$$

Eigenfrequencies of this system of equations satisfy the equation

$$\left[-m\omega^2 + k\left(1 - \frac{l_0}{l_{\rm eq}}\frac{1 - \cos\,(2\theta)}{2}\right)\right]^2$$
$$-\,k^2\left(\cos\,(2\theta) + \frac{l_0}{l_{\rm eq}}\frac{1 - \cos\,(2\theta)}{2}\right)^2 = 0 \tag{3.7.37}$$

or

$$\left[-m\omega^2 + k\left(1+\cos(2\theta)\right)\right]\left[-m\omega^2 + k\left(1-\frac{l_0}{l_{\text{eq}}}\right)\left(1-\cos(2\theta)\right)\right] = 0 \tag{3.7.38}$$

or

$$\left[-m\omega^2 + 2k\cos^2\theta\right]\left[-m\omega^2 + 2k\left(1-\frac{l_0}{l_{\text{eq}}}\right)\sin^2\theta\right] = 0. \tag{3.7.39}$$

Thus, the two eigenfrequencies are given by

$$\omega_1^2 = 2\cos^2\theta\frac{k}{m}, \quad \omega_2^2 = 2\sin^2\theta\frac{k}{m}\left(1-\frac{l_0}{l_{\text{eq}}}\right). \tag{3.7.40}$$

One can see that the first of the oscillation modes is purely elastic. The second oscillation frequency with the help of Eq. (3.7.33) can be rewritten as

$$\omega_2^2 = 2\sin^2\theta\tan\theta\frac{g}{l_{\text{eq}}}, \tag{3.7.41}$$

i.e., the second oscillation mode is a pure gravity mode. This mode survives in the limit $k \to \infty$.

3.7.2 *Double pendulum*

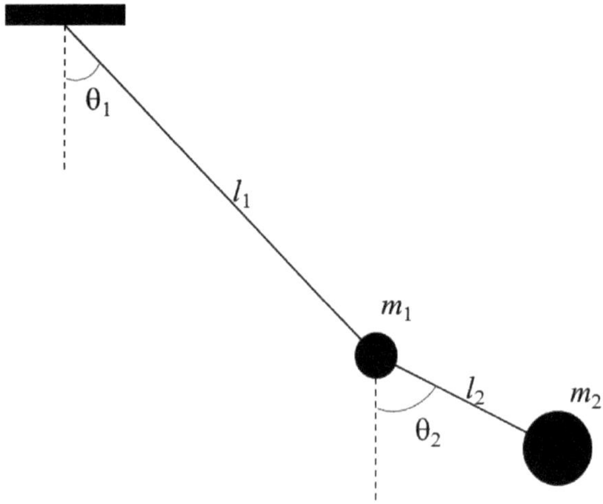

Consider a double pendulum that consists of pendulum 1 attached in a standard way and pendulum 2 attached to the end of pendulum 1. For simplicity, consider the case $m_1 = m_2 = m$ and $l_1 = l_2 = l$. Find eigenmodes and eigenfrequencies of this system.

Solution: Describing the pendula with the angles θ_1 and θ_2, one writes

$$x_1 = l\sin\theta_1, \quad x_2 = l\sin\theta_1 + l\sin\theta_2,$$
$$y_1 = -l\cos\theta_1, \quad y_2 = -l\cos\theta_1 - l\cos\theta_2. \tag{3.7.42}$$

Potential energy reads

$$U = mg\,(y_1 + y_2) = -mgl\,(2\cos\theta_2 + \cos\theta_1)\,, \tag{3.7.43}$$

and the kinetic energy is given by

$$\begin{aligned} E_k &= \frac{m}{2}\left[\dot{x}_1^2 + \dot{y}_1^2 + \dot{x}_2^2 + \dot{y}_2^2\right] \\ &= \frac{ml^2}{2}\left[\dot{\theta}_1^2 + \left(\cos\theta_1\dot{\theta}_1 + \cos\theta_2\dot{\theta}_2\right)^2 + \left(\sin\theta_1\dot{\theta}_1 + \sin\theta_2\dot{\theta}_2\right)^2\right] \\ &= \frac{ml^2}{2}[2\dot{\theta}_1^2 + \dot{\theta}_2^2 + 2(\cos\theta_1\cos\theta_2 + \sin\theta_1\sin\theta_2)\dot{\theta}_1\dot{\theta}_2]. \end{aligned} \tag{3.7.44}$$

Near the ground state $\theta_1 = \theta_2 = 0$, one leaves only quadratic terms that results in

$$\mathcal{L} = E_k - U = \frac{ml^2}{2}\left(2\dot{\theta}_1^2 + \dot{\theta}_2^2 + 2\dot{\theta}_1\dot{\theta}_2\right) - \frac{mgl}{2}(2\theta_1^2 + \theta_2^2). \tag{3.7.45}$$

Identifying this with Eq. (3.6.2) yields

$$\mathbb{M} = ml^2\begin{pmatrix} 2 & 1 \\ 1 & 1 \end{pmatrix}, \quad \mathbb{K} = mgl\begin{pmatrix} 2 & 0 \\ 0 & 1 \end{pmatrix}. \tag{3.7.46}$$

The generalized eigenvalue problem, Eq. (3.6.6), has the form

$$\begin{pmatrix} 2\left(-\omega^2 + \omega_0^2\right) & -\omega^2 \\ -\omega^2 & -\omega^2 + \omega_0^2 \end{pmatrix}\begin{pmatrix} a_1 \\ a_2 \end{pmatrix} = 0, \tag{3.7.47}$$

where we have introduced

$$\omega_0^2 = g/l. \tag{3.7.48}$$

Secular equation corresponding to Eq. (3.7.47) has the form

$$\begin{aligned} 0 &= 2\left(-\omega^2 + \omega_0^2\right)^2 - \omega^4 = \left[\sqrt{2}\left(\omega^2 - \omega_0^2\right) + \omega^2\right]\left[\sqrt{2}\left(\omega^2 - \omega_0^2\right) - \omega^2\right] \\ &= \left[\left(\sqrt{2}+1\right)\omega^2 - \sqrt{2}\omega_0^2\right]\left[\left(\sqrt{2}-1\right)\omega^2 - \sqrt{2}\omega_0^2\right]. \end{aligned} \tag{3.7.49}$$

Thus, the eigenfrequencies are given by

$$\omega_1^2 = \frac{\sqrt{2}\omega_0^2}{\sqrt{2}+1} = \left(2-\sqrt{2}\right)\omega_0^2, \quad \omega_2^2 = \frac{\sqrt{2}\omega_0^2}{\sqrt{2}-1} = \left(2+\sqrt{2}\right)\omega_0^2. \tag{3.7.50}$$

Let us now find generalized eigenvectors and normal modes of the system. Equation (3.7.47) becomes

$$\begin{aligned} 2\left(\omega_\alpha^2 - \omega_0^2\right) a_{1\alpha} + \omega_\alpha^2 a_{2\alpha} &= 0, \\ \omega_\alpha^2 a_{1\alpha} + \left(\omega_\alpha^2 - \omega_0^2\right) a_{2\alpha} &= 0. \end{aligned} \tag{3.7.51}$$

For instance, from the second equation, one obtains

$$\mathbf{a}_\alpha = \mu_\alpha \begin{pmatrix} -\omega_\alpha^2 + \omega_0^2 \\ \omega_\alpha^2 \end{pmatrix}, \tag{3.7.52}$$

where μ_α are normalization coefficients. In particular,

$$\begin{aligned} \mathbf{a}_1 &= \mu_1 \begin{pmatrix} -\omega_1^2 + \omega_0^2 \\ \omega_1^2 \end{pmatrix} = \mu_1\omega_0^2 \begin{pmatrix} \sqrt{2}-1 \\ 2-\sqrt{2} \end{pmatrix} = \mu_1\omega_0^2\left(\sqrt{2}-1\right)\begin{pmatrix} 1 \\ \sqrt{2} \end{pmatrix}, \\ \mathbf{a}_2 &= \mu_2 \begin{pmatrix} -\omega_2^2 + \omega_0^2 \\ \omega_2^2 \end{pmatrix} = \mu_2\omega_0^2 \begin{pmatrix} -\sqrt{2}-1 \\ 2+\sqrt{2} \end{pmatrix} = \mu_2\omega_0^2\left(\sqrt{2}+1\right)\begin{pmatrix} -1 \\ \sqrt{2} \end{pmatrix}. \end{aligned} \tag{3.7.53}$$

Let us check the orthogonality of these eigenvectors, Eq. (3.6.14):

$$\mathbf{a}_1^T \cdot \mathbb{M} \cdot \mathbf{a}_2 = \mu_1\mu_2 ml^2\omega_0^4 \begin{pmatrix} 1 & \sqrt{2} \end{pmatrix} \cdot \begin{pmatrix} 2 & 1 \\ 1 & 1 \end{pmatrix} \cdot \begin{pmatrix} -1 \\ \sqrt{2} \end{pmatrix} = 0, \tag{3.7.54}$$

as expected. Now, we calculate the normalization factors:

$$\begin{aligned} 1 = \mathbf{a}_1^T \cdot \mathbb{M} \cdot \mathbf{a}_1 &= \mu_1^2 ml^2\omega_0^4\left(\sqrt{2}-1\right)^2 \begin{pmatrix} 1 & \sqrt{2} \end{pmatrix} \cdot \begin{pmatrix} 2 & 1 \\ 1 & 1 \end{pmatrix} \cdot \begin{pmatrix} 1 \\ \sqrt{2} \end{pmatrix} \\ &= \mu_1^2 ml^2\omega_0^4\left(\sqrt{2}-1\right)2\sqrt{2}, \\ 1 = \mathbf{a}_2^T \cdot \mathbb{M} \cdot \mathbf{a}_2 &= \mu_2^2 ml^2\omega_0^4\left(\sqrt{2}+1\right)^2 \begin{pmatrix} -1 & \sqrt{2} \end{pmatrix} \cdot \begin{pmatrix} 2 & 1 \\ 1 & 1 \end{pmatrix} \cdot \begin{pmatrix} -1 \\ \sqrt{2} \end{pmatrix} \\ &= \mu_2^2 ml^2\omega_0^4\left(\sqrt{2}+1\right)2\sqrt{2}. \end{aligned} \tag{3.7.55}$$

Thus,

$$\mu_1\omega_0^2 = \frac{1}{\sqrt{ml^2}}\left(\frac{1}{(\sqrt{2}-1)\,2\sqrt{2}}\right)^{1/2} = \frac{1}{\sqrt{ml^2}}\left(\frac{\sqrt{2}+1}{2\sqrt{2}}\right)^{1/2},$$

$$\mu_2\omega_0^2 = \frac{1}{\sqrt{ml^2}}\left(\frac{1}{(\sqrt{2}+1)\,2\sqrt{2}}\right)^{1/2} = \frac{1}{\sqrt{ml^2}}\left(\frac{\sqrt{2}-1}{2\sqrt{2}}\right)^{1/2}, \quad (3.7.56)$$

and, finally,

$$\mathbf{a}_1 = \frac{1}{\sqrt{ml^2}}\left(\frac{\sqrt{2}-1}{2\sqrt{2}}\right)^{1/2}\begin{pmatrix}1\\ \sqrt{2}\end{pmatrix},$$

$$\mathbf{a}_2 = \frac{1}{\sqrt{ml^2}}\left(\frac{\sqrt{2}+1}{2\sqrt{2}}\right)^{1/2}\begin{pmatrix}-1\\ \sqrt{2}\end{pmatrix}. \quad (3.7.57)$$

Now, the angles expressed through the eigenmodes are given by Eq. (3.6.16) in matrix form or by Eq. (3.6.21) in components. The latter yields

$$\theta_1 = a_{11}\zeta_1 + a_{12}\zeta_2 = \frac{1}{\sqrt{ml^2}}\left(\frac{\sqrt{2}-1}{2\sqrt{2}}\right)^{1/2}\zeta_1 - \frac{1}{\sqrt{ml^2}}\left(\frac{\sqrt{2}+1}{2\sqrt{2}}\right)^{1/2}\zeta_2,$$

$$\theta_2 = a_{21}\zeta_1 + a_{22}\zeta_2 = \frac{\sqrt{2}}{\sqrt{ml^2}}\left(\frac{\sqrt{2}-1}{2\sqrt{2}}\right)^{1/2}\zeta_1 + \frac{\sqrt{2}}{\sqrt{ml^2}}\left(\frac{\sqrt{2}+1}{2\sqrt{2}}\right)^{1/2}\zeta_2.$$

$$(3.7.58)$$

Inverse transformation with the help of

$$\det \equiv a_{11}a_{22} - a_{21}a_{12} = \frac{1}{ml^2} \quad (3.7.59)$$

takes the form

$$\zeta_1 = \frac{\det_1}{\det} = (\theta_1 a_{22} - \theta_2 a_{12})\,ml^2 = \sqrt{ml^2}\left(\frac{\sqrt{2}+1}{2\sqrt{2}}\right)^{1/2}\left(\sqrt{2}\theta_1 + \theta_2\right),$$

$$\zeta_2 = \frac{\det_2}{\det} = (a_{11}\theta_2 - a_{21}\theta_1)\,ml^2 = \sqrt{ml^2}\left(\frac{\sqrt{2}-1}{2\sqrt{2}}\right)^{1/2}\left(-\sqrt{2}\theta_1 + \theta_2\right).$$

$$(3.7.60)$$

One can see that in normal mode 1, both pendula are swinging in-phase and in mode 2, they are swinging in antiphase. The frequency of mode 1 is lower.

3.7.3 *Two coupled oscillators*

Consider the model that illustrates general principles and has an interesting physical behavior—the model of two coupled oscillators with the Lagrangian

$$\mathcal{L} = \frac{m}{2}(\dot{x}_1^2 - \omega_1^2 x_1^2 + \dot{x}_2^2 - \omega_2^2 x_2^2 - \Delta^2 x_1 x_2). \tag{3.7.61}$$

Find eigenmodes and eigenfrequencies of this system. Investigate the resonance case, in which the frequencies ω_1 and ω_2 are close to each other. Find the time dependences of the oscillators' energies at resonance, $\omega_1 = \omega_2 = \omega_0$, if at the beginning one of the oscillators was excited and another one was in its ground state.

Solution: We will show that if $|\omega_1 - \omega_2| \gg \Delta$ (off-resonance case), the oscillators behave nearly independently of each other and oscillate with their own frequencies $\omega_{1,2}$, whereas for $|\omega_1 - \omega_2| \sim \Delta$ (resonance case), they strongly hybridize and each normal mode involves both coordinates. Here, the matrices $\mathbb{M}$ and $\mathbb{K}$ are given by

$$\mathbb{M} = m \begin{pmatrix} 1 & 0 \\ 0 & 1 \end{pmatrix}, \quad \mathbb{K} = m \begin{pmatrix} \omega_1^2 & \Delta^2/2 \\ \Delta^2/2 & \omega_2^2 \end{pmatrix} \tag{3.7.62}$$

and the eigenvalue problem, Eq. (3.6.6), reads

$$\begin{pmatrix} -\omega^2 + \omega_1^2 & \Delta^2/2 \\ \Delta^2/2 & -\omega^2 + \omega_2^2 \end{pmatrix} \cdot \begin{pmatrix} a_1 \\ a_2 \end{pmatrix} = 0. \tag{3.7.63}$$

Note that this is a standard eigenvalue problem, not a generalized one. The secular equation has the form

$$\left(\omega^2 - \omega_1^2\right)\left(\omega^2 - \omega_2^2\right) - \frac{\Delta^4}{4} = 0, \tag{3.7.64}$$

and eigenfrequencies are given by

$$\omega_\pm^2 = \frac{1}{2}\left(\omega_1^2 + \omega_2^2 \pm \sqrt{(\omega_1^2 - \omega_2^2) + \Delta^4}\right). \tag{3.7.65}$$

One can see that, indeed, for $|\omega_1 - \omega_2| \gg \Delta$, the eigenfrequencies $\omega_\pm$ coincide with $\omega_{1,2}$. On the other hand, at resonance, $\omega_1 = \omega_2 = \omega_0$, one obtains frequencies of two modes split by coupling:

$$\omega_\pm^2 = \omega_0^2 \pm \frac{1}{2}\Delta^2. \tag{3.7.66}$$

Let us now find the eigenvectors and normal modes. From the equation

$$\left(-\omega_{\pm}^{2}+\omega_{1}^{2}\right) a_{1\pm}+\frac{\Delta^{2}}{2} a_{2\pm}=0, \tag{3.7.67}$$

one obtains

$$\mathbf{a}_{\pm}=\mu_{\pm}\begin{pmatrix}\Delta^{2}/2\\ \omega_{\pm}^{2}-\omega_{1}^{2}\end{pmatrix}, \tag{3.7.68}$$

where μ_α $(\alpha = \pm)$ are normalization coefficients. This form is non-symmetric, and it is convenient to symmetrize it. To this end, consider the case $\omega_1 < \omega_2$. One can see from Eq. (3.7.65) that the frequencies are ordered as $\omega_- < \omega_1 < \omega_2 < \omega_+$. Keeping these relations in mind, one can symmetrize $\mathbf{a}_\pm$ as

$$\mathbf{a}_{+}=\mu_{+}\begin{pmatrix}\sqrt{\frac{\Delta^{2}}{2\left(\omega_{+}^{2}-\omega_{1}^{2}\right)}}\\ \sqrt{\frac{2\left(\omega_{+}^{2}-\omega_{1}^{2}\right)}{\Delta^{2}}}\end{pmatrix}, \quad \mathbf{a}_{-}=\mu_{-}\begin{pmatrix}\sqrt{\frac{\Delta^{2}}{2\left(\omega_{1}^{2}-\omega_{-}^{2}\right)}}\\ -\sqrt{\frac{2\left(\omega_{1}^{2}-\omega_{-}^{2}\right)}{\Delta^{2}}}\end{pmatrix}, \tag{3.7.69}$$

where μ_α is another set of normalization coefficients. One can check orthogonality of these eigenvectors, $\mathbf{a}_+^T \cdot \mathbf{a}_- = 0$. From the normalization condition, $\mathbf{a}_\alpha^T \cdot \mathbf{a}_\alpha = 1$, one finds μ_α that turn out to be the same:

$$\mu_{\alpha} \equiv \mu_{\pm}=\sqrt{\frac{\Delta^{2}}{2 \sqrt{\left(\omega_{1}^{2}-\omega_{2}^{2}\right)+\Delta^{4}}}}. \tag{3.7.70}$$

Substituting this into the formulas above and simplifying, one finally obtains a symmetric expression

$$\mathbf{a}_{\pm}=\frac{1}{\sqrt{2}}\begin{pmatrix}\sqrt{1 \pm \frac{\omega_{1}^{2}-\omega_{2}^{2}}{\sqrt{\left(\omega_{1}^{2}-\omega_{2}^{2}\right)+\Delta^{4}}}}\\ +\sqrt{1 \mp \frac{\omega_{1}^{2}-\omega_{2}^{2}}{\sqrt{\left(\omega_{1}^{2}-\omega_{2}^{2}\right)+\Delta^{4}}}}\end{pmatrix}. \tag{3.7.71}$$

Actually, this result is valid for an arbitrary relation between $\omega_1$and ω_2. Off resonance for $\omega_1 < \omega_2$, one has

$$\mathbf{a}_{-} \cong\begin{pmatrix}1\\ 0\end{pmatrix}, \quad \mathbf{a}_{+} \cong\begin{pmatrix}0\\ 1\end{pmatrix}, \tag{3.7.72}$$

which means that $\omega_- \cong \omega_1$ corresponds to the first oscillator and $\omega_+ \cong \omega_2$ corresponds to the second oscillator. Off resonance, for $\omega_1 > \omega_2$, it is vice versa:

$$\mathbf{a}_{-} \cong\begin{pmatrix}0\\ 1\end{pmatrix}, \quad \mathbf{a}_{+} \cong\begin{pmatrix}1\\ 0\end{pmatrix}, \tag{3.7.73}$$

thus $\omega_- \cong \omega_2$ corresponds to the second oscillator and $\omega_+ \cong \omega_1$ corresponds to the first one. At resonance, $\omega_1 = \omega_2 = \omega_0$, one obtains

$$\mathbf{a}_\pm = \frac{1}{\sqrt{2}} \begin{pmatrix} 1 \\ \pm 1 \end{pmatrix}, \tag{3.7.74}$$

which means that the modes are strongly mixed.

The original coordinates expressed through the normal modes are given by the formula similar to Eq. (3.6.21), i.e.,

$$x_1 = a_{1+}\zeta_+ + a_{1-}\zeta_-, \quad x_2 = a_{2+}\zeta_+ + a_{2-}\zeta_-. \tag{3.7.75}$$

Normal modes behave as independent harmonic oscillators with frequencies $\omega_\pm$. Thus, the time dependence of both x_1 and x_2 is a superposition of these two oscillations.

Now, consider, as an important illustration, the resonance case $\omega_1 = \omega_2 = \omega_0$ with a small coupling $\Delta \ll \omega_0$. Here,

$$x_1 = \frac{1}{\sqrt{2}} \left(\zeta_+ + \zeta_-\right), \quad x_2 = \frac{1}{\sqrt{2}} \left(\zeta_+ - \zeta_-\right), \tag{3.7.76}$$

and inversely,

$$\zeta_+ = \frac{1}{\sqrt{2}} \left(x_1 + x_2\right), \quad \zeta_- = \frac{1}{\sqrt{2}} (x_1 - x_2). \tag{3.7.77}$$

One can see that ζ_+ is in-phase oscillation, while ζ_- is anti-phase oscillation. The latter has a lower frequency that can be inferred from Eq. (3.7.61). Suppose at $t = 0$, oscillator 1 was in a general state, while oscillator 2 was in its ground state. Then the initial condition for the normal modes is

$$\zeta_\pm(0) = \frac{1}{\sqrt{2}} x_1(0), \quad \dot{\zeta}_\pm(0) = \frac{1}{\sqrt{2}} \dot{x}_1(0). \tag{3.7.78}$$

Thus, the time dependence of the normal-mode coordinates is given by

$$\zeta_\pm(t) = \frac{1}{\sqrt{2}} x_1(0) \cos(\omega_\pm t) + \frac{1}{\sqrt{2}} \dot{x}_1(0) \frac{\sin(\omega_\pm t)}{\omega_\pm}. \tag{3.7.79}$$

Inserting this into Eq. (3.7.76), one obtains

$$\begin{aligned} x_1(t) &= \frac{1}{2} \left\{ x_1(0) \left[\cos(\omega_+ t) + \cos(\omega_- t)\right] + \dot{x}_1(0) \left[\frac{\sin(\omega_+ t)}{\omega_+} + \frac{\sin(\omega_- t)}{\omega_-}\right] \right\}, \\ x_2(t) &= \frac{1}{2} \left\{ x_1(0) \left[\cos(\omega_+ t) - \cos(\omega_- t)\right] + \dot{x}_1(0) \left[\frac{\sin(\omega_+ t)}{\omega_+} - \frac{\sin(\omega_- t)}{\omega_-}\right] \right\}. \end{aligned} \tag{3.7.80}$$

In the small-coupling case, the eigenfrequencies are close to each other, so that one can approximate

$$\frac{1}{\omega_+} \cong \frac{1}{\omega_-} \cong \frac{1}{\omega_0}. \tag{3.7.81}$$

However, one cannot make this approximation in the arguments of sin and cos since it will break down at large times. From Eq. (3.7.66) in the weak-coupling case follows

$$\omega_\pm = \omega_0 \sqrt{1 \pm \frac{\Delta^2}{2\omega_0^2}} \cong \omega_0 \pm \frac{\Delta^2}{4\omega_0}. \tag{3.7.82}$$

Using this and trigonometric relations, one obtains

$$\begin{aligned} x_1(t) &= \left[x_1(0)\cos(\omega_0 t) + \frac{\dot{x}_1(0)}{\omega_0}\sin(\omega_0 t)\right]\cos\left(\frac{\Delta^2}{4\omega_0}t\right), \\ x_2(t) &= \left[-x_1(0)\sin(\omega_0 t) + \frac{\dot{x}_1(0)}{\omega_0}\cos(\omega_0 t)\right]\sin\left(\frac{\Delta^2}{4\omega_0}t\right). \end{aligned} \tag{3.7.83}$$

This is a standard time dependence for a harmonic oscillator (square brackets) multiplied by a slowly oscillating function of time. Using

$$\dot{x}_1(t) = [-x_1(0)\omega_0 \sin(\omega_0 t) + \dot{x}_1(0)\cos(\omega_0 t)]\cos\left(\frac{\Delta^2}{4\omega_0}t\right) \tag{3.7.84}$$

(we do not differentiate the slow function), one can compute the time dependence of the energy of the first oscillator as

$$E_1(t) = \frac{m}{2}\left(\dot{x}_1^2(t) + \omega_0^2 x_1^2(0)\right) = E_1(0)\cos^2\left(\frac{\Delta^2}{4\omega_0}t\right). \tag{3.7.85}$$

Similarly, for the second oscillator, one obtains

$$E_2(t) = E_1(0)\sin^2\left(\frac{\Delta^2}{4\omega_0}t\right). \tag{3.7.86}$$

One can see that the total energy $E(t) = E_1(t) + E_2(t)$ is conserved and slowly migrating between the two oscillators.

Chapter 4

Rotational Motion of Rigid Bodies

4.1 Rotational kinematics

4.1.1 *Large and small rotations*

Rotations of rigid bodies are described with respect to their center of mass (CM) located at O or with respect to any other point O', in particular, a support point or a point on the so-called "instantaneous axis of rotation". In the latter case, the point O' is not connected to the body and moves with respect to the body's particles.

The orientation of a rigid body is defined by the directions of its own embedded Descarte axes $\mathbf{e}^{(1)}$, $\mathbf{e}^{(2)}$, $\mathbf{e}^{(3)}$ with respect to the laboratory Descarte axes $\mathbf{e}_x$, $\mathbf{e}_y$, $\mathbf{e}_z$. Any rotation of a body out of the reference orientation $\mathbf{e}^{(\alpha)} = \mathbf{e}_\alpha$, $\alpha = x, y, z = 1, 2, 3$, to an arbitrary orientation can be described by a rotation by a finite angle χ around some axis $\boldsymbol{\nu}$. This is the Euler theorem obtained by geometrical arguments.[1] As the direction of $\boldsymbol{\nu}$ is specified by two angles ϑ and φ of the spherical coordinate system, this finite rotation is specified by three angles: ϑ, φ, and χ.

In the finite rotation described above, any embedded vector $\mathbf{r}$, including $\mathbf{e}^{(\alpha)}$, is rotated as

$$\begin{aligned}\mathbf{r} \rightarrow \mathbf{r}' &= (\boldsymbol{\nu} \cdot \mathbf{r})\,\boldsymbol{\nu} + [\mathbf{r} - \boldsymbol{\nu}\,(\boldsymbol{\nu} \cdot \mathbf{r})] \cos\chi + [\boldsymbol{\nu} \times \mathbf{r}] \sin\chi \\ &= \mathbf{r}\cos\chi + \boldsymbol{\nu}\,(\boldsymbol{\nu} \cdot \mathbf{r})\,(1 - \cos\chi) + [\boldsymbol{\nu} \times \mathbf{r}] \sin\chi. \qquad (4.1.1)\end{aligned}$$

The first line of this equation tells the story of how it was obtained. Vector $\boldsymbol{\nu}\,(\boldsymbol{\nu} \cdot \mathbf{r})$ is the projection of $\mathbf{r}$ on $\boldsymbol{\nu}$ (the longitudinal component of $\mathbf{r}$) that remains invariant under rotation. Vectors $\mathbf{r} - \boldsymbol{\nu}\,(\boldsymbol{\nu} \cdot \mathbf{r}) \equiv -\,[\boldsymbol{\nu} \times [\boldsymbol{\nu} \times \mathbf{r}]]$ and $[\boldsymbol{\nu} \times \mathbf{r}]$ are perpendicular to $\boldsymbol{\nu}$ and to each other, thus they form a basis

[1] L. Euler, Novi Comment. Petrop. XX, 189 (1776).

in the plane perpendicular to $\boldsymbol{\nu}$. The first of them is just the transverse component of $\mathbf{r}$. The transverse component of $\mathbf{r}'$ that is rotated by χ is projected onto these two basis vectors. Note that double vector products can be simplified as

$$[\mathbf{a}\times[\mathbf{b}\times\mathbf{c}]] = \mathbf{b}\,(\mathbf{a}\cdot\mathbf{c}) - \mathbf{c}\,(\mathbf{a}\cdot\mathbf{b}). \tag{4.1.2}$$

For small rotation angles $\delta\chi$ at linear order in $\delta\chi$, this transformation simplifies to

$$\mathbf{r}\to\mathbf{r}'\cong\mathbf{r}+[\boldsymbol{\nu}\times\mathbf{r}]\,\delta\chi = \mathbf{r}+[\delta\boldsymbol{\chi}\times\mathbf{r}], \tag{4.1.3}$$

where the vector rotation angle $\delta\boldsymbol{\chi}\equiv\boldsymbol{\nu}\delta\chi$ was introduced. Note that the latter is possible only for small rotation angles. First-order equation (4.1.3) is sufficient to introduce the angular velocity in the following section. Note that, as the body rotates, there are individual $\boldsymbol{\nu}$ and χ for each moment of time.

The result of two and more successive rotations by large angles using Eq. (4.1.1) depend on the order in which rotations are performed. Indeed, Eq. (4.1.1) is a linear transformation that can be represented by a matrix, and matrices in general do not commute. However, for small rotations at linear order in $\delta\chi$, the result does not depend on the order of rotations. As an example, consider the two successive rotations

$$\mathbf{r}^{(1)}\cong\mathbf{r}+\left[\delta\boldsymbol{\chi}^{(1)}\times\mathbf{r}\right],\quad \mathbf{r}^{(2)}\cong\mathbf{r}^{(1)}+\left[\delta\boldsymbol{\chi}^{(2)}\times\mathbf{r}^{(1)}\right]. \tag{4.1.4}$$

Combining these two formulas and keeping only terms linear in rotation angles, one obtains

$$\begin{aligned}\mathbf{r}^{(2)} &\cong \mathbf{r}+\left[\delta\boldsymbol{\chi}^{(1)}\times\mathbf{r}\right]+\left[\delta\boldsymbol{\chi}^{(2)}\times\left(\mathbf{r}+\left[\delta\boldsymbol{\chi}^{(1)}\times\mathbf{r}\right]\right)\right]\\ &\cong \mathbf{r}+\left[\left(\delta\boldsymbol{\chi}^{(1)}+\delta\boldsymbol{\chi}^{(2)}\right)\times\mathbf{r}\right].\end{aligned} \tag{4.1.5}$$

This result is very important because one can consider parts of $\delta\boldsymbol{\chi}$ without taking care of their order and one can treat more complicated rotations (such as rolling of a cone) as a superposition of two or more small rotations.

4.1.2 *Angular velocity*

The change of any vector $\mathbf{r}$ embedded in a rigid body due to an infinitesimal rotation and the corresponding velocity follow from Eq. (4.1.3):

$$\delta\mathbf{r} = \mathbf{r}' - \mathbf{r} \cong [\delta\boldsymbol{\chi}\times\mathbf{r}] \tag{4.1.6}$$

and

$$\mathbf{v} = \dot{\mathbf{r}} = [\boldsymbol{\omega} \times \mathbf{r}], \tag{4.1.7}$$

where the angular velocity $\boldsymbol{\omega}$ is defined by

$$\boldsymbol{\omega} = \frac{\delta \boldsymbol{\chi}}{\delta t}. \tag{4.1.8}$$

The definition of ω is written in this form (and not as $\boldsymbol{\omega} = d\boldsymbol{\chi}/dt$) since the function $\boldsymbol{\chi}(t)$ that could be differentiated over time does not exist in general. Time-continuous rotations are not described by a single rotation vector because as the body rotates, the direction of the rotation axis changes too. This makes rotational dynamics complicated.

Equation (4.1.7) is valid in the case of the origin of the coordinate system for $\mathbf{r}$ fixed in space, such as the case of rotation around a point of support (fulcrum). If the origin is the center of mass of the body or any other point that is moving with velocity $\mathbf{V}$, one has to add this motion to the rotational motion and write a more general formula

$$\mathbf{v} = \mathbf{V} + \dot{\mathbf{r}} = \mathbf{V} + [\boldsymbol{\omega} \times \mathbf{r}], \tag{4.1.9}$$

where $\mathbf{r}$ is defined with respect to the rotation center.

One can find the acceleration of any point of a body due to its rotation. Differentiating Eq. (4.1.9) and using $\dot{\mathbf{r}} \equiv \mathbf{v} - \mathbf{V}$ and Eq. (4.1.2) yield

$$\begin{aligned}\dot{\mathbf{v}} &= \dot{\mathbf{V}} + [\dot{\boldsymbol{\omega}} \times \mathbf{r}] + [\boldsymbol{\omega} \times \dot{\mathbf{r}}] = \dot{\mathbf{V}} + [\dot{\boldsymbol{\omega}} \times \mathbf{r}] + [\boldsymbol{\omega} \times [\boldsymbol{\omega} \times \mathbf{r}]] \\ &= \dot{\mathbf{V}} + [\dot{\boldsymbol{\omega}} \times \mathbf{r}] + \boldsymbol{\omega}\,(\boldsymbol{\omega} \cdot \mathbf{r}) - \mathbf{r}\omega^2 = \dot{\mathbf{V}} + [\dot{\boldsymbol{\omega}} \times \mathbf{r}] - \mathbf{r}_\perp \omega^2, \end{aligned} \tag{4.1.10}$$

where $\mathbf{r}_\perp$ is the component of $\mathbf{r}$ in the direction perpendicular to the rotation axis $\boldsymbol{\omega}$ since the component of $\mathbf{r}$ parallel to $\boldsymbol{\omega}$ was subtracted. Thus, the vector $-\mathbf{r}_\perp$ is directed toward the rotation axis (in practical cases, going through the CM), and because of that, the last term in Eq. (4.1.10) is called *centripetal acceleration.*

4.1.3 *Rolling constraint, instantaneous axis of rotation*

Now, consider Eq. (4.1.9) with the origin of rotations O at the CM and introduce another origin O' such that O is located by the vector $\mathbf{b}$ away from O'. The position vectors $\mathbf{r}$ and $\mathbf{r}'$ in both coordinate systems are related by

$$\mathbf{r} = \mathbf{r}' - \mathbf{b}. \tag{4.1.11}$$

Thus, the velocity of the same physical point of the body, Eq. (4.1.9), expressed via $\mathbf{r}'$ is

$$\mathbf{v} = \mathbf{V} + [\boldsymbol{\omega} \times \mathbf{r}'] - [\boldsymbol{\omega} \times \mathbf{b}]. \tag{4.1.12}$$

If a body of a round shape is rolling on a surface without skidding, it is convenient to place O' at the contact point between the body and the surface. The velocity of the point of the body that is in contact with the surface ($\mathbf{r}' = 0$) is zero. This constraint condition can be written as $\mathbf{0} = \mathbf{V} - [\boldsymbol{\omega} \times \mathbf{b}]$, i.e.,

$$\mathbf{V} = [\boldsymbol{\omega} \times \mathbf{b}]. \tag{4.1.13}$$

Now, from Eq. (4.1.12) for any point of the body, one obtains

$$\mathbf{v} = [\boldsymbol{\omega} \times \mathbf{r}']. \tag{4.1.14}$$

This is similar to Eq. (4.1.7) and describes the rotation around an instantaneous axis that goes through the contact point. It is called instantaneous not only because it can change its direction with time (for sphere or disc but not for cylinder) but also because different points of the surface of the sphere are in contact with the support surface at different moments of time.

If, for instance, a cylinder or a sphere of radius R is rolling on a surface without skidding, one has $\mathbf{b} = \mathbf{n}R$, where $\mathbf{n}$ is the unit vector normal to the surface and directed up toward the center of the body. In this case, the constraint equation has the form

$$\mathbf{V} = R\,[\boldsymbol{\omega} \times \mathbf{n}]. \tag{4.1.15}$$

For a disk rolling on a plane, the constraint equation is different and has the form

$$\mathbf{V} = R\,[\boldsymbol{\omega} \times \mathbf{e}_O], \tag{4.1.16}$$

where $\mathbf{e}_O$ is a unit vector pointing from O' to the center of the disk O. This vector is not necessarily perpendicular to the plane and depends on the orientation of the disc.

There are other types of constraints describing the contact of two rigid bodies. For a body on a plane without friction, the constraint requires that only the vertical component of the velocity $\mathbf{v}$ at the contact point is zero:

$$\mathbf{V} \cdot \mathbf{n} = [\boldsymbol{\omega} \times \mathbf{b}] \cdot \mathbf{n}. \tag{4.1.17}$$

If $\mathbf{b} \parallel \mathbf{n}$, as for the cylinder of a sphere, then the rhs vanishes and the constraint becomes simply $\mathbf{V} \cdot \mathbf{n} \equiv V_z = 0$. For the disk, the second term does not vanish.

An interesting case is a round body with a sharp edge, such as disk, moving on the ice. In this case, the constraint allows the rotation of the disk around its symmetry axis $\mathbf{e}^{(3)}$, whereas the vertical and sideways motion is prohibited. Thus, the ice constraint relations can be written as

$$\mathbf{V} \cdot \mathbf{n} = R\left[\boldsymbol{\omega} \times \mathbf{e}_O\right] \cdot \mathbf{n}, \quad \mathbf{V} \cdot \mathbf{e}^{(3)} = R\left[\boldsymbol{\omega} \times \mathbf{e}_O\right] \cdot \mathbf{e}^{(3)}. \tag{4.1.18}$$

In the case of a body of a general round shape, such as an ellipsoid, rolling on a plane, the constraint equation becomes more complicated and includes the equation of the surface of the body, $F(r_1, r_2, r_3) = 0$, where r_k are coordinates in the frame embedded in the body. The perpendicular to the body's surface, which is the gradient of this function, should also be perpendicular to the supporting plane at the contact point. This is the definition of the contact point, which constitutes one of the equations. To this, the no-skidding condition, Eq. (4.1.15), should be added.

Let us calculate the acceleration of the points of the body using Eq. (4.1.14). Note that one cannot use $\dot{\mathbf{r}}' = [\boldsymbol{\omega} \times \mathbf{r}']$ because $\dot{\mathbf{r}}' = \dot{\mathbf{r}} + \dot{\mathbf{b}} = \mathbf{v} - \mathbf{V} + \dot{\mathbf{b}} \neq \mathbf{v}$. Instead, one should use $\dot{\mathbf{r}} = [\boldsymbol{\omega} \times \mathbf{r}]$, see Eq. (4.1.9). One obtains

$$\begin{aligned}
\dot{\mathbf{v}} &= [\dot{\boldsymbol{\omega}} \times \mathbf{r}'] + \left[\boldsymbol{\omega} \times \dot{\mathbf{r}}'\right] = [\dot{\boldsymbol{\omega}} \times \mathbf{r}'] + \left[\boldsymbol{\omega} \times \left(\dot{\mathbf{r}} + \dot{\mathbf{b}}\right)\right] \\
&= [\dot{\boldsymbol{\omega}} \times \mathbf{r}'] + \left[\boldsymbol{\omega} \times \dot{\mathbf{b}}\right] + [\boldsymbol{\omega} \times [\boldsymbol{\omega} \times \mathbf{r}]] \\
&= [\dot{\boldsymbol{\omega}} \times \mathbf{r}'] + \left[\boldsymbol{\omega} \times \dot{\mathbf{b}}\right] - \mathbf{r}_{\perp}\omega^2,
\end{aligned} \tag{4.1.19}$$

where $-\mathbf{r}_{\perp}$ is directed toward true rotation axis defined by Eq. (4.1.9). For $\dot{\boldsymbol{\omega}} = \dot{\mathbf{b}} = \mathbf{0}$, the acceleration $\dot{\mathbf{v}}$ is directed in the direction of $-\mathbf{r}_{\perp}$ despite the rotation being represented as rotation around the instantaneous rotation axis.

Let us now calculate the acceleration in a standard way. Returning to the preceding section and substituting the rolling constraint, Eq. (4.1.13), into Eq. (4.1.10) and using Eq. (4.1.11), one obtains

$$\dot{\mathbf{v}} = \frac{d}{dt}[\boldsymbol{\omega} \times \mathbf{b}] + [\dot{\boldsymbol{\omega}} \times \mathbf{r}] - \mathbf{r}_\perp \omega^2$$

$$[\dot{\boldsymbol{\omega}} \times \mathbf{b}] + \left[\boldsymbol{\omega} \times \dot{\mathbf{b}}\right] + [\dot{\boldsymbol{\omega}} \times \mathbf{r}] - \mathbf{r}_\perp \omega^2$$

$$= [\dot{\boldsymbol{\omega}} \times \mathbf{r}'] + \left[\boldsymbol{\omega} \times \dot{\mathbf{b}}\right] - \mathbf{r}_\perp \omega^2, \tag{4.1.20}$$

which coincides with the expression above. In particular, for a cylinder or sphere rolling on a plane ($\mathbf{b} = -\mathbf{n}R = \mathbf{const}$) with a constant angular velocity, the acceleration of the points of the body is directed toward the center of the body, whether one considers the rotation around the geometrical center or around the instantaneous axis of rotation.

One can ask what is the difference between rolling described as a rotation around the instantaneous axis going through the contact point and the rotation around a fixed axis. In the latter case, $\mathbf{v} = [\boldsymbol{\omega} \times \mathbf{r}']$ and $\dot{\mathbf{v}} = [\dot{\boldsymbol{\omega}} \times \mathbf{r}'] - \mathbf{r}'_\perp \omega^2$, i.e., the centripetal acceleration is directed toward the fixed axis.

The rolling constraints above for the sphere and the disk are non-holonomic because they cannot be integrated, eliminating generalized velocities and resulting in a constraint for generalized coordinates. Only for a cylinder that is described by a single dynamical variable φ, this can be done, and thus, the rolling constraint is holonomic.

4.1.4 *Euler angles*

In Sec. 4.1.1, we have seen that any orientation of the embedded coordinate system of a rigid body $\mathbf{e}^{(1)}$, $\mathbf{e}^{(2)}$, $\mathbf{e}^{(3)}$ with respect to the laboratory frame $\mathbf{e}_x$, $\mathbf{e}_y$, $\mathbf{e}_z$ can be described by a single finite rotation specified by three angles. However, in mechanics of rigid bodies, it is more convenient to describe orientations as a result of three successive elementary rotations by Euler angles θ, ϕ, and ψ shown in Fig. 4.1.1. The detailed description of the three rotation is as follows:

1. Rotation around $\mathbf{e}_x$ by the angle θ (nutation). After that, $\mathbf{e}^{(1)} = \mathbf{e}_x$.
2. Rotation around $\mathbf{e}_z$ by the angle ϕ (precession). After that, $\mathbf{e}^{(1)} = \mathbf{e}_N$ and $\mathbf{e}^{(2)} = \mathbf{e}_A$.
3. Rotation around $\mathbf{e}^{(3)}$ by the angle ψ (spin for prolate and wobble for oblate bodies).

Rotation of the body gives rise to the three constituents of the angular velocity:

$$\boldsymbol{\omega} = \dot{\theta}\mathbf{e}_N + \dot{\phi}\mathbf{e}_z + \dot{\psi}\mathbf{e}^{(3)}, \tag{4.1.21}$$

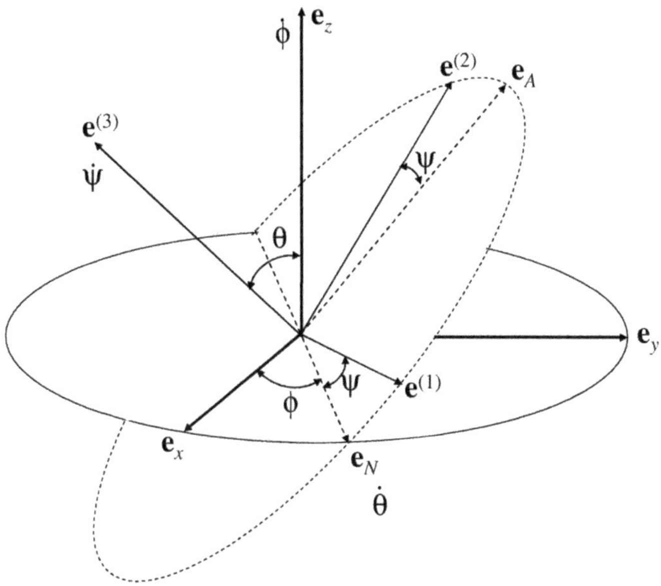

Figure 4.1.1. Euler angles.

where $\mathbf{e}_N$ is the unit vector along the line of nodes shown in Fig. 4.1.1. In addition to the auxiliary node vector $\mathbf{e}_N$, we will need the "antinode" vector $\mathbf{e}_A$ that lies in the plane spanned by $\mathbf{e}^{(1)}$ and $\mathbf{e}^{(2)}$ and is perpendicular to both $\mathbf{e}_N$ and $\mathbf{e}^{(3)}$. Using

$$\begin{aligned}
\mathbf{e}_N &= \mathbf{e}^{(1)} \cos\psi - \mathbf{e}^{(2)} \sin\psi, \\
\mathbf{e}_z &= \mathbf{e}_A \sin\theta + \mathbf{e}^{(3)} \cos\theta, \\
\mathbf{e}_A &= \mathbf{e}^{(1)} \sin\psi + \mathbf{e}^{(2)} \cos\psi,
\end{aligned} \tag{4.1.22}$$

one can project $\boldsymbol{\omega}$ onto the internal frame:

$$\begin{aligned}
\omega_1 &= \dot{\phi} \sin\theta \sin\psi + \dot{\theta} \cos\psi, \\
\omega_2 &= \dot{\phi} \sin\theta \cos\psi - \dot{\theta} \sin\psi, \\
\omega_3 &= \dot{\phi} \cos\theta + \dot{\psi}.
\end{aligned} \tag{4.1.23}$$

One can also project $\boldsymbol{\omega}$ of Eq. (4.1.21) onto the laboratory axes:

$$\begin{aligned}
\omega_x &= \dot{\psi} \sin\theta \sin\phi + \dot{\theta} \cos\phi, \\
\omega_y &= -\dot{\psi} \sin\theta \cos\phi + \dot{\theta} \sin\phi, \\
\omega_z &= \dot{\psi} \cos\theta + \dot{\phi}.
\end{aligned} \tag{4.1.24}$$

This can be done using $\mathbf{e}_N$ in Eq. (4.1.25) and $\mathbf{e}^{(3)}$ in Eq. (4.1.27) in the following section.

4.1.5 *Rotation matrices**

First, let us now find the complete relationship between the two sets of vectors $\mathbf{e}^{(1)}$, $\mathbf{e}^{(2)}$, $\mathbf{e}^{(3)}$ and $\mathbf{e}_x$, $\mathbf{e}_y$, $\mathbf{e}_z$ in terms of Euler's angles. The three vectors, $\mathbf{e}^{(3)}$, $\mathbf{e}_A$, and $\mathbf{e}_z$ lie in the same plane. The auxiliary vectors can be projected onto $\mathbf{e}_x$, $\mathbf{e}_y$, $\mathbf{e}_z$ as follows:

$$\begin{aligned} \mathbf{e}_N &= \cos\phi \mathbf{e}_x + \sin\phi \mathbf{e}_y, \\ \mathbf{e}_A &= -\cos\theta \sin\phi \mathbf{e}_x + \cos\theta\cos\phi \mathbf{e}_y + \sin\theta \mathbf{e}_z. \end{aligned} \tag{4.1.25}$$

To project $\mathbf{e}^{(1)}$, $\mathbf{e}^{(2)}$, $\mathbf{e}^{(3)}$ onto $\mathbf{e}_x$, $\mathbf{e}_y$, $\mathbf{e}_z$, one can at first project $\mathbf{e}^{(1)}$ and $\mathbf{e}^{(2)}$ onto $\mathbf{e}_N$ and $\mathbf{e}_A$:

$$\begin{aligned} \mathbf{e}^{(1)} &= \cos\psi \mathbf{e}_N + \sin\psi \mathbf{e}_A, \\ \mathbf{e}^{(2)} &= -\sin\psi \mathbf{e}_N + \cos\psi \mathbf{e}_A, \end{aligned} \tag{4.1.26}$$

and then use Eq. (4.1.25). Vector $\mathbf{e}^{(3)}$ can be projected onto $\mathbf{e}_x$, $\mathbf{e}_y$, $\mathbf{e}_z$ directly. As the result, one obtains

$$\begin{aligned} \mathbf{e}^{(1)} &= (\cos\phi\cos\psi - \cos\theta\sin\phi\sin\psi)\,\mathbf{e}_x \\ &\quad + (\sin\phi\cos\psi + \cos\theta\cos\phi\sin\psi)\,\mathbf{e}_y + \sin\theta\sin\psi \mathbf{e}_z, \\ \mathbf{e}^{(2)} &= (-\cos\phi\sin\psi - \cos\theta\sin\phi\cos\psi)\,\mathbf{e}_x \\ &\quad + (-\sin\phi\sin\psi + \cos\theta\cos\phi\cos\psi)\,\mathbf{e}_y + \sin\theta\cos\psi \mathbf{e}_z, \\ \mathbf{e}^{(3)} &= \sin\theta\sin\phi \mathbf{e}_x - \sin\theta\cos\phi \mathbf{e}_y + \cos\theta \mathbf{e}_z. \end{aligned} \tag{4.1.27}$$

Coefficient in these equations form a 3×3 rotation matrix, properties of which will be studied in a more general context in the following section.

4.1.5.1 *General*

A number of properties of the rotation matrices can be proven. In particular, these matrices are orthogonal:

$$\mathbb{A}^T \cdot \mathbb{A} \equiv \mathbb{A} \cdot \mathbb{A}^T = \mathbb{I}, \quad \mathbb{A}^T = \mathbb{A}^{-1}, \tag{4.1.28}$$

where $\mathbb{A}^T$ is the matrix transposed to $\mathbb{A}$. Transformations by orthogonal matrices (i.e., rotations) preserve the length of the vectors, $\mathbf{r}'^2 = \mathbf{r}^2$.

Indeed, considering vectors as columns ($\mathbf{r}$) and rows ($\mathbf{r}^T$) of components, one obtains

$$\mathbf{r}'^2 = \mathbf{r}' \cdot \mathbf{r}' = \mathbf{r}'^T \cdot \mathbf{r}' = \mathbf{r}^T \cdot \mathbb{A}^T \cdot \mathbb{A} \cdot \mathbf{r}. \tag{4.1.29}$$

Requiring $\mathbb{A}^T \cdot \mathbb{A} = \mathbb{I}$, one secures $\mathbf{r}'^2 = \mathbf{r}^2$.

4.1.5.2 *Active and passive rotations*

One has to distinguish between active and passive rotations. Active rotations are actual rotations of vectors such as the rotation described by Eq. (4.1.1). Active rotations are described with respect to the unchanged coordinate system and are represented by

$$r'_\alpha = A_{\alpha\beta} r_\beta, \quad \mathbf{r}' = \mathbb{A} \cdot \mathbf{r} \tag{4.1.30}$$

(with summation over repeated indices), where $\mathbb{A}$ is a matrix of elements $A_{\alpha\beta}$ and $\mathbf{r}$ is a column of components r_β. Passive rotations are rotations of the frame axes, whereas $\mathbf{r}$ remains unchanged in the original frame. With respect to the new rotated frame, the components of $\mathbf{r}$ are rotated. However, this *passive* rotation of $\mathbf{r}$ is in the opposite direction with respect to the active rotation of the frame vectors. Passive rotation is used to relate components of a vector in different frames with each other.

Let $\mathbf{e}_\gamma$ be the original frame vectors, such as $\mathbf{e}_x$, $\mathbf{e}_y$, and $\mathbf{e}_z$ of the laboratory system and $\mathbf{e}^{(\gamma)}$ are rotated vectors such as the body-frame vectors $\mathbf{e}^{(1)}$, $\mathbf{e}^{(2)}$, and $\mathbf{e}^{(3)}$. In the initial state before rotation, $\mathbf{e}^{(\gamma)} = \mathbf{e}_\gamma$. The rotation of the vectors $\mathbf{e}^{(\gamma)} \equiv \mathbf{e}'_\gamma$ is the active rotation defined by Eq. (4.1.30) in the vector form

$$\mathbf{e}^{(\gamma)} = \mathbb{A} \cdot \mathbf{e}_\gamma \tag{4.1.31}$$

($\mathbf{e}_\gamma$ is a column of its components) or

$$e_\alpha^{(\gamma)} = A_{\alpha\beta} \left(\mathbf{e}_\gamma\right)_\beta = A_{\alpha\beta} \delta_{\beta\gamma} = A_{\alpha\gamma}. \tag{4.1.32}$$

The whole vector $\mathbf{e}^{(\gamma)}$ is thus given by

$$\mathbf{e}^{(\gamma)} = e_\alpha^{(\gamma)} \mathbf{e}_\alpha = A_{\alpha\gamma} \mathbf{e}_\alpha = \left(\mathbb{A}^T\right)_{\gamma\alpha} \mathbf{e}_\alpha. \tag{4.1.33}$$

There is a subtle difference with Eq. (4.1.31): In the former, $\mathbf{e}_\gamma$ is a single vector, the initial value of $\mathbf{e}^{(\gamma)}$, whereas in the latter, there are three different vectors $\mathbf{e}_\alpha$, on which $\mathbf{e}^{(\gamma)}$ is projected. In particular, the coefficients in Eq. (4.1.27) directly form the matrix $\mathbb{A}^T$.

Let us consider passive rotations now. The vector $\mathbf{r}$ remains the same and it can be projected on both frames as follows:

$$\mathbf{r} = r_\gamma \mathbf{e}_\gamma = r^{(\gamma)} \mathbf{e}^{(\gamma)}. \tag{4.1.34}$$

Using Eq. (4.1.33), one can relate the components of $\mathbf{r}$ in both frames:

$$\mathbf{r} = r^{(\gamma)} \mathbf{e}^{(\gamma)} = r^{(\gamma)} A_{\alpha\gamma} \mathbf{e}_\alpha = A_{\alpha\gamma} r^{(\gamma)} \mathbf{e}_\alpha, \tag{4.1.35}$$

i.e.,

$$r_\alpha = A_{\alpha\gamma} r^{(\gamma)}, \quad r^{(\gamma)} = \left(\mathbb{A}^T\right)_{\gamma\alpha} r_\alpha, \tag{4.1.36}$$

where we used the orthogonality of $\mathbb{A}$, i.e., Eq. (4.1.28) or $\left(\mathbb{A}^T\right)_{\gamma'\alpha} A_{\alpha\gamma} = \delta_{\gamma\gamma'}$. One can see that since $r^{(\gamma)} = r'_\gamma$, the passive rotation is done by the inverse rotation matrix,

$$r'_\alpha = \left(\mathbb{A}^T\right)_{\alpha\beta} r_\beta, \quad \mathbf{r}' = \mathbb{A}^T \cdot \mathbf{r}, \tag{4.1.37}$$

i.e., in the opposite direction compared to the active rotation, c.f. Eq. (4.1.30). Difference between active and passive rotations is a subtle point ignored in many publications, leading to uncertainties and errors.

4.1.5.3 *Time-dependent rotation and angular velocity*

Let us consider orientations of the body at time t as obtained from the reference orientation $\mathbf{e}^{(\gamma)} = \mathbf{e}_\gamma$ at $t = 0$ by a rotations matrix that depends on time. Then any vector $\mathbf{r}$ rotating with the body is given by

$$\mathbf{r}(t) = \mathbb{A}(t) \cdot \mathbf{r}_0, \tag{4.1.38}$$

where $\mathbf{r}_0 = \mathbf{r}(0)$ and we neglect the trivial translational motion. The velocity of the point represented by $\mathbf{r}$ obeys the equation

$$\mathbf{v} = \dot{\mathbf{r}} = \dot{\mathbb{A}} \cdot \mathbf{r}_0 = \boldsymbol{\Omega} \cdot \mathbf{r}, \tag{4.1.39}$$

where the matrix $\boldsymbol{\Omega}$ is defined by

$$\boldsymbol{\Omega} \equiv \dot{\mathbb{A}} \cdot \mathbb{A}^T. \tag{4.1.40}$$

Note that we are performing an active rotation, while passive rotation is unsuitable for this purpose. The matrix $\boldsymbol{\Omega}$ is antisymmetric. Indeed, from Eq. (4.1.28), one obtains

$$\dot{\mathbb{A}} \cdot \mathbb{A}^T + \mathbb{A}^T \cdot \dot{\mathbb{A}} = 0 \quad \Rightarrow \quad \dot{\mathbb{A}} \cdot \mathbb{A}^T = -\mathbb{A}^T \cdot \dot{\mathbb{A}}. \tag{4.1.41}$$

Now, transposing $\boldsymbol{\Omega}$ gives

$$\boldsymbol{\Omega}^T = \left(\dot{\mathbb{A}} \cdot \mathbb{A}^T\right)^T = \mathbb{A} \cdot \dot{\mathbb{A}}^T = -\dot{\mathbb{A}} \cdot \mathbb{A}^T = -\boldsymbol{\Omega}. \tag{4.1.42}$$

Thus, all diagonal elements of $\boldsymbol{\Omega}$ are zero and non-diagonal elements are antisymmetric and can be represented by only three different numbers as follows:

$$\boldsymbol{\Omega} = \begin{pmatrix} 0 & -\omega_z & \omega_y \\ \omega_z & 0 & -\omega_x \\ -\omega_y & \omega_x & 0 \end{pmatrix}, \quad \Omega_{\alpha\beta} = -\epsilon_{\alpha\beta\gamma}\omega_\gamma, \tag{4.1.43}$$

where $\epsilon_{\alpha\beta\gamma}$ is a fully antisymmetric. Thus, Eq. (4.1.39) takes the form

$$v_\alpha = -\epsilon_{\alpha\beta\gamma}\omega_\gamma r_\beta = \epsilon_{\alpha\beta\gamma}\omega_\beta r_\gamma = [\boldsymbol{\omega} \times \mathbf{r}]_\alpha, \tag{4.1.44}$$

which coincides with Eq. (4.1.7). One can resolve Eq. (4.1.43) for ω_γ by multiplying it by $\epsilon_{\alpha\beta\gamma'}$ and summing over α and β:

$$\epsilon_{\alpha\beta\gamma'}\Omega_{\alpha\beta} = -\epsilon_{\alpha\beta\gamma'}\epsilon_{\alpha\beta\gamma}\omega_\gamma = -2\omega_{\gamma'}, \tag{4.1.45}$$

which yields

$$\omega_\gamma = -\frac{1}{2}\Omega_{\alpha\beta}\epsilon_{\alpha\beta\gamma}. \tag{4.1.46}$$

These are the components of the angular velocity $\boldsymbol{\omega}$ in the laboratory frame. The components of this same vector in the body frame can be obtained by the *passive* rotation. Putting the component indices up ($\omega^{(1)} \equiv \omega_1$, etc.) to avoid confusion and using the second equation in Eq. (4.1.36), one obtains

$$\omega^{(\gamma)} = \left(\mathbb{A}^T\right)_{\gamma\delta}\omega_\delta = \omega_\delta A_{\delta\gamma} = -\frac{1}{2}\Omega_{\alpha\beta}\epsilon_{\alpha\beta\delta}A_{\delta\gamma}. \tag{4.1.47}$$

The formalism in this section does not imply any particular parametrization of $\mathbb{A}$ through rotation angles. The latter can be Euler angles, the angles that specify a single rotation, or any other parameters. Examples will be considered in the following section.

4.1.5.4 *Euler angles*

As follows from Eqs. (4.1.27) and (4.1.33),

$$\mathbb{A}^T = \begin{pmatrix} \cos\phi\cos\psi - \cos\theta\sin\phi\sin\psi & \sin\phi\cos\psi + \cos\theta\cos\phi\sin\psi & \sin\theta\sin\psi \\ -\cos\phi\sin\psi - \cos\theta\sin\phi\cos\psi & -\sin\phi\sin\psi + \cos\theta\cos\phi\cos\psi & \sin\theta\cos\psi \\ \sin\theta\sin\phi & -\sin\theta\cos\phi & \cos\theta \end{pmatrix}, \tag{4.1.48}$$

hence

$$\mathbb{A} = \begin{pmatrix} \cos\phi\cos\psi - \cos\theta\sin\phi\sin\psi & -\cos\phi\sin\psi - \cos\theta\sin\phi\cos\psi & \sin\theta\sin\phi \\ \sin\phi\cos\psi + \cos\theta\cos\phi\sin\psi & -\sin\phi\sin\psi + \cos\theta\cos\phi\cos\psi & -\sin\theta\cos\phi \\ \sin\theta\sin\psi & \sin\theta\cos\psi & \cos\theta \end{pmatrix}. \tag{4.1.49}$$

(H. Goldstein in his *Classical Mechanics* considers *passive* rotations as the basis, thus his passive rotation $\mathbb{A}$ is the same as our $\mathbb{A}^T$, while our $\mathbb{A}$ is the active rotation.) For the rotation matrix above, Eq. (4.1.43) takes the form

$$\mathbf{\Omega} = \begin{pmatrix} 0 & -\left(\dot\psi\cos\theta + \dot\phi\right) & -\dot\psi\sin\theta\cos\phi + \dot\theta\sin\phi \\ \dot\psi\cos\theta + \dot\phi & 0 & -\left(\dot\psi\sin\theta\sin\phi + \dot\theta\cos\phi\right) \\ -\left(-\dot\psi\sin\theta\cos\phi + \dot\theta\sin\phi\right) & \dot\psi\sin\theta\sin\phi + \dot\theta\cos\phi & 0 \end{pmatrix}, \tag{4.1.50}$$

from which the laboratory-frame components of $\boldsymbol{\omega}$ can be read out. One can see that they coincide with those given by Eq. (4.1.24). Equation (4.1.47) yields components of $\boldsymbol{\omega}$ in the body frame that are given by Eq. (4.1.23).

4.1.5.5 *Single rotation*

The finite rotation described by Eq. (4.1.1) can be represented as acting by a matrix on the vector $\mathbf{r}$. To work out the form of this matrix, we represent Eq. (4.1.1) via components with respect to the laboratory axes $\mathbf{e}_\alpha$:

$$r'_\alpha = r_\alpha\cos\chi + \nu_\alpha\nu_\beta r_\beta\left(1 - \cos\chi\right) + \epsilon_{\alpha\beta\gamma}\nu_\beta r_\gamma\sin\chi. \tag{4.1.51}$$

This can be rewritten in the form of Eq. (4.1.30) where the elements of the rotation matrix $\mathbf{A}$ are given by

$$A_{\alpha\beta} = \delta_{\alpha\beta}\cos\chi + \nu_\alpha\nu_\beta\left(1 - \cos\chi\right) - \epsilon_{\alpha\beta\gamma}\nu_\gamma\sin\chi. \tag{4.1.52}$$

The matrix of small rotations at linear order in $\delta\chi$ becomes

$$A_{\alpha\beta} = \delta_{\alpha\beta} - \epsilon_{\alpha\beta\gamma}\delta\chi_\gamma, \quad \delta\chi_\gamma \equiv \nu_\gamma\delta\chi. \tag{4.1.53}$$

The components of ν are explicitly given by

$$\nu_x = \sin\vartheta\cos\varphi, \quad \nu_y = \sin\vartheta\sin\varphi, \quad \nu_z = \cos\vartheta. \tag{4.1.54}$$

Substitution yields the single-rotation matrix

$$\mathbb{A} = (\mathbf{A}_1, \mathbf{A}_2, \mathbf{A}_3), \tag{4.1.55}$$

where columns $\mathbf{A}_i$ are given by

$$\mathbf{A}_1 = \begin{pmatrix} \cos\chi + (1-\cos\chi)\sin^2\vartheta\cos^2\varphi \\ (1-\cos\chi)\sin^2\vartheta\cos\varphi\sin\varphi + \sin\chi\cos\vartheta \\ \cos\vartheta\cos\varphi(1-\cos\chi)\sin\vartheta - \sin\vartheta\sin\varphi\sin\chi \end{pmatrix},$$

$$\mathbf{A}_2 = \begin{pmatrix} (1-\cos\chi)\sin^2\vartheta\sin\varphi\cos\varphi - \sin\chi\cos\vartheta \\ \cos\chi + (1-\cos\chi)\sin^2\vartheta\sin^2\varphi \\ (1-\cos\chi)\sin\vartheta\cos\vartheta\sin\varphi + \sin\chi\cos\varphi\sin\vartheta \end{pmatrix},$$

$$\mathbf{A}_3 = \begin{pmatrix} (1-\cos\chi)\cos\vartheta\sin\vartheta\cos\varphi + \sin\chi\sin\vartheta\sin\varphi \\ \cos\vartheta(1-\cos\chi)\sin\vartheta\sin\varphi - \sin\chi\cos\varphi\sin\vartheta \\ (1-\cos\chi)\cos^2\vartheta + \cos\chi \end{pmatrix}. \tag{4.1.56}$$

In the special cases of rotation around the axes x, y, and z, this matrix simplifies to

$$\mathbb{A}^{(x)} = \begin{pmatrix} 1 & 0 & 0 \\ 0 & \cos\chi & -\sin\chi \\ 0 & \sin\chi & \cos\chi \end{pmatrix}, \quad \mathbb{A}^{(y)} = \begin{pmatrix} \cos\chi & 0 & \sin\chi \\ 0 & 1 & 0 \\ -\sin\chi & 0 & \cos\chi \end{pmatrix},$$

$$\mathbb{A}^{(z)} = \begin{pmatrix} \cos\chi & -\sin\chi & 0 \\ \sin\chi & \cos\chi & 0 \\ 0 & 0 & 1 \end{pmatrix}. \tag{4.1.57}$$

In the general case, however, this matrix is much more cumbersome than the rotation matrix in terms of three successive rotations by Euler angles, Eq. (4.1.49). Also, components of ω in both the laboratory and body frames have a complicated form. This is the reason why the single-rotation matrix is never used.

The matrices $\mathbb{A}(\theta, \phi, \psi)$ of Eq. (4.1.49) and $\mathbb{A}(\vartheta, \varphi, \chi)$ of Eq. (4.1.55) should coincide with a proper choice of the angles. Indeed, according to the Euler theorem, the three rotations by Euler angles should be equivalent to a single rotation. This means that there is a vector collinear with the rotation axis that is invariant under rotation. This is $\boldsymbol{\nu}$ given by Eq. (4.1.54). The same vector should be the eigenvector of $\mathbb{A}(\theta, \phi, \psi)$ of Eq. (4.1.49) with eigenvalue $\lambda = 1$. Thus, if the rotation matrix $\mathbb{A}(\theta, \phi, \psi)$ is given, one can find its eigenvector of eigenvalue 1 and equate it to $\boldsymbol{\nu}$, thus obtaining the direction of the single-rotation axis in terms of Euler's angles. However, it is difficult to do analytically in the general case. On the other hand, equating the traces of both matrices, one easily obtains

$$\cos\theta + (1+\cos\theta)\cos(\phi+\psi) = 1 + 2\cos\chi, \tag{4.1.58}$$

which defines the single-rotation angle χ in terms of Euler's angles. Using computer algebra on Eqs. (4.1.46) and (4.1.47), one can calculate the components of ω in the laboratory and body frames in terms of ϑ, φ, χ and their first derivatives turn out to be cumbersome. Although the parametrization of the body orientations through single rotations out of a reference state is a valid approach that could be used instead of the parametrization based on Euler's angles, at least for numerical work, it is never practically applied.

4.2 Rotational dynamics

4.2.1 *Kinetic energy of a rotating body: Moments of inertia*

Inserting Eq. (4.1.9) into the general expression for the kinetic energy, one proceeds as

$$\begin{aligned} E_k &= \frac{1}{2}\sum_i m_i \mathbf{v}_i^2 = \frac{1}{2}\sum_i m_i \left(\mathbf{V} + [\boldsymbol{\omega}\times\mathbf{r}_i]\right)^2 \\ &= \sum_i m_i\left(\frac{1}{2}\mathbf{V}^2 + \mathbf{V}\cdot[\boldsymbol{\omega}\times\mathbf{r}_i] + \frac{1}{2}[\boldsymbol{\omega}\times\mathbf{r}_i]^2\right) \\ &= \frac{1}{2}M\mathbf{V}^2 + M\mathbf{V}\cdot[\boldsymbol{\omega}\times\mathbf{R}] + \frac{1}{2}\sum_i m_i\left\{\omega^2 r_i^2 - (\boldsymbol{\omega}\cdot\mathbf{r}_i)^2\right\}. \end{aligned} \tag{4.2.1}$$

To transform the last term, the identity

$$[\mathbf{a}\times\mathbf{b}]\cdot[\mathbf{c}\times\mathbf{d}] = (\mathbf{a}\cdot\mathbf{c})(\mathbf{b}\cdot\mathbf{d}) - (\mathbf{a}\cdot\mathbf{d})(\mathbf{b}\cdot\mathbf{c}) \tag{4.2.2}$$

was used. If the body rotates around a point of support or around an instantaneous axis of rotation, the velocities are given by Eq. (4.1.7) or Eq. (4.1.14), so that the first and second terms in Eq. (4.2.1) do not occur. In the second term, $\mathbf{R}$ is the position of the CM with respect to the origin O, moving with velocity $\mathbf{V}$. If the latter is put in the CM, then $\mathbf{R} = \mathbf{0}$ and the second term is absent. In this case, the first term in Eq. (4.2.1) is the kinetic energy of the translational motion of the center of mass. The last term in Eq. (4.2.1), the rotational energy, can be rewritten in components as follows:

$$E_{\mathrm{rot}} = \frac{1}{2}\sum_i m_i \left\{\omega_\alpha\omega_\beta\delta_{\alpha\beta}r_{i\gamma}^2 - \omega_\alpha r_{i\alpha}\omega_\beta r_{i\beta}\right\} = \frac{1}{2}I_{\alpha\beta}\omega_\alpha\omega_\beta, \qquad (4.2.3)$$

where $I_{\alpha\beta}$ is the tensor of inertia defined by

$$I_{\alpha\beta} = \sum_i m_i \left(\delta_{\alpha\beta}r_{i\gamma}^2 - r_{i\alpha}r_{i\beta}\right). \qquad (4.2.4)$$

Equation (4.2.4) works out to

$$I_{xx} = \sum_i m_i \left(y_i^2 + z_i^2\right), \quad I_{xy} = -\sum_i m_i x_i y_i, \qquad (4.2.5)$$

etc. In the diagonal components of $I_{\alpha\beta}$, the squares of the distances from the corresponding axes enter, like $y_i^2 + z_i^2$ being the square of the distance from the x-axis. For solid bodies, moments of inertia can be obtained by replacing summation by integration. In some cases, symmetry can be used to simplify the calculation.

If $I_{\alpha\beta}$ is defined with respect to the CM, the tensor of inertia $I'_{\alpha\beta}$ with respect to another origin O' shifted by $\mathbf{a}$ from the CM can be expressed via $I_{\alpha\beta}$ by using $\mathbf{r}' = \mathbf{r} - \mathbf{a}$. One obtains

$$\begin{aligned} I'_{\alpha\beta} &= \sum_i m_i \left\{\delta_{\alpha\beta}r_{i\gamma}'^2 - r'_{i\alpha}r'_{i\beta}\right\} \\ &= \sum_i m_i \left\{\delta_{\alpha\beta}\left(r_{i\gamma} - a_\gamma\right)\left(r_{i\gamma} - a_\gamma\right) - \left(r_{i\alpha} - a_\alpha\right)\left(r_{i\beta} - a_\beta\right)\right\} \\ &= I_{\alpha\beta} + M\left\{\delta_{\alpha\beta}a_\gamma a_\gamma - a_\alpha a_\beta\right\}, \end{aligned} \qquad (4.2.6)$$

where we used $\sum_i m_i r_{i\gamma} = 0$. This relation is known as the Steiner theorem. Equation (4.2.6) yields

$$I'_{xx} = I_{xx} + M\left(a_y^2 + a_z^2\right), \qquad (4.2.7)$$

etc. Here, $a_y^2 + a_z^2$ is the square of the distance between the x-axis and the CM. One can see that diagonal moments of inertia are minimal with respect to the CM and the shifted moments of inertia are greater than CM-centered ones.

In Eqs. (4.2.4)–(4.2.6), the moments of inertia are written with respect to an arbitrarily oriented coordinate system. If the laboratory coordinate system is used then, in general, components of the tensor of inertia $I_{\alpha\beta}$ change with time, as the body rotates. This is very inconvenient and makes writing down a closed system of equations of motion hardly possible. A much better choice is to use the body's own embedded coordinate system with respect to which $I_{\alpha\beta}$ are independent of the body's orientation. Moreover, for a body of any shape, the tensor of inertia can be diagonalized by an appropriate choice of the orientation of the embedded Descarte axes. Such axes are called *principal axes.* Then the rotational energy has the simple form

$$E_{\text{rot}} = \frac{1}{2}\left(I_1\omega_1^2 + I_2\omega_2^2 + I_3\omega_3^2\right). \tag{4.2.8}$$

In most practical cases, bodies of a symmetric shape are considered, for which the choice of the internal coordinate system is obvious.

Substituting Eq. (4.1.23) into Eq. (4.2.8) gives the expression of the rotational kinetic energy in terms of Euler angles:

$$\begin{aligned}E_{\text{rot}} &= \frac{1}{2}I_1\left(\dot\phi\sin\theta\sin\psi + \dot\theta\cos\psi\right)^2 + \frac{1}{2}I_2\left(\dot\phi\sin\theta\cos\psi - \dot\theta\sin\psi\right)^2\\ &\quad + \frac{1}{2}I_3\left(\dot\phi\cos\theta + \dot\psi\right)^2.\end{aligned} \tag{4.2.9}$$

In the case of symmetric top ($I_1 = I_2$), simplification yields

$$E_{\text{rot}} = \frac{1}{2}I_1\left(\dot\phi^2\sin^2\theta + \dot\theta^2\right) + \frac{1}{2}I_3\left(\dot\phi\cos\theta + \dot\psi\right)^2. \tag{4.2.10}$$

Since E_{rot} contains terms beyond quadratic in generalized coordinates and velocities, rotational motion of even a free body can be rather complicated.

4.2.2 *Sliding principal axes*

If the two moments of inertia are equal, one can simplify the formalism by using the so-called *sliding principal axes.* In the case $I_1 = I_2$, the orientation of the vectors $\mathbf{e}^{(1)}$ and $\mathbf{e}^{(2)}$ is not fixed by any condition. One can choose, instead of the embedded vectors, the vectors $\mathbf{e}^{(1)} = \mathbf{e}_N$ and $\mathbf{e}^{(2)} = \mathbf{e}_A$, see Fig. 4.1.1, which "slide" with respect to the spinning body. Strictly speaking, the embedded vectors are still used in the background to define the spin $\dot\psi$, but all dynamical vectors are projected onto the sliding axes. The set of the sliding principal axes can be obtained from the general formula, Eq. (4.1.27), by setting $\psi = 0$:

$$\begin{aligned}\mathbf{e}^{(1)} &= \cos\phi\mathbf{e}_x + \sin\phi\mathbf{e}_y,\\ \mathbf{e}^{(2)} &= -\cos\theta\sin\phi\mathbf{e}_x + \cos\theta\cos\phi\mathbf{e}_y + \sin\theta\mathbf{e}_z,\\ \mathbf{e}^{(3)} &= \sin\theta\sin\phi\mathbf{e}_x - \sin\theta\cos\phi\mathbf{e}_y + \cos\theta\mathbf{e}_z\end{aligned} \tag{4.2.11}$$

(note that $\mathbf{e}^{(3)}$ does not depend on ψ and remains the same). The formulas for the components of the angular velocity, Eq. (4.1.23), after setting $\psi = 0$ simplify to

$$\omega_1 = \dot{\theta}, \quad \omega_2 = \dot{\phi}\sin\theta, \quad \omega_3 = \dot{\phi}\cos\theta + \dot{\psi} \tag{4.2.12}$$

(note that we are keeping $\dot{\psi}$). One can test this formalism by inserting the above expressions into Eq. (4.2.8), which yields the same in Eq. (4.2.10).

4.2.3 *Angular momentum in laboratory and body frames*

Let us now express the angular momentum $\mathbf{L}$ through the components of the angular velocity $\boldsymbol{\omega}$. Using Eqs. (4.1.9) and (4.1.2), one proceeds as

$$\begin{aligned}\mathbf{L} &= \sum_i m_i \left[\mathbf{r}_i \times \mathbf{v}_i\right] = \sum_i m_i \left[\mathbf{r}_i \times \left(\mathbf{V} + \left[\boldsymbol{\omega} \times \mathbf{r}_i\right]\right)\right] \\ &= M\left[\mathbf{R} \times \mathbf{V}\right] + \sum_i m_i \left\{\boldsymbol{\omega} r_i^2 - \mathbf{r}_i\left(\boldsymbol{\omega}\cdot\mathbf{r}_i\right)\right\}.\end{aligned} \tag{4.2.13}$$

The first term of this result is the angular momentum associated with the translation motion of the center of mass. This contribution can be removed by choosing the frame that is moving with the CM. If the rotation around a point of support or rolling without skidding is considered, Eqs. (4.1.7) or (4.1.14) are used and this term does not arise. Henceforth, we will consider only the last term in Eq. (4.2.13) that is due to the rotation of the body. In components, it has the form

$$L_\alpha = I_{\alpha\beta}\omega_\beta, \tag{4.2.14}$$

where the tensor of inertia is given by Eq. (4.2.4). One can also define $\mathbf{L}$ with respect to the instantaneous rotation axis, see Sec. 4.1.3. In this case, one has to replace $I_{\alpha\beta} \Rightarrow I'_{\alpha\beta}$, where $I'_{\alpha\beta}$ is defined with respect to the instantaneous axis, see Eq. (4.2.6).

The angular momentum $\mathbf{L}$ in the absence of torques acting on the system is conserved, $\mathbf{L} = \mathbf{const}$. Thus, all components L_α with respect to the laboratory coordinate system are constants. As the body rotates, $I_{\alpha\beta}$ with respect to the laboratory frame in general is changing, apart from special cases such as spherical body. Thus, the vector ω is changing too, so that the orientation of the body changes in time in a complicated way. The relation between the components of $\mathbf{L}$ and ω simplifies in the principal coordinate system:

$$\mathbf{L} = I_1\omega_1\mathbf{e}^{(1)} + I_2\omega_2\mathbf{e}^{(2)} + I_3\omega_3\mathbf{e}^{(3)}. \tag{4.2.15}$$

The components of the angular momentum in the laboratory system can be obtained from this with the help of Eqs. (4.1.27). In particular,

$$L_z = I_1\omega_1 \sin\theta \sin\psi + I_2\omega_2 \sin\theta \cos\psi + I_3\omega_3 \cos\theta. \tag{4.2.16}$$

Now, using Eqs. (4.1.23), one obtains the full expression for L_z in terms of Euler's angles:

$$\begin{aligned} L_z &= I_1 \sin\theta \sin\psi \left(\dot{\phi} \sin\theta \sin\psi + \dot{\theta} \cos\psi\right) \\ &\quad + I_2 \sin\theta \cos\psi \left(\dot{\phi} \sin\theta \cos\psi - \dot{\theta} \sin\psi\right) + I_3 \cos\theta \left(\dot{\phi} \cos\theta + \dot{\psi}\right). \end{aligned} \tag{4.2.17}$$

For a symmetric top, $I_1 = I_2$, this simplifies to

$$L_z = I_2\dot{\phi} \sin^2\theta + I_3 \left(\dot{\phi} \cos\theta + \dot{\psi}\right) \cos\theta = I_2\omega_2 \sin\theta + I_3\omega_3 \cos\theta. \tag{4.2.18}$$

With the help of Eq. (4.2.15), the rotational kinetic energy of Eq. (4.2.8) can be rewritten in the form

$$E_{\text{rot}} = \frac{1}{2}\left(\frac{L_1^2}{I_1} + \frac{L_2^2}{I_2} + \frac{L_3^2}{I_3}\right). \tag{4.2.19}$$

Comparing Eqs. (4.2.3) and (4.2.14), one can easily obtain the relations

$$\mathbf{L} = \frac{\partial E_{\text{rot}}}{\partial \boldsymbol{\omega}}, \quad L_k = \frac{\partial E_{\text{rot}}}{\partial \omega_k}, \tag{4.2.20}$$

where $k = 1, 2, 3$. This means that the angular momentum plays the role of a generalized momentum corresponding to the angular velocity within the Lagrangian formalism.

4.2.4 *Equation of motion for the angular momentum*

The equation of motion for the angular momentum $\mathbf{L}$ is given by Eq. (1.3.11). For the rotating solid body, the angular momentum can be expressed via the moments of inertia and and the components of the angular velocity with respect to the principal axes, as it was done in Sec. 4.2.3.

The case of rolling without skidding (see Sec. 4.1.3) requires special consideration. Here, one can define the angular momentum with respect to the instantaneous axis of rotation, $\mathbf{L}' = \sum_i m_i \left[\mathbf{r}'_i \times \mathbf{v}_i\right]$, and use Eq. (4.1.11) to connect to the velocities as $\mathbf{v}_i = \mathbf{V} + \dot{\mathbf{r}}_i = \mathbf{V} - \dot{\mathbf{b}} + \dot{\mathbf{r}}'$.

Differentiating $\mathbf{L}'$, one obtains

$$\dot{\mathbf{L}}' = \frac{d}{dt}\sum_i m_i\left[\mathbf{r}_i' \times \mathbf{v}_i\right] = \sum_i m_i\left\{\left[\left(\mathbf{v}_i - \mathbf{V} + \dot{\mathbf{b}}\right) \times \mathbf{v}_i\right] + \left[\mathbf{r}_i' \times \dot{\mathbf{v}}_i\right]\right\}$$
$$= \left(\mathbf{V} + \dot{\mathbf{b}}\right) \times M\mathbf{V} + \sum_i \mathbf{r}_i' \times \mathbf{F}_i = \dot{\mathbf{b}} \times \mathbf{P} + \boldsymbol{\tau}', \tag{4.2.21}$$

where $\boldsymbol{\tau}'$ is the torque with respect to the contact point. The first term in the rhs of this equation resembles the kinematic term in Eq. (1.3.11) and it has a similar origin. Using the rolling constraint, Eq. (4.1.13), one can write the rotational dynamical equation as

$$\dot{\mathbf{L}}' = \tilde{\boldsymbol{\tau}}' \equiv M\dot{\mathbf{b}} \times [\boldsymbol{\omega} \times \mathbf{b}] + \boldsymbol{\tau}'. \tag{4.2.22}$$

For the cylinder or sphere rolling on a plane, $\mathbf{b} = \mathbf{n} = \mathbf{const}.$, and the constraint term disappears.

A torque created by potential forces is related to the dependence of the potential energy U on the body's orientation. Infinitesimal change of U due to rotations can be represented as

$$\delta U = -\sum_i \mathbf{f}_i \cdot \delta\mathbf{r}_i = -\sum_i \mathbf{f}_i \cdot [\delta\boldsymbol{\chi} \times \mathbf{r}_i] = -\sum_i \delta\boldsymbol{\chi} \cdot [\mathbf{r}_i \times \mathbf{f}_i] = -\boldsymbol{\tau} \cdot \delta\boldsymbol{\chi}, \tag{4.2.23}$$

thus

$$\boldsymbol{\tau} = -\frac{\delta U}{\delta\boldsymbol{\chi}}. \tag{4.2.24}$$

Here, we avoid writing $\boldsymbol{\tau} = -\partial U(\boldsymbol{\chi})/\partial\boldsymbol{\chi}$ since the function $U(\boldsymbol{\chi})$ that could be differentiated does not exist. On the other hand, using Euler angles, one obtains the potential energy as $U = U(\theta, \phi, \psi)$, which can be differentiated. In particular, one obtains

$$\tau_N = -\frac{\partial U}{\partial\theta}, \quad \tau_z = -\frac{\partial U}{\partial\phi}, \quad \tau_3 = -\frac{\partial U}{\partial\psi}, \tag{4.2.25}$$

where τ_N is the projection of the torque on the line of nodes.

4.2.5 *Free symmetric top*

The tensor relation between $\mathbf{L}$ and $\boldsymbol{\omega}$, Eq. (4.2.14), makes the rotational dynamics complicated. To illustrate this, consider a free symmetric top with $I_1 = I_2 \neq I_3$ and having the angular momentum $\mathbf{L}$, as shown in Fig. 4.2.1. For a free body $\mathbf{L} = \mathbf{const}.$, it is thus convenient to choose $\mathbf{e}_z$ along $\mathbf{L}$.

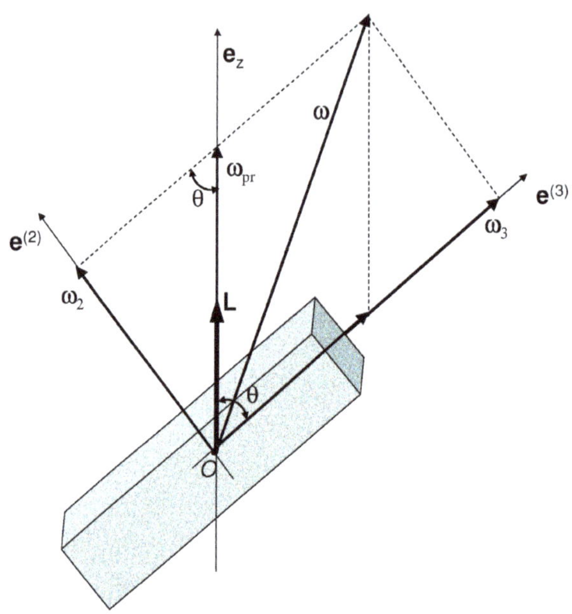

Figure 4.2.1. Free symmetric top.

In general, $\mathbf{L}$ does not coincide with the principal axes $\mathbf{e}^{(\alpha)}$. Thus, $\boldsymbol{\omega}$ is not collinear to $\mathbf{L}$. The body-frame components of $\mathbf{L}$ and $\boldsymbol{\omega}$ are given by

$$L_1 = 0, \quad L_2 = L\sin\theta, \quad L_3 = L\cos\theta,$$
$$\omega_1 = 0, \quad \omega_2 = \frac{L}{I_1}\sin\theta, \quad \omega_3 = \frac{L}{I_3}\cos\theta, \tag{4.2.26}$$

where we have chosen the sliding frame with $\mathbf{e}^{(1)} = \mathbf{e}_N$ and $\mathbf{e}^{(2)} = \mathbf{e}_A$ using the freedom of choice of $\mathbf{e}^{(1)}$ and $\mathbf{e}^{(2)}$ for an axially symmetric body. One can see that $\boldsymbol{\omega}$ is in the same plane as $\mathbf{L}$ and $\mathbf{e}^{(3)}$. From $\omega_1 = 0$ and Eq. (4.2.12) follows $\dot{\theta} = 0$, and thus, $\theta = \text{const}$. Then, from the second equation of Eq. (4.2.12), one obtains

$$\dot{\phi} = \frac{\omega_2}{\sin\theta} = \frac{L}{I_1}. \tag{4.2.27}$$

Then, from the last equation of Eq. (4.2.12) and the last of Eqs. (4.2.26), one obtains

$$\dot{\psi} = \omega_3 - \dot{\phi}\cos\theta = \left(\frac{1}{I_3} - \frac{1}{I_1}\right) L\cos\theta. \tag{4.2.28}$$

This formula can be rewritten as

$$\dot{\psi} = \left(1 - \frac{I_3}{I_1}\right)\omega_3, \tag{4.2.29}$$

which nicely illustrates that $\dot{\psi}$ and ω_3 are not the same. One can also relate precession (wobble) and spin using the two equations above as

$$\dot{\psi} = \left(\frac{I_1}{I_3} - 1\right)\dot{\phi}\cos\theta. \tag{4.2.30}$$

The Earth is very close to a fully symmetric top,

$$1 - \frac{I_3}{I_1} \simeq -\frac{1}{300}, \tag{4.2.31}$$

and the angle θ is small. The small deviation from the full symmetry must be due to the centrifugal forces caused by the Earth's rotation and the Earth's incomplete rigidity. Thus, the apparent rate of rotation of the Earth ω_3 that leads to the change of day and night is in fact mostly precession $\dot{\phi}$ of the Earth's symmetry axis (wobble), $\omega_3 = \dot{\phi}\cos\theta + \dot{\psi} \cong \dot{\phi}$. The actual Earth's spin $\dot{\psi}$ is very small, and the period of this rotation should be 300 days, according to Eqs. (4.2.29) and (4.2.31). If $\dot{\psi}$ were zero, the vector $\boldsymbol{\omega}$ would cross the Earth's surface at a fixed point (about 10 m from the North Pole). Due to of the rotation $\dot{\psi}$, the crossing point should be non-stationary, making circles around the North Pole with a period of 300 days. Observations show that a similar effect exists, and it is called the Chandler wobbling. However, the period of the Chandler wobbling is about 427 days and the trajectory of the crossing point has a large irregular component. The latter can be blamed on the Earth's incomplete rigidity, whereas the theoretical result of 300 days was obtained for a rigid body (already by Leonhard Euler). The fact that the Earth's rate of rotation around its symmetry axis is very small seems to contradict our everyday experience. However, since θ is small, and thus the axes $\mathbf{e}_z$ and $\mathbf{e}^{(3)}$ are close to each other, it makes little difference for an observer whether the body rotates around $\mathbf{e}_z$ (rate $\dot{\phi}$) or around $\mathbf{e}^{(3)}$ (rate $\dot{\psi}$).

Another example is a disc that rotates freely aroud the axis nearly parallel to the symmetry axis, $\theta \ll 1$. Since for a disc, $I_3 = 2I_1$, the relation between the wobble and the spin, Eq. (4.2.30), becomes

$$\dot{\phi} \cong -2\dot{\psi}. \tag{4.2.32}$$

If $I_3 \to 0$ like for an infinitely thin rod, Eq. (4.2.30) yields $\dot{\psi}$ diverging as $\dot{\psi} \cong (I_1/I_3)\,\dot{\phi}\cos\theta$. According to Eq. (4.1.23), this yields the diverging

$\omega_3 \cong \dot{\psi} \cong (I_1/I_3)\,\dot{\phi}\cos\theta$ and further the diverging kinetic energy $E \cong (1/2)\,I_3\omega_3^2 \cong (1/2)\left(I_1^2/I_3\right)\dot{\phi}^2\cos^2\theta$. For $\dot{\phi} \neq 0$, the only chance to avoid this divergence is to have $\theta = \pi/2$. The same can be seen on a more basic level. The kinetic-energy contribution $L_3^2/(2I_3)$ in Eq. (4.2.19) should not diverge for $I_3 \to 0$, thus L_3 must vanish. According to Eq. (4.2.37), it requires $\cos\theta \to 0$. Thus, a rod-like body with $I_3 \to 0$ would freely rotate with $\mathbf{L}$ perpendicular to its length. Deviations from this orientation would induce very large $\dot{\psi}$ that cost much kinetic energy.

Let us now consider the dynamics of $\boldsymbol{\omega}$ in the laboratory frame. It is precessing at rate $\dot{\phi}$ around the constant vector $\mathbf{L} \parallel \mathbf{e}_z$. One can use Eqs. (4.1.24) with $\dot{\theta} = 0$ that yield

$$\begin{aligned} \omega_x &= \dot{\psi}\sin\theta\sin\phi, \\ \omega_y &= -\dot{\psi}\sin\theta\cos\phi, \\ \omega_z &= \dot{\psi}\cos\theta + \dot{\phi}. \end{aligned} \tag{4.2.33}$$

From Eqs. (4.2.27) and (4.2.28), one obtains

$$\omega_z = \frac{L}{I_1} + \left(\frac{1}{I_3} - \frac{1}{I_1}\right) L\cos^2\theta = \text{const.} \tag{4.2.34}$$

and

$$\begin{Bmatrix} \omega_x \\ \omega_y \end{Bmatrix} = \left(\frac{1}{I_3} - \frac{1}{I_1}\right) L\cos\theta\sin\theta \begin{Bmatrix} \sin\phi \\ -\cos\phi \end{Bmatrix}, \tag{4.2.35}$$

where the angle ϕ depends on time as $\phi(t) = (L/I_1)\,t + \text{const}$. Equation (4.2.35) describes precession of the vector $\boldsymbol{\omega}$ around the constant vector $\mathbf{L}$. For a nearly fully symmetric top such as the Earth, this precession has a very small amplitude, although it is not slow. For $\cos\theta = 0$ or $\sin\theta = 0$, the vectors $\boldsymbol{\omega}$ and $\mathbf{L}$ are collinear and there is no precession.

4.2.6 *Free asymmetric top*

Let us now consider a more general case of an asymmetric top having all moments of inertia different. From Eqs. (4.2.15) and (4.1.23), one obtains

$$\begin{aligned} L_1 &= I_1\left(\dot{\phi}\sin\theta\sin\psi + \dot{\theta}\cos\psi\right), \\ L_2 &= I_2\left(\dot{\phi}\sin\theta\cos\psi - \dot{\theta}\sin\psi\right), \\ L_3 &= I_3\left(\dot{\phi}\cos\theta + \dot{\psi}\right). \end{aligned} \tag{4.2.36}$$

Let us now choose $\mathbf{e}_z$ along $\mathbf{L} = \mathbf{const}$. and project $\mathbf{L}$ onto the body frame. Using Eqs. (4.1.27) relating the principal axes to the axes of the laboratory system, one obtains

$$L_1 = L\sin\theta\sin\psi, \quad L_2 = L\sin\theta\cos\psi, \quad L_3 = L\cos\theta. \tag{4.2.37}$$

Substituting here Eq. (4.2.36), after simple algebra, one arrives at the system of equations of motion for Euler's angles of a free asymmetric top:

$$\begin{aligned}
\dot{\theta} &= \left(\frac{1}{I_1} - \frac{1}{I_2}\right) L\sin\theta\sin\psi\cos\psi, \\
\dot{\phi} &= \left(\frac{\sin^2\psi}{I_1} + \frac{\cos^2\psi}{I_2}\right) L, \\
\dot{\psi} &= \left(\frac{1}{I_3} - \frac{\sin^2\psi}{I_1} - \frac{\cos^2\psi}{I_2}\right) L\cos\theta.
\end{aligned} \tag{4.2.38}$$

Note that equations for $\dot{\theta}$ (nutation) and $\dot{\psi}$ (spin) form an autonomous system of equations that can be solved as the first step. After that, the equation for the precession $\dot{\phi}$ can be integrated using previously found $\psi(t)$. It is difficult to believe that something can be unknown in mechanics of rigid bodies. Still, this very important and beautiful system of equations is difficult to find in mechanics textbooks. For a free symmetric top, $I_1 = I_2 \neq I_3$, these equations simplify to

$$\dot{\theta} = 0, \quad \dot{\phi} = \frac{L}{I_1} \quad \dot{\psi} = \left(\frac{1}{I_3} - \frac{1}{I_1}\right) L\cos\theta, \tag{4.2.39}$$

which coincides with the solution found in Sec. 4.2.5.

There is a huge contrast with the translational dynamics of a free body, for which the conservation of the momentum $\mathbf{P}$ yields the three component equations

$$\dot{r}_\alpha = P_\alpha/M, \quad \alpha = x, y, z, \tag{4.2.40}$$

which can be easily integrated. Equation (4.2.38) is the rotational analog of the trivial equations above. However, rotational equations are much more complicated because of the intrinsic nonlinearity and coupling of different rotational variables.

One can substitute Eq. (4.2.37) into the rotational kinetic energy of Eq. (4.2.19), obtaining

$$E_{\text{rot}} = \frac{L^2}{2}\left[\frac{\sin^2\theta\sin^2\psi}{I_1} + \frac{\sin^2\theta\cos^2\psi}{I_2} + \frac{\cos^2\theta}{I_3}\right] = \text{const.}, \tag{4.2.41}$$

which relates θ and ψ under conservation of energy and angular momentum. Examples of the landscape of this function are shown in Fig. 4.2.2. In fact, one could restrict the region of ψ to $0 \leq \psi \leq \pi$ because of periodicity. The minimal value of E_{rot} corresponds to the rotation around the principal axis with the maximal moment of inertia. Using Eq. (4.1.27), one can see that if I_3 is maximal, this is the orientation with $\cos^2\theta = 1$, which corresponds to $\mathbf{e}^{(3)} = \mathbf{e}_z \| \mathbf{L}$. If I_2 is maximal, the energy minimum corresponds to $\sin^2\theta\cos^2\psi = 1$, i.e., $\mathbf{e}^{(2)} = \mathbf{e}_z$. If I_1 is minimal, it is $\sin^2\theta\sin^2\psi = 1$, i.e., $\mathbf{e}^{(1)} = \mathbf{e}_z$. Relaxation processes within an isolated system (e.g., due to plastic deformations) lead to the decrease in the mechanical energy, while $L = \text{const}$. In this case, the system ends up rotating around the axis with the maximal moment of inertia.

Expressing ψ via θ from Eq. (4.2.41) and substituting the result into the first equation of Eq. (4.2.38) one can obtain an isolated equation for $\dot{\theta}$ that can be solved in terms of Jacobian elliptic functions. In a similar way, one can obtain an isolated equation for $\dot{\psi}$. Much easier, however, is to solve Eq. (4.2.38) numerically. The application of the above equations to the asymmetric top will be discussed in the following.

The expression for the angular velocity ω_z can be obtained from Eqs. (4.1.24). Substituting the time derivatives from Eqs. (4.2.38), one obtains

$$\omega_z = \dot{\psi}\cos\theta + \dot{\phi} = 2E_{\text{rot}}/L = \text{const}. \tag{4.2.42}$$

There is a "magic" stationary value of the angle $\psi = \tilde{\psi}$ for which the expression in the brackets in the last of Eq. (4.2.38) is zero, and thus, $\dot{\psi} = 0$. One can find the value of $\tilde{\psi}$ by changing to the double angle:

$$\frac{1}{I_3} - \frac{\sin^2\psi}{I_1} - \frac{\cos^2\psi}{I_2} = \frac{1}{I_3} - \frac{1}{2I_1} - \frac{1}{2I_2} + \left(\frac{1}{I_1} - \frac{1}{I_2}\right)\frac{\cos 2\psi}{2} = 0, \tag{4.2.43}$$

which yields

$$\cos 2\tilde{\psi} = \frac{\frac{2}{I_3} - \frac{1}{I_1} - \frac{1}{I_2}}{\frac{1}{I_2} - \frac{1}{I_1}}. \tag{4.2.44}$$

The absolute value of the expression on the rhs of this formula has the absolute value smaller than one if I_3 is the intermediate moment of inertia, $I_1 < I_3 < I_2$ or $I_2 < I_3 < I_1$. This is the condition for the existence of the magic angle $\tilde{\psi}$. The latter is clearly visible in the rotational-energy landscape in the case when I_3 is the intermediate moment of inertia, see Fig. 4.2.3.

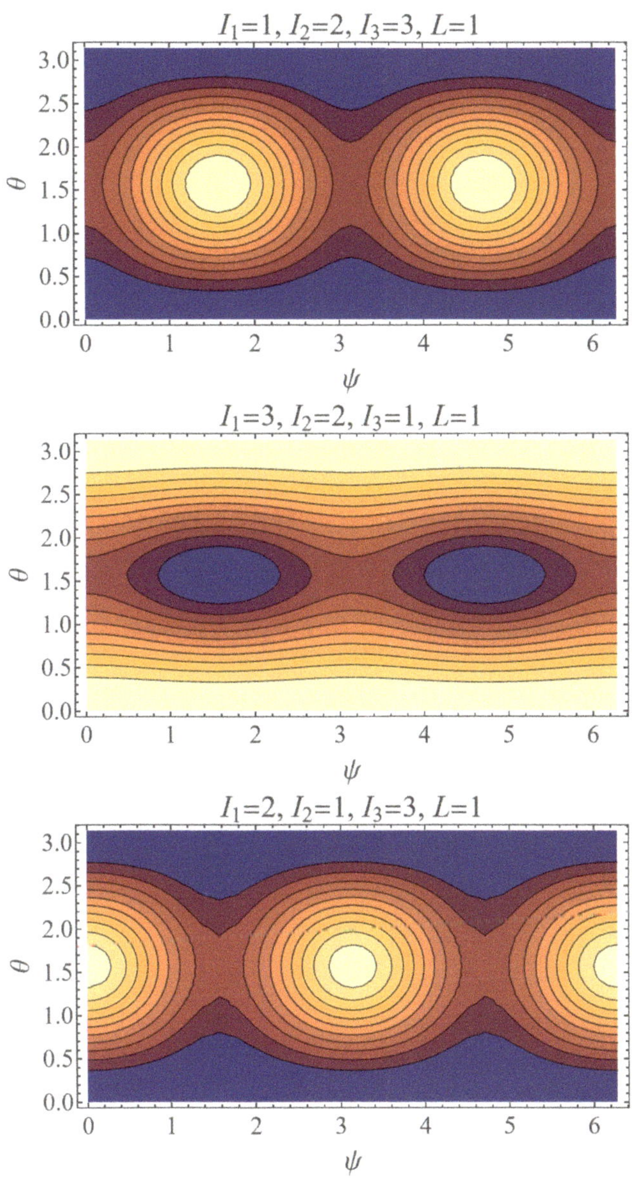

Figure 4.2.2. Examples of the energy landscape of the free asymmetric top ($0 \leq \theta \leq \pi$, $0 \leq \psi \leq 2\pi$). Brighter regions correspond to higher energy, darker regions correspond to lower energy. The top is moving along the lines of the constant energy clockwise around the energy maxima and counterclockwise around the energy minima. If the values of the highest and lowest moments of inertia interchange, energy maxima interchange with energy minima.

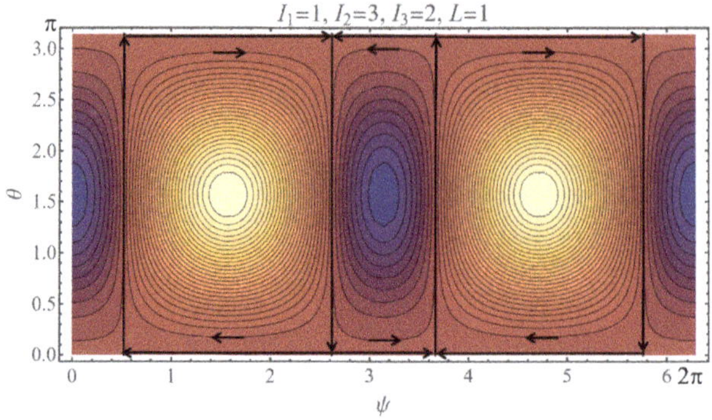

Figure 4.2.3. Energy landscape of the free asymmetric top in the case of I_3 being the intermediate moment of inertia. The vertical lines with arrows correspond to the "magic" values of the spin angle, $\psi = \tilde{\psi}$, see Eq. (4.2.44). The arrows show the direction of the motion of the system along the lines of constant energy.

If $\psi = \tilde{\psi}$, the body does not spin but precesses at the constant rate

$$\dot{\phi} = \frac{L}{I_3}, \tag{4.2.45}$$

according to the second of Eq. (4.2.38). This results in the apparent rotation of the body. Substituting $\psi = \tilde{\psi}$ into the first of Eqs. (4.2.38) and using $\sin\psi\cos\psi = \frac{1}{2}\sin 2\psi = \pm\frac{1}{2}\sqrt{1-\cos^2 2\psi}$, one obtains the equation

$$\dot{\theta} = \pm K \sin\theta, \tag{4.2.46}$$

where

$$K = \sqrt{\left(\frac{1}{I_3} - \frac{1}{I_1}\right)\left(\frac{1}{I_2} - \frac{1}{I_3}\right)}L = \sqrt{\left(1 - \frac{I_3}{I_1}\right)\left(\frac{I_3}{I_2} - 1\right)}\omega, \tag{4.2.47}$$

where $\omega = L/I_3$. The quantity K is real if I_3 is the intermediate moment of inertia. The solution of the differential equation above is

$$\cos\theta = \pm\tanh\left[K(t - t_0)\right], \tag{4.2.48}$$

where t_0 is the integration constant. It describes the relaxation toward the states with $\cos\theta = \pm 1$ corresponding to the rotation around the axis $\mathbf{e}^{(3)}$. The switching-width time is $\tau \equiv 1/K$. It shortens for large L (fast rotation) and small moments of inertia. If the medium moment of inertia I_3 is close to I_1 or I_2, K becomes small and switching becomes slow.

The solution found above is related to the spectacular Dzhanibekov's effect considered in the following section. It should be noted that this solution is unstable. If there is a small deviation of ψ from the magic value $\tilde{\psi}$, it initially increases with time and the evolution of the system becomes periodic, as can be seen in Fig. 4.2.2. If the intermediate moment of inertia is I_1 or I_2, the "magic" solution also exists but has a different form in terms of Euler angles. Nevertheless, the time dependences of the projections of the principal axes onto the laboratory axes is independent of the definition of Euler's angles and should behave similarly.

4.2.7 *Stability of rotations of the free asymmetric top*

Equation (4.2.38) provides a direct access to the time-dependent orientation of a free asymmetric top, bypassing Euler's equations (see the following). Let us investigate the stability of rotation around principal axes. In these cases, one of the principal axes, $\mathbf{e}^{(1)}$, $\mathbf{e}^{(2)}$, or $\mathbf{e}^{(3)}$, nearly coincides with $\mathbf{e}_z \| \mathbf{L}$. As it is done in the stability analysis, we will linearize the equations and investigate whether the initial deviation of the rotation (principal) axis from $\mathbf{e}_z$ remains small or increases with time. In the latter case, the motion is unstable. To investigate the stability, it is sufficient to study only one of the three orientations, as the principal axes of the body can be renamed.

The first idea would be to consider the rotation with $\mathbf{e}^{(3)} \approx \mathbf{e}_z$. However, Eq. (4.2.38) cannot be linearized in the vicinity of this orientation corresponding to $\theta = 0$. This is the manifestation of the well-known singularity of the spherical coordinate system. Linear analysis of Eq. (4.2.38) is possible for rotations around $\mathbf{e}^{(1)}$ and $\mathbf{e}^{(2)}$.

Consider, for instance, rotation around $\mathbf{e}^{(1)} = \mathbf{e}_z$. This orientation is described by $\theta = \psi = \pi/2$. (After the nutation to $\theta = \pi/2$, one has $\mathbf{e}^{(1)} = \mathbf{e}_x$, $\mathbf{e}^{(2)} = \mathbf{e}_z$, and $\mathbf{e}^{(3)} = -\mathbf{e}_y$. After a subsequent spin by $\psi = \pi/2$, one has $\mathbf{e}^{(1)} = \mathbf{e}_z$, $\mathbf{e}^{(2)} = -\mathbf{e}_x$, and $\mathbf{e}^{(3)} = -\mathbf{e}_y$). Consider small deviations from this orientation:

$$\theta = \pi/2 + \delta\theta, \quad \psi = \pi/2 + \delta\psi, \tag{4.2.49}$$

with small $\delta\theta$ and $\delta\psi$. Linearizing the equations for $\dot{\theta}$ and $\dot{\psi}$ yields

$$\begin{aligned} \frac{d}{dt}\delta\theta &= -\left(\frac{1}{I_1} - \frac{1}{I_2}\right) L\delta\psi, \\ \frac{d}{dt}\delta\psi &= \left(\frac{1}{I_1} - \frac{1}{I_3}\right) L\delta\theta. \end{aligned} \tag{4.2.50}$$

These equations describe a small precession of $\mathbf{e}^{(1)}$ around $\mathbf{e}_z$ with the frequency

$$\Omega = \sqrt{\left(\frac{1}{I_1} - \frac{1}{I_2}\right)\left(\frac{1}{I_1} - \frac{1}{I_3}\right)} L = \sqrt{\left(1 - \frac{I_1}{I_2}\right)\left(1 - \frac{I_1}{I_3}\right)}\omega_1, \qquad (4.2.51)$$

similar to Eq. (4.4.19). If I_1 is the maximal or minimal of the three moments of inertia, Ω is real and rotation is stable. This situation corresponds to the maximum and minimum of the energy, see the first and second panels of Fig. 4.2.2. If I_1 is intermediate, Ω is imaginary and small angular deviations exponentially increase. This situation corresponds to the saddle point of the energy, see the third panel of Fig. 4.2.2.

The precession rate $\dot{\phi}$ for small $\delta\theta$ and $\delta\psi$ in the lowest approximation is given by

$$\dot{\phi} = \frac{L}{I_1} = \omega_1, \qquad (4.2.52)$$

corresponding to $\psi = \pi/2$. One can see that the top performs a double motion with incommensurate frequencies Ω and ω_1. Small projections of $\mathbf{e}^{(2)}$ and $\mathbf{e}^{(3)}$ onto $\mathbf{e}_z$, proportional to $\delta\theta$ and $\delta\psi$, depend additionally on ϕ and evolve with combinational frequencies $\Omega \pm \omega_1$. Physically, the top rotates (precesses) with the frequency $\dot{\phi} = \omega_1$ around $\mathbf{e}_z$, while its principal axis $\mathbf{e}^{(1)}$ rotates around the same axis with the frequency Ω. The latter is a small-amplitude motion, thus the eye sees only the former.

One can do a similar analysis for the rotation around the orientation $\theta = \pi/2$ and $\psi = 0$, i.e., $\mathbf{e}^{(2)} = \mathbf{e}_z$ and $\mathbf{e}^{(3)}$, $\mathbf{e}^{(1)}$ in the x–y-plane. For $I_1 < I_2 < I_3$, rotation around $\mathbf{e}^{(2)}$ is unstable.

4.2.8 *Dzhanibekov's effect*

It was known since long ago that the rotation of the body around the axis with the intermediate moment of inertia is unstable. However, only in 1985, the Soviet astronaut Vladimir Dzhanibekov, working on a space station in the condition of weightlessness, observed the full dynamics on this unstable motion. It turned out that the body spends some time rotating around the unstable axis, then this axis quickly turns over and the body rotates around it for some time in the opposite direction (in the same direction around $\mathbf{e}_z \| \mathbf{L}$ in the laboratory frame). Then the unstable axis quickly turns over into its original orientation, and the process repeats infinitely. One can easily find the information about Dzhanibekov's effect online, including videos from experiments in space and numerical simulations using the conventional dynamics of the rigid body based on Euler's equations (see the following). However, Dzhanibekov's effect can be described in a simpler way using the equations of motion directly for Euler's angles, Eqs. (4.2.38).

Let us start with the numerical solution of Eqs. (4.2.38). It does not matter which principal axis, 1, 2, or 3, is initially close to $\mathbf{e}_z \| \mathbf{L}$. For Dzhanibekov's effect, it is needed that this axis has the intermediate moment of inertia. Since we are not performing the linear analysis here, we can make the natural choice as $\mathbf{e}^{(3)} \approx \mathbf{e}_z$, so that I_3 is the intermediate moment of inertia. The numerical solution for the model with $I_1 = 1$, $I_2 = 3$, $I_3 = 2$ and $L = 1$ with the initial conditions $\theta_0 = 10^{-6}$ (in radians) and $\psi_0 = 0$ is shown in Fig. 4.2.4. The top panel of this figure shows the projections of the medium axis $\mathbf{e}^{(3)}$ onto the axes of the laboratory systems. The most important is $\mathbf{e}^{(3)} \cdot \mathbf{e}_z = \cos\theta$ that shows periodic switching between the two opposite orientations. The projection $\mathbf{e}^{(3)} \cdot \mathbf{e}_x = \sin\theta \sin\phi$ is visible only in the switching regions, and $\mathbf{e}^{(3)} \cdot \mathbf{e}_y = -\sin\theta\cos\phi$ (not shown) behaves similarly. Projections of the other principal axes on the laboratory axes (not shown) demonstrate the precession of the body with the frequency $\omega \approx L/I_3$, which is perturbed in the switching regions.

One could also plot the time dependences of ω_3 and ω_z. The former follows from Eqs. (4.2.37), $\omega_3 = (L/I_3)\cos\theta$, and has the same time dependence as the middle-axis orientation shown in Fig. 4.2.4. The latter is constant according to Eq. (4.2.42) and retains its initial value $\omega_z = L/I_3$.

The spin angle ψ in the bottom of Fig. 4.2.4 evolves in the antiphase to $\theta(t)$, which follows from Fig. 4.2.3. The evolution shown in Fig. 4.2.4 corresponds to the trajectories in the energy landscape that are close to the rectangles made of thick black lines in Fig. 4.2.3. Near the corners of the rectangles, the motion becomes very slow. The analytical solution found above, Eq. (4.2.48), corresponds to the trajectories coinciding exactly with the thick black vertical lines. In this case, the motion takes an infinite time. For the trajectories close to the black rectangles, there is multiple switching observed in Dzhanibekov's effect. In this case (and in general), an analytical solution at all times should be available in terms of the Jacoby elliptic functions.

4.2.9 *Precession of a rapidly spinning top*

Let us consider a body rapidly rotating around one of its principal axes, say, the 3-axis, and having a fixed point of support O' on this axis at the

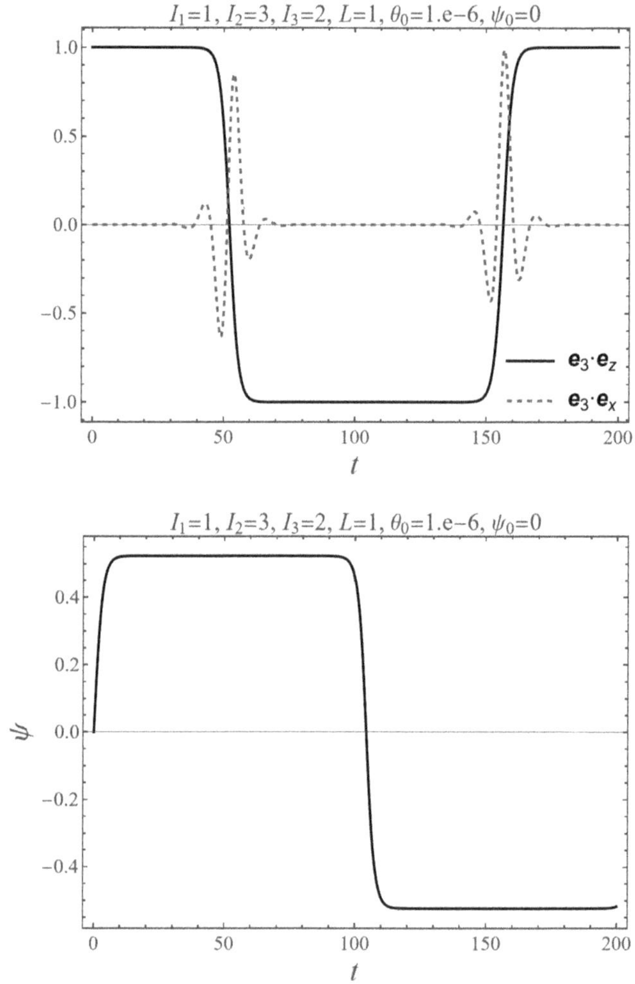

Figure 4.2.4. Dzhanibekov's effect: Time dependences of the orientation of the medium axis $\mathbf{e}_3$ and of the spin angle ψ.

distance a from the center of mass O. We direct the vector $\mathbf{e}^{(3)}$ from O' to O. With respect to O', the gravity force $\mathbf{F} = -Mg\mathbf{e}_z$ produces a torque

$$\boldsymbol{\tau} = \left[a\mathbf{e}^{(3)} \times (-Mg\mathbf{e}_z)\right] = Mga\left[\mathbf{e}_z \times \mathbf{e}^{(3)}\right]. \tag{4.2.53}$$

Due to the gravity, the scalar product $\left(\mathbf{e}_z \cdot \mathbf{e}^{(3)}\right)$ can change and the top initially spinning around the 3-axis can acquire rotations around the 1- and

2-axes as well. In other words, if there is spin $\dot{\psi} \neq 0$ in the initial state, then also precession $\dot{\phi}$ and nutation $\dot{\theta}$ will arise. The full dynamics of the heavy top will be considered in Sec. 4.3.3, and here, we assume that for a fast spinning top, these contributions are small and, to a good approximation, the angular momentum is due to the spin only and thus, $\mathbf{L}$ is directed in one of the two directions, collinear or anticollinear to $\mathbf{e}^{(3)}$:

$$\mathbf{L} = I_3\omega_3\mathbf{e}^{(3)} = L\mathbf{e}^{(3)}, \tag{4.2.54}$$

where L can have positive and negative values depending on the direction of the spin. Now, the torque τ can be expressed via $\mathbf{L}$ using $\mathbf{e}^{(3)} = \mathbf{L}/L$, and the dynamical equation for $\mathbf{L}$, Eq. (1.3.11), becomes

$$\dot{\mathbf{L}} = [\mathbf{L} \times \boldsymbol{\Omega}], \quad \boldsymbol{\Omega} \equiv -\frac{Mga}{L}\mathbf{e}_z. \tag{4.2.55}$$

This is the equation describing the precession of $\mathbf{L}$ around the vector $\boldsymbol{\Omega}$ in the anticlockwise direction with the frequency $\Omega = Mga/L$, and the vector $\boldsymbol{\Omega}$ is directed up or down, depending on the sign of L. A similar equation can be written for $\mathbf{e}^{(3)}$, which describes the orientation of the top:

$$\dot{\mathbf{e}}^{(3)} = \left[\mathbf{e}^{(3)} \times \boldsymbol{\Omega}\right]. \tag{4.2.56}$$

Note that this precession is in the positive direction, $\dot{\phi} = \Omega > 0$, and it does not change the potential energy of the body

$$U = Mga\left(\mathbf{e}^{(3)} \cdot \mathbf{e}_z\right) = -\left(\mathbf{L} \cdot \boldsymbol{\Omega}\right). \tag{4.2.57}$$

$\boldsymbol{\Omega}$ can be interpreted as a generalized force corresponding to the variable $\mathbf{L}$:

$$\boldsymbol{\Omega} = -\partial U/\partial \mathbf{L}. \tag{4.2.58}$$

For this particular problem, a more natural choice of $\mathbf{e}_z$ would be down in the direction of the gravity force. (The same is done in magnetism, directing $\mathbf{e}_z$ parallel to the magnetic field.) With this choice, in the formulas above, one should replace $g \Rightarrow -g$, which leads to $\boldsymbol{\Omega} = (Mga/L)\,\mathbf{e}_z$.

One can see that for a rapidly spinning top, the precession frequency is small:

$$\frac{\Omega}{\omega_3} = \frac{Mga}{I_3\omega_3^2} \ll 1, \tag{4.2.59}$$

so that our method is self-consistent.

4.3 Lagrangian formalism

4.3.1 *General scheme*

Lagrangian formalism for rotational dynamics is the most efficient in the case of holonomic constraints. For non-holonomic constraints, the more physically appealing Newtonian formalism is preferred (see the following). To set up the Lagrangian formalism, one has to express the kinetic and potential energies in terms of the generalized coordinates that in this case are Euler angles θ, φ, and ψ. The first step is to express the kinetic energy via $\mathbf{V}$ and $\boldsymbol{\omega}$ in Eq. (4.1.9).

Expressing the potential energy in terms of Euler's angles usually poses no problem. If the body has no point of support, translational and rotational motions separate in most practical cases. If the body has a point of support, one can consider rotations around this point without any translational motion. The Lagrangian describing the rotational motion can be written as

$$\mathcal{L} = E_{\text{rot}}(\theta, \phi, \psi, \dot{\theta}, \dot{\phi}, \dot{\psi}) - U(\theta, \phi, \psi),$$

and the Lagrange equations read

$$\frac{d}{dt}\frac{\partial \mathcal{L}}{\partial \dot{\theta}} = \frac{\partial \mathcal{L}}{\partial \theta}, \quad \frac{d}{dt}\frac{\partial \mathcal{L}}{\partial \dot{\phi}} = \frac{\partial \mathcal{L}}{\partial \phi}, \quad \frac{d}{dt}\frac{\partial \mathcal{L}}{\partial \dot{\psi}} = \frac{\partial \mathcal{L}}{\partial \psi}. \tag{4.3.1}$$

Let us at first consider the generalized momenta

$$p_\theta \equiv \frac{\partial \mathcal{L}}{\partial \dot{\theta}}, \quad p_\phi \equiv \frac{\partial \mathcal{L}}{\partial \dot{\phi}}, \quad p_\psi \equiv \frac{\partial \mathcal{L}}{\partial \dot{\psi}}. \tag{4.3.2}$$

Comparison with Eq. (4.2.20) suggests that they must be related to the components of the angular momentum $\mathbf{L}$. Indeed, as ϕ is responsible for rotations around the z-axis and ψ is responsible for rotations around the 3-axis, it should be

$$p_\phi \equiv \frac{\partial \mathcal{L}}{\partial \dot{\phi}} = L_z, \quad p_\psi \equiv \frac{\partial \mathcal{L}}{\partial \dot{\psi}} = L_3. \tag{4.3.3}$$

These formulas can be checked by the differentiation of the rotational energy in terms of Euler angles, Eq. (4.2.9). The resulting p_ϕ coincides with L_z given by Eq. (4.2.17). The resulting p_ψ coincides with $L_3 = I_3\omega_3$, where ω_3 is given by Eq. (4.1.23). If ϕ and ψ are cyclic, $p_\phi = L_z$ and $p_\psi = L_3$ are integrals of motion, and we know that integrals of motion following from rotational symmetry are components of $\mathbf{L}$. The generalized momentum p_θ is the projection of $\mathbf{L}$ on $\mathbf{e}_N$, thus it is a combination of different components of $\mathbf{L}$.

Lagrangian equations are second-order differential equations for all three Euler angles. Thus, the initial conditions for their angles and their time derivatives are needed for its solution.

4.3.2 *Free symmetric top by the Lagrangian formalism*

Consider, as an illustration, a free symmetric top, having $\mathcal{L} = E_{\mathrm{rot}}$ of Eq. (4.2.10). This system was already considered in Sec. 4.2.1. Nevertheless, it makes sense to give it another consideration within the Largangian formalism because the same method will be applied in the following to the less trivial problem of a heavy symmetric top. The variables φ and ψ are cyclic, so that one obtains

$$\frac{\partial \mathcal{L}}{\partial \dot{\phi}} = I_1 \dot{\phi} \sin^2\theta + I_3 \left(\dot{\phi}\cos\theta + \dot{\psi}\right)\cos\theta = L_z = \text{const.},$$

$$\frac{\partial \mathcal{L}}{\partial \dot{\psi}} = I_3 \left(\dot{\phi}\cos\theta + \dot{\psi}\right) = L_3 = \text{const.} \tag{4.3.4}$$

Using these results, one can express $\dot{\phi}$ and $\dot{\psi}$ through the integrals of motion,

$$\dot{\phi} = \frac{L_z - L_3\cos\theta}{I_1 \sin^2\theta},$$

$$\dot{\psi} = \frac{L_3}{I_3} - \frac{L_z - L_3\cos\theta}{I_1\sin^2\theta}\cos\theta. \tag{4.3.5}$$

The Lagrange equation for θ has the form

$$I_1\ddot{\theta} = (I_1 - I_3)\,\dot{\phi}^2 \sin\theta\cos\theta - I_3\dot{\psi}\dot{\phi}\sin\theta. \tag{4.3.6}$$

Here, $\dot{\phi}$ and $\dot{\psi}$ can be eliminated using Eq. (4.3.5):

$$I_1\ddot{\theta} = \frac{(L_z\cos\theta - L_3)\,(L_z - L_3\cos\theta)}{I_1\sin^3\theta}. \tag{4.3.7}$$

This equation can be written in the form

$$I_1\ddot{\theta} = -\frac{dU_{\mathrm{eff}}(\theta)}{d\theta}, \tag{4.3.8}$$

where $U_{\mathrm{eff}}(\theta)$ is an effective potential energy for θ. Equation (4.3.8) has an integral of motion

$$\frac{1}{2}I_1\dot{\theta}^2 + U_{\mathrm{eff}}(\theta) = E_{\mathrm{eff}}, \tag{4.3.9}$$

where $E_{\text{eff}} = \text{const.}$ The effective energy $U_{\text{eff}}(\theta)$ can be found from Eq. (4.3.7) or from the energy conservation $E_{\text{rot}} = E = \text{const.}$ using Eqs. (4.2.10) and (4.3.5). The result has the form

$$U_{\text{eff}}(\theta) = \frac{(L_z - L_3 \cos\theta)^2}{2I_1 \sin^2\theta}, \tag{4.3.10}$$

up to an irrelevant constant. The three integrals of motion, L_z, L_3, and E_{eff}, completely define the motion of the system. The time dependence of θ can be found from Eq. (4.3.9). It is a motion between the two turning points at which $U_{\text{eff}}(\theta) = E_{\text{eff}}$. Then time dependences of ϕ and ψ can be found from Eq. (4.3.5). This procedure seems to be complicated and it does not promise a simple analytical solution.

At this moment, one can note that Descarte axes, with respect to which Euler angles are defined, can be chosen in different ways. The best way is to choose z-axis along $\mathbf{L}$. This yields $L_x = L_y = 0$, $L_z = L$ and $L_3 = L\cos\theta$. Now, everything simplifies, the right side of Eq. (4.3.7) disappears, and one obtains $\dot{\theta} = \text{const.}$ This constant should be zero since non-zero $\dot{\theta}$ would create a component of the angular momentum along $\mathbf{e}_N$ perpendicular to the z-axis that contradicts the choice of the z axis made above. In this way, all previously obtained results for a free symmetric top are reproduced.

The solution outlined above is methodologically questionable since one has to go beyond the Lagrangian formalism and use other considerations. The problem is that the Lagrangian formalism readily yields two conserved components of the angular momentum, whereas, in fact, three components are conserved. Thus, there is an additional hidden integral of motion. Without physical idea of conservation of the whole vector $\mathbf{L}$, straightforwardly applying Lagrange formalism, one would not find the best choice of the laboratory axes. On the other hand, for a heavy symmetric top considered in the following section, Largangian formalism is fully adequate since the choice of the laboratory axes is predefined by the direction gravity and there is no hidden integral of motion.

4.3.3 *Heavy symmetric top*

For a heavy symmetric top with a point of support O' on the symmetry axis at the distance a from the CM at O, one has to replace $I_1 \Rightarrow I_1' = I_1 + Ma^2$ and add the potential of gravity to the effective potential of Eq. (4.3.10). This yields

$$U_{\text{eff}}(\theta) = \frac{(L_z - L_3 \cos\theta)^2}{2I_1' \sin^2\theta} + Mga\cos\theta, \tag{4.3.11}$$

whereas Eq. (4.3.5) remains the same with the replacement $I_1 \Rightarrow I_1'$. Motion of the heavy symmetric top described by the primary equation (4.3.8) with $U_{\text{eff}}(\theta)$ above and by the secondary equations (4.3.5) is more complicated. While $\mathbf{L}$ is precessing around the vertical axis due to gravity, the symmetry axis $\mathbf{e}^{(3)}$ is precessing around $\mathbf{L}$, similar to that of the free symmetric top. Both these precessions are not simple precessions with a constant projection angle and precession rate. According to the first of Eqs. (4.3.5), $\dot{\phi}$ can change the sign. If the *average* rate of the precession of $\mathbf{L}$ is smaller than the rate of nutation (weak gravity), $\mathbf{e}^{(3)}$ makes loops and its trajectory crosses itself repeatedly. In the other case (strong gravity), the trajectory of $\mathbf{e}^{(3)}$ does not cross itself. In this case, $\dot{\phi}$ given by Eqs. (4.3.5) does not change the sign.

Nutations are described by Eq. (4.3.9) with $U_{\text{eff}}(\theta)$ above that has upward curvature as the function of $\cos\theta$. Indeed, $(L_z - L_3\cos\theta)^2$ is a parabola with the upward curvature and $1/\sin^2\theta = 1/\left(1-\cos^2\theta\right)$ has upward curvature and diverges at the boundaries of the interval, $\cos\theta = \pm 1$. The potential energy is a straight line on $\cos\theta$. Thus $U_{\text{eff}}(\theta)$ has one minimum in the interval $-1 \le \cos\theta \le 1$. Let us consider some particular cases of the motion of a heavy symmetric top.

4.3.3.1 *"Sleeping" top*

If the top is oriented vertically up and rotating around z-axis (the so-called "sleeping top"), then $\theta = 0$ and $L_3 = L_z = L$, see Eq. (4.2.26). Analyzing $U_{\text{eff}}(\theta)$ of Eq. (4.3.11) allows us to find the stability region of this state. Expanding $U_{\text{eff}}(\theta)$ up to θ^2, one obtains

$$U_{\text{eff}}(\theta) \cong \left(\frac{L^2}{4I_1'} - Mga\right)\frac{\theta^2}{2}. \tag{4.3.12}$$

Thus, the up state is stable if the rotation is fast enough:

$$L^2 > 4I_1'Mga. \tag{4.3.13}$$

Similar analysis shows that the rotation with the downward orientation, $L_3 = -L_z$ and $\theta = \pi$, is always stable. In all other cases, $L_3 \neq \pm L_z$ and $U_{\text{eff}}(\theta)$ diverges for $\theta \to 0, \pi$. Thus, $U_{\text{eff}}(\theta)$ has a single minimum inside the interval $\theta \in (0, \pi)$.

4.3.3.2 *Nutationless top*

Let us find the condition for the top to move with $\theta = \text{const}$. Obviously, this requires that $\theta = \theta_{\min}$ corresponding to the minimum of $U_{\text{eff}}(\theta)$, i.e., $U_{\text{eff}}(\theta)/d\theta = 0$. The latter entails the relation between the spin and precession that can be obtained by expressing L_z and L_3 via $\dot{\psi}$ and $\dot{\phi}$ using Eq. (4.2.18) and $L_3 = I_3\omega_3$, with ω_3 given by Eq. (4.1.23). The stationarity condition for θ defines the required spin for a given θ and the precession rate $\dot{\phi}$:

$$\dot{\psi} = \left(\frac{I_1'}{I_3} - 1\right)\dot{\phi}\cos\theta + \frac{Mga}{I_3\dot{\phi}}. \tag{4.3.14}$$

For a free top, $Mga = 0$, this reduces to Eq. (4.2.30) with $I_1 \Rightarrow I_1'$. It is interesting to resolve this relation for $\dot{\phi}$ that yields two solutions:

$$\dot{\phi} = \frac{I_3\dot{\psi} \pm \sqrt{\left(I_3\dot{\psi}\right)^2 - 4Mga\,(I_1' - I_3)\cos\theta}}{2\,(I_1' - I_3)\cos\theta}. \tag{4.3.15}$$

One can see that the nutationless regime exists if the spin is fast enough:

$$\dot{\psi}^2 \geq 4Mga\frac{I_1' - I_3}{I_3^2}\cos\theta. \tag{4.3.16}$$

In the fast-spinning limit, the solutions for $\dot{\phi}$ simplify to

$$\dot{\phi} \cong \frac{\dot{\psi}}{(I_1'/I_3 - 1)\cos\theta} - \frac{Mga}{I_3\dot{\psi}} \tag{4.3.17}$$

(fast precession) and

$$\dot{\phi} \cong \frac{Mga}{I_3\dot{\psi}} \tag{4.3.18}$$

(slow precession). The fast precession is the free-top precession studied in Sec. 4.2.5, here slightly modified by gravity. The slow precession was studied in Sec. 4.2.5. In this case, $\mathbf{L} \cong L\mathbf{e}^{(3)}$ and $L \cong I_3\dot{\psi}$, so that the rate of precession $\dot{\phi}$ is the frequency $\Omega = Mga/L$, see Eq. (4.2.55).

For a fast spinning top, both L_3 and L_z are large, thus the minimum of $U_{\text{eff}}(\theta)$ is deep and narrow. If Eq. (4.3.14) is not satisfied exactly, θ deviates from its value in the energy minimum, but these deviations remain small. This justifies the more intuitive derivation of the rapidly rotating top equation in Sec. 4.2.4

4.3.3.3 *Spinless top*

Let us now consider the spinless regime $\dot{\psi} = 0$ with $\dot{\theta} = 0$. According to Eq. (4.3.14), this regime is realized if

$$\dot{\phi} = \pm\sqrt{\frac{Mga}{(I_3 - I_1')\cos\theta}}, \quad (I_3 - I_1')\cos\theta > 0. \tag{4.3.19}$$

If I_3 is small, then Eq. (4.3.14) yields $\dot{\psi} \cong \left(I_1'\dot{\phi}\cos\theta + Mga/\dot{\phi}\right)/I_3$. To avoid divergence of the kinetic energy, Eq. (4.2.19), for $I_3 \to 0$, the expression in brackets must be zero and precession rate $\dot{\phi}$ must be given by

$$\dot{\phi} = \pm\sqrt{\frac{Mga}{-I_1'\cos\theta}}, \quad \pi/2 < \theta < \pi. \tag{4.3.20}$$

This looks like a particular case of the preceding formula, although its origin is different. Equation (4.3.20) describes the stationary precession of a thin-rod pendulum that does not rotate around its symmetry axis and is fully described by θ and ϕ. Of course, for the pendulum, Eq. (4.3.20) can be obtained in a shorter way.

Results for $I_3 \to 0$ can be obtained in a more general way. According to Eq. (4.2.19), $I_3 \to 0$ must be $L_3 \to 0$ to avoid divergence of the kinetic energy. Thus, one should drop L_3 in Eq. (4.3.11) and in the first line of Eq. (4.3.5). This yields the expressions for the pendulum

$$U_{\text{eff}}(\theta) = \frac{L_z^2}{2I_1'\sin^2\theta} + Mga\cos\theta, \quad \dot{\phi} = \frac{L_z}{I_1\sin^2\theta}, \tag{4.3.21}$$

whereas $\dot{\psi}$ becomes irrelevant and should be discarded. From here, the requirement $\theta = \text{const.}$, (i.e., $\partial U_{\text{eff}}(\theta)/\partial\theta = 0$) leads to Eq. (4.3.20). Note that for the pendulum, the reference orientation is conventionally chosen down, so that in standard notations, one should replace $\theta \to \pi - \theta$.

4.4 Newtonian formalism

Whereas in many important cases, Lagrangian formalism for rotational dynamics is sufficient and leads to the result in the shortest way, the Newtonian formalism remains the most versatile. It allows us to consider systems with non-holonomic constraints in a more physically appealing way and it can be easily extended to include forces of new kinds.

4.4.1 *Euler equations*

The implementation of the Newtonian approach to rotational dynamics leads to Euler equations, which are projections of the rotational dynamical equation (1.3.11) onto the principal axes of the body. One could think that one could start with Eqs. (4.2.36), differentiate them over time and write $\dot{L'}_k = \tau_k$ with $k = 1, 2, 3$, obtaining equations of motion for Euler's angles. However, this would be wrong because the vector $\mathbf{L}$ also depends on the time-dependent directions of the principal axes via Eq. (4.2.15). Differentiating this equation over time, one obtains six terms:

$$\dot{\mathbf{L}}' = \dot{L}'_1\mathbf{e}^{(1)} + \dot{L}'_2\mathbf{e}^{(2)} + \dot{L}'_3\mathbf{e}^{(3)} + L'_1\dot{\mathbf{e}}^{(1)} + L'_2\dot{\mathbf{e}}^{(2)} + L'_3\dot{\mathbf{e}}^{(3)}. \tag{4.4.1}$$

If the principal axes are embedded in the body, they rotate according to

$$\dot{\mathbf{e}}^{(k)} = \left[\boldsymbol{\omega} \times \mathbf{e}^{(k)}\right] \tag{4.4.2}$$

as any embedded vector, see Eq. (4.1.7). Using this, one can cast the equation of motion for $\mathbf{L}$ into a compact form

$$\frac{d\mathbf{L}'}{dt} = \frac{\tilde{d}\mathbf{L}'}{dt} + [\boldsymbol{\omega} \times \mathbf{L}'] = [\mathbf{P} \times \mathbf{v}_O] + \boldsymbol{\tau}', \tag{4.4.3}$$

where the term $\tilde{d}\mathbf{L}'/dt$ comprises the first three terms of Eq. (4.4.1), which yield the time derivative of $\mathbf{L}$ as seen by the observer rotating with the body.

For a practical solution, one needs the principal-axes components of this vector equation. Using $\boldsymbol{\omega} = \omega_1\mathbf{e}^{(1)} + \omega_2\mathbf{e}^{(2)} + \omega_3\mathbf{e}^{(3)}$ and $\left[\mathbf{e}^{(1)} \times \mathbf{e}^{(2)}\right] = \mathbf{e}^{(3)}$, etc., one obtains

$$\begin{aligned}
\left[\boldsymbol{\omega} \times \mathbf{e}^{(1)}\right] &= \mathbf{e}^{(2)}\omega_3 - \mathbf{e}^{(3)}\omega_2, \\
\left[\boldsymbol{\omega} \times \mathbf{e}^{(2)}\right] &= \mathbf{e}^{(3)}\omega_1 - \mathbf{e}^{(1)}\omega_3, \\
\left[\boldsymbol{\omega} \times \mathbf{e}^{(3)}\right] &= \mathbf{e}^{(1)}\omega_2 - \mathbf{e}^{(2)}\omega_1.
\end{aligned} \tag{4.4.4}$$

Then one proceeds by substituting this into Eq. (4.4.1), expressing L'_k as $L'_k = I_k\omega_k$ and collecting the terms with the same $\mathbf{e}^{(k)}$:

$$\begin{aligned}
\dot{\mathbf{L}}' &= \left\{\dot{L}'_1 - L'_2\omega_3 + L'_3\omega_2\right\}\mathbf{e}^{(1)} + \left\{\dot{L}'_2 - L'_3\omega_1 + L'_1\omega_3\right\}\mathbf{e}^{(2)} \\
&\quad + \left\{\dot{L}'_3 - L'_1\omega_2 + L'_2\omega_1\right\}\mathbf{e}^{(3)}, \\
&= \left\{I_1\dot{\omega}_1 - (I_2 - I_3)\omega_2\omega_3\right\}\mathbf{e}^{(1)} + \left\{I_2\dot{\omega}_2 - (I_3 - I_1)\omega_3\omega_1\right\}\mathbf{e}^{(2)} \\
&\quad + \left\{I_3\dot{\omega}_3 - (I_1 - I_2)\omega_1\omega_2\right\}\mathbf{e}^{(3)}.
\end{aligned} \tag{4.4.5}$$

Finally, adding the components of the torque in Eq. (1.3.11) projected onto $\mathbf{e}^{(k)}$, one arrives at the set of the famous Euler equations

$$
\begin{aligned}
I_1\dot{\omega}_1 &= (I_2 - I_3)\omega_2\omega_3 + [\mathbf{P}\times\mathbf{v}_O]_1 + \tau_1,\\
I_2\dot{\omega}_2 &= (I_3 - I_1)\omega_3\omega_1 + [\mathbf{P}\times\mathbf{v}_O]_2 + \tau_2,\\
I_3\dot{\omega}_3 &= (I_1 - I_2)\omega_1\omega_2 + [\mathbf{P}\times\mathbf{v}_O]_3 + \tau_3.
\end{aligned} \tag{4.4.6}
$$

If rotations around the center of mass or around an instantaneous axis of rotation are considered, the kinematic term $[\mathbf{P}\times\mathbf{v}_O]$ disappears.

These equations can also be written for the components of $\mathbf{L}$:

$$
\begin{aligned}
\dot{L}'_1 + \left(\frac{1}{I_2} - \frac{1}{I_3}\right) L'_2 L'_3 &= [\mathbf{P}\times\mathbf{v}_O]_1 + \tau_1,\\
\dot{L}'_2 + \left(\frac{1}{I_3} - \frac{1}{I_1}\right) L'_3 L'_1 &= [\mathbf{P}\times\mathbf{v}_O]_2 + \tau_2,\\
\dot{L}'_3 + \left(\frac{1}{I_1} - \frac{1}{I_2}\right) L'_1 L'_2 &= [\mathbf{P}\times\mathbf{v}_O]_3 + \tau_3.
\end{aligned} \tag{4.4.7}
$$

One can see that there are nonlinear terms in the lhs of these equations in contrast to the set of naive equations considered at the beginning of this section. This makes rotational dynamics complicated even in the case of a free body. In the absence of the kinematic term and the torque, these equations conserve $\mathbf{L}^2 = L_1^2 + L_2^2 + L_3^2$:

$$
\partial_t\mathbf{L}^2 = 2\mathbf{L}\cdot\dot{\mathbf{L}} = 2\left(L_1\dot{L}_1 + L_2\dot{L}_2 + L_3\dot{L}_3\right) = 0, \tag{4.4.8}
$$

where primes have been dropped for convenience.

For a free body, $\tau_k = 0$, Euler's equations are closed, which lends them a special beauty. One can at first solve the equations of motion for ω_k and then substitute the solution into Eq. (4.1.23), thus obtaining a set of equations for Euler's angles that can be solved as the second step. This is how the analytical solution (in terms of the Jacoby elliptic functions, See *Mechanics* by Landau and Lifshitz) for a free asymmetric $I_1 \neq I_2 \neq I_3$ top can be obtained. However, for a torque-free body, one can use Eq. (4.2.38) that gives a one-step solution of this problem directly in terms of Euler's angles. Thus, the free-body Euler equations are practically of little use.

If a torque is applied to the body, the beauty of Euler's equations gets killed. In general, the torque depends on Euler's angles θ, ϕ, and ψ. In this case, one cannot solve the equations in terms of the components ω_k or L_k, as for the free body, and has to solve it in terms of Euler's angles. For this, one has to substitute ω_k by their expressions in terms of θ, ϕ, and ψ,

Eq. (4.1.23). In fact, one should not regret that Euler's equations cannot be solved in two steps in the presence of a torque. The practical way to solve these equations is numerical, and here, it is simpler to solve the equations in one step.

Euler equations can also be written for rolling bodies with respect to the contact point, and the non-holonomic constraint can be worked out to yield a purely rotational dynamical equation, see Eq. (4.2.22). This technique can be applied in the case of a sphere or disc rolling on a surface without skidding. Within this approach, one at first solves a purely rotational problem and at the second stage finds the motion of the center of mass, as is done for the wheel rolling on a plane in Chapter 8. The purely rotational equations are easier to analyze analytically than the big system of equations containing rotational equations, translational equation, and a non-holonomic constraint.

4.4.2 *Euler equations with sliding principal axes*

If instead of the embedded principal axes, the sliding axes for the symmetric body with $I_1 = I_2$ are used, see Sec. 4.4.2, then Eq. (4.4.2) becomes invalid and one should work with Euler's angles. Differentiating Eq. (4.2.11), one obtains

$$\begin{aligned}\dot{\mathbf{e}}^{(1)} &= (-\sin\phi\mathbf{e}_x + \cos\phi\mathbf{e}_y)\,\dot{\phi} = \left(\mathbf{e}^{(2)}\cos\theta - \mathbf{e}^{(3)}\sin\theta\right)\dot{\phi},\\ \dot{\mathbf{e}}^{(2)} &= \mathbf{e}^{(3)}\dot{\theta} - \mathbf{e}^{(1)}\cos\theta\dot{\phi},\\ \dot{\mathbf{e}}^{(3)} &= -\mathbf{e}^{(2)}\dot{\theta} + \mathbf{e}^{(1)}\sin\theta\dot{\phi}.\end{aligned} \tag{4.4.9}$$

Substituting it into Eq. (4.4.1) and collecting the terms with the same $\mathbf{e}^{(k)}$, one obtains

$$\begin{aligned}\dot{\mathbf{L}}' = &\left\{\dot{L}'_1 - L'_2\dot{\phi}\cos\theta + L'_3\dot{\phi}\sin\theta\right\}\mathbf{e}^{(1)}\\ &+\left\{\dot{L}'_2 + L'_1\dot{\phi}\cos\theta - L'_3\dot{\theta}\right\}\mathbf{e}^{(2)}\\ &+\left\{\dot{L}'_3 - L'_1\dot{\phi}\sin\theta + L'_2\dot{\theta}\right\}\mathbf{e}^{(3)}.\end{aligned} \tag{4.4.10}$$

Using $L'_k = I_k\omega_k$ and adding the components of the torque, one obtains the equations

$$\begin{aligned}I_1\dot{\omega}_1 &= I_2\omega_2\dot{\phi}\cos\theta - I_3\omega_3\dot{\phi}\sin\theta + [\mathbf{P}\times\mathbf{v}_O]_1 + \tau_1,\\ I_2\dot{\omega}_2 &= -I_1\omega_1\dot{\phi}\cos\theta + I_3\omega_3\dot{\theta} + [\mathbf{P}\times\mathbf{v}_O]_2 + \tau_2,\\ I_3\dot{\omega}_3 &= I_1\omega_1\dot{\phi}\sin\theta - I_2\omega_2\dot{\theta} + [\mathbf{P}\times\mathbf{v}_O]_3 + \tau_3.\end{aligned} \tag{4.4.11}$$

Here, one can use Eq. (4.2.12) to replace $\dot{\theta} \Rightarrow \omega_1$, $\dot{\phi}\sin\theta \Rightarrow \omega_2$, as well as $\dot{\phi}\cos\theta \Rightarrow \omega_2\cot\theta$ to bring the equations to the form close to Eqs. (4.4.6). This yields

$$\begin{aligned} I_1\dot{\omega}_1 &= (I_2\omega_2\cot\theta - I_3\omega_3)\,\omega_2 + [\mathbf{P}\times\mathbf{v}_O]_1 + \tau_1, \\ I_2\dot{\omega}_2 &= (-I_1\omega_2\cot\theta + I_3\omega_3)\,\omega_1 + [\mathbf{P}\times\mathbf{v}_O]_2 + \tau_2, \\ I_3\dot{\omega}_3 &= (I_1 - I_2)\,\omega_1\omega_2 + [\mathbf{P}\times\mathbf{v}_O]_3 + \tau_3. \end{aligned} \tag{4.4.12}$$

If the torque is a function of θ and ϕ, independent of ψ, one can use ω_3 as one of the dynamic variables and express ω_1 and ω_2 via Euler's angles. This yields

$$\begin{aligned} I_1\ddot{\theta} &= \left(I_2\dot{\phi}\cos\theta - I_3\omega_3\right)\dot{\phi}\sin\theta + [\mathbf{P}\times\mathbf{v}_O]_1 + \tau_1, \\ I_2\ddot{\phi}\sin\theta &= \left[-(I_1 + I_2)\,\dot{\phi}\cos\theta + I_3\omega_3\right]\dot{\theta} + [\mathbf{P}\times\mathbf{v}_O]_2 + \tau_2, \\ I_3\dot{\omega}_3 &= (I_1 - I_2)\,\dot{\theta}\dot{\phi}\sin\theta + [\mathbf{P}\times\mathbf{v}_O]_3 + \tau_3. \end{aligned} \tag{4.4.13}$$

4.4.3 *Free asymmetric top revisited*

Alternative to Eqs. (4.2.38), insights in the motion of the free asymmetric top can be gained already from geometrical considerations based on the conservation of the angular momentum $\mathbf{L}^2 = L^2 = \text{const.}$ and energy $E_{\text{rot}} = E = \text{const.}$ From Eqs. (4.2.15) and (4.2.19) one obtains

$$\begin{aligned} \frac{L_1^2}{L^2} + \frac{L_2^2}{L^2} + \frac{L_3^2}{L^2} &= 1, \\ \frac{L_1^2}{2EI_1} + \frac{L_2^2}{2EI_2} + \frac{L_3^2}{2EI_3} &= 1. \end{aligned} \tag{4.4.14}$$

The first of these equations represents a sphere of radius L, and the second equation represents an ellipsoid with half-axes $\sqrt{2EI_1}$, $\sqrt{2EI_2}$, and $\sqrt{2EI_3}$. Vector $\mathbf{L}$ (or its terminus) can move along the lines in space that are defined by the intersection of the sphere and the ellipsoid. Let us assume, for a moment, $I_1 < I_2 < I_3$. Then, for $L^2 < 2EI_1$ and $L^2 > 2EI_3$, there is no intersection, so that such values of L and E are unphysical. If L^2 slightly exceeds $2EI_1$ or is slightly below $2EI_3$, the intersection lines are small contours encircling $\mathbf{e}^{(1)}$ or $\mathbf{e}^{(3)}$. Thus, the system prepared in such states will stay forever with directions of $\mathbf{L}$ near $\mathbf{e}^{(1)}$ or $\mathbf{e}^{(3)}$. One concludes that rotations around the principal axis with minimal and maximal moments of inertia are stable. This was also confirmed by careful experiments with a chalk eraser at the Graduate Center and Lehman College of the CUNY.

To obtain these small contours, one can subtract the two equations from each other and replace the appropriate component of $\mathbf{L}$ by L. For instance, if L^2 is slightly below $2EI_3$, one obtains

$$\begin{aligned} 0 &= \left(\frac{1}{L^2} - \frac{1}{2EI_1}\right) L_1^2 + \left(\frac{1}{L^2} - \frac{1}{2EI_2}\right) L_2^2 + \left(\frac{1}{L^2} - \frac{1}{2EI_3}\right) L_3^2 \\ &\cong \left(\frac{1}{2EI_3} - \frac{1}{2EI_1}\right) L_1^2 + \left(\frac{1}{2EI_3} - \frac{1}{2EI_2}\right) L_2^2 + \left(\frac{1}{L^2} - \frac{1}{2EI_3}\right) L^2 \end{aligned} \tag{4.4.15}$$

or

$$\left(\frac{I_3}{I_1} - 1\right) L_1^2 + \left(\frac{I_3}{I_2} - 1\right) L_2^2 = 2EI_3 - L^2. \tag{4.4.16}$$

This is an equation of a small (L_1, L_2) ellipse with half-axes

$$\sqrt{(2EI_3 - L^2)\,\frac{I_1}{I_3 - I_1}}, \quad \sqrt{(2EI_3 - L^2)\,\frac{I_2}{I_3 - I_2}}, \tag{4.4.17}$$

whereas $L_3 \cong L$. If I_3 is the intermediate moment of inertia, say, $I_1 < I_3 < I_2$, Eq. (4.4.16) describes hyperbolas since the coefficients in front of L_1^2 and L_2^2 have different signs. Thus, with time, the top strongly deviates from the initial direction of rotation and Eq. (4.4.16) loses its validity.

The dynamics of the above phenomena can be studied with the help of Euler's equations for $\boldsymbol{\omega}$ or $\mathbf{L}$. Let us take Eq. (4.4.6) with $\boldsymbol{\tau} = 0$ and consider rotations around the directions close to $\mathbf{e}^{(3)}$, so that ω_1 and ω_2 are small and $\omega_1\omega_2$ in the third equation can be neglected. Then $\omega_3 \cong$ const. and with Eq. (4.4.2), one obtains

$$\dot{\omega}_1 = \frac{I_2 - I_3}{I_1}\omega_2\omega_3, \quad \dot{\omega}_2 = \frac{I_3 - I_1}{I_2}\omega_1\omega_3. \tag{4.4.18}$$

Combining the two equations yields

$$\ddot{\omega}_1 + \Omega^2\omega_1 = 0, \quad \Omega = \sqrt{\left(\frac{I_3}{I_1} - 1\right)\left(\frac{I_3}{I_2} - 1\right)}\,\omega_3, \tag{4.4.19}$$

and the same equation for ω_2. If I_3 is the maximal or minimal moment of inertia, Ω is real and both ω_1 and ω_2 perform oscillations with frequency Ω. In fact, $\boldsymbol{\omega}$ performs elliptic precession around $\mathbf{e}^{(3)}$, and $\mathbf{L}$ does the similarly. Elliptic character of precession is clear from Eq. (4.4.16). If, however, I_3 is the intermediate moment of inertia, Ω is imaginary and both ω_1 and ω_2

exponentially grow with time from small initial values and soon become comparable with ω_3, so that the approximation breaks down.

4.5 Problems on the angular motion of rigid bodies

The most complicated case of rotational motion in this book is that of a wheel rolling on a plane. This non-holonomic problem is considered in Chapter 8.

4.5.1 *Moments of inertia*

Calculate tensors of inertia with respect to the principal axes of the following bodies:

1. hollow sphere of mass M and radius R;
2. cone of the height h and radius of the base R, both with respect to the apex and to the center of mass;
3. body of a box shape with sides a, b, and c. Consider the limit of a thin rod.

Solution:
(1) We calculate tensor of inertia with respect to the center of mass that is located in the geometrical center. Tensor of inertia is diagonal and all its diagonal elements are the same, $I_{xx} = I_{yy} = I_{zz} = I$. Let us calculate, for instance, I as I_{zz}. Introducing the surface mass density

$$\sigma \equiv \frac{M}{S} = \frac{M}{4\pi R^2}, \tag{4.5.1}$$

one defines $I = I_{zz}$ as

$$I = \sigma \int\limits_S dS\,(x^2 + y^2) = \sigma R^2 \int\limits_0^\pi d\theta \sin\theta \int\limits_0^{2\pi} d\varphi\,(x^2 + y^2). \tag{4.5.2}$$

In this and all similar cases, it is convenient to use $u = \cos\theta$ as an integration variable. Using $x^2 + y^2 = R^2 \sin^2\theta = R^2\left(1 - u^2\right)$, one obtains

$$\begin{aligned} I &= \sigma R^2 2\pi R^2 \int\limits_{-1}^{1} du\,(1 - u^2) = 4\pi R^2 \sigma R^2 \int\limits_0^1 du\,(1 - u^2) \\ &= MR^2\left(1 - \frac{1}{3}\right) = \frac{2}{3} MR^2. \end{aligned} \tag{4.5.3}$$

One can calculate moments of inertia using the symmetry, $I_{xx} = I_{yy} = I_{zz}$, and writing

$$I = \frac{1}{3}\left(I_{xx} + I_{yy} + I_{zz}\right) = \frac{1}{3}\sigma \int\limits_S dS\left(y^2 + z^2 + z^2 + x^2 + x^2 + y^2\right)$$

$$= \frac{2}{3}\sigma R^2 \int\limits_S dS = \frac{2}{3}MR^2. \qquad (4.5.4)$$

In this calculation, one can replace integration by summation, then it becomes suitable for a non-calculus-based physics course.

(2) Let us first calculate moments of inertia of the cone with respect to its apex and then obtain those with respect to CM with the help of the Steiner theorem. The volume of the cone is

$$V = \int\limits_0^h dz S(z) = \int\limits_0^h dz \pi r^2(z) = \int\limits_0^h dz \pi \left(R\frac{z}{h}\right)^2 = \frac{1}{3}\pi R^2 h. \qquad (4.5.5)$$

Introducing the mass density

$$\rho \equiv \frac{M}{V}, \qquad (4.5.6)$$

one proceeds

$$I_{zz} = \rho \int\limits_V dV\left(x^2 + y^2\right) = \rho \int\limits_0^h dz \int\limits_0^{Rz/h} 2\pi r dr\, r^2 = \rho \int\limits_0^h dz \frac{\pi}{2}\left(R\frac{z}{h}\right)^4$$

$$= \frac{1}{10}\rho\pi R^4 h = \frac{3}{10}MR^2. \qquad (4.5.7)$$

Similarly,

$$I'_{xx} = \rho \int\limits_V dV\left(y^2 + z^2\right), \quad I'_{yy} = \rho \int\limits_V dV\left(z^2 + x^2\right). \qquad (4.5.8)$$

Since, by symmetry, $I'_{xx} = I'_{yy} \equiv I_\perp$, one can simplify the calculation,

$$I_\perp = \frac{1}{2}\left(I'_{xx} + I'_{yy}\right) = \frac{1}{2}\rho \int\limits_V dV\left(x^2 + y^2 + 2z^2\right) = \frac{1}{2}I_{zz} + I''_\perp, \qquad (4.5.9)$$

where

$$I''_\perp \equiv \rho \int\limits_V dV z^2 = \rho \int\limits_0^h dz \pi \left(R\frac{z}{h}\right)^2 z^2 = \frac{1}{5}\rho\pi R^2 h^3 = \frac{3}{5}Mh^2. \qquad (4.5.10)$$

Thus,

$$I'_\perp = \frac{3}{20}MR^2 + \frac{3}{5}Mh^2. \tag{4.5.11}$$

According to the Steiner theorem, transverse moment of inertia with respect to CM is given by

$$I_\perp = I'_\perp - Ma^2, \tag{4.5.12}$$

where a is the distance from the apex to CM. The latter can be calculated as

$$a = \frac{1}{M}\rho \int_V dV z = \frac{1}{M}\rho \int_0^h dz\pi \left(R\frac{z}{h}\right)^2 z = \frac{1}{M}\rho\frac{1}{4}\pi R^2 h^2 = \frac{3}{4}h. \tag{4.5.13}$$

Using this, one obtains

$$I_\perp = \frac{3}{20}MR^2 + \frac{3}{80}Mh^2. \tag{4.5.14}$$

I_{zz} with respect to CM is the same.

(3) Calculate moments of inertia with respect to CM.

$$I_{zz} = \frac{M}{abc}\int_{-a/2}^{a/2} dx \int_{-b/2}^{b/2} dy \int_{-c/2}^{c/2} dx \left(x^2 + y^2\right) = \frac{M}{abc}\left(\frac{1}{3}a^3 b + \frac{1}{3}ab^3\right)$$

$$c = \frac{1}{3}M\left(a^2 + b^2\right). \tag{4.5.15}$$

Similarly, one obtains

$$I_{xx} = \frac{1}{3}M\left(b^2 + c^2\right), \quad I_{yy} = \frac{1}{3}M\left(c^2 + a^2\right). \tag{4.5.16}$$

The limit of a thin rod can be obtained from this by setting $a = b = 0$ and $c = L$. This yields

$$I_{xx} = I_{yy} = \frac{1}{3}ML^2, \quad I_{zz} = 0. \tag{4.5.17}$$

4.5.2 *Rotational dynamics with respect to the instantaneous axis of rotation*

A hollow cylinder of mass M and radius R is lying on a horizontal plane with sufficient friction, so that the cylinder can roll without skidding. A horizontal force F is applied to the cylinder's center of mass in the direction of the perspective rolling. Find the linear and angular acceleration by considering the rotational dynamics with respect to the CM. Check whether rotational dynamics with respect to the contact point (cylinder plane) yields the same result.

Solution with respect to the CM: The angular acceleration is caused by the friction force F_{fr} directed opposite to F, while the acceleration of the CM is caused by the total horizontal force $F - F_{fr}$. The dynamical equations are

$$\begin{aligned} \dot{\omega} I &= F_{fr} R, \\ M\dot{v} &= F - F_{fr}. \end{aligned} \tag{4.5.18}$$

where the moment of inertia is given by $I = MR^2$ and the linear and angular velocities are related by the no-skidding constraint $v = \omega R$. Substituting I and v, one obtains the system of equations

$$\begin{aligned} \dot{\omega} &= F_{fr} R/I = F_{fr}/(MR), \\ \dot{\omega} &= (F - F_{fr})/(MR). \end{aligned} \tag{4.5.19}$$

Adding these equations cancels the friction force, and one obtains

$$\dot{\omega} = F/(2MR), \quad \dot{v} = F/(2M). \tag{4.5.20}$$

Solution with respect to the instantaneous rotation axis going through the contact point: Here, there is only rotational motion described by Eq. (4.2.22) with $\dot{\mathbf{a}} = 0$:

$$\dot{\omega} I' = FR, \tag{4.5.21}$$

where $I' = I + MR^2 = 2MR^2$ is the shifted moment of inertia with respect to the axis going through the contact point. The solution is $\dot{\omega} = F/(2MR)$, the same as above.

4.5.3 *Rolling half-cylinder*

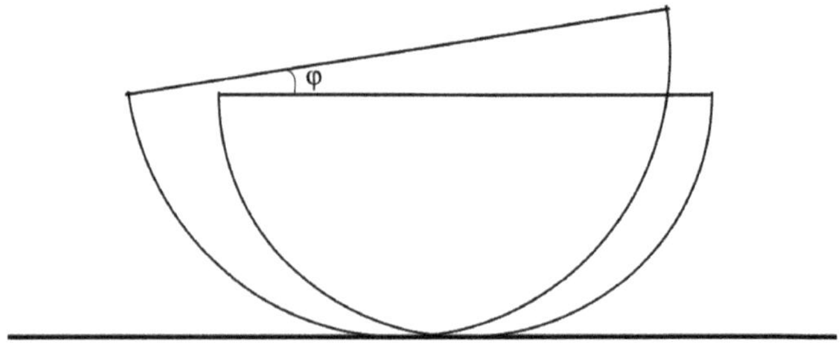

Consider a half-cylinder of mass M and radius R on a horizontal plane.

1. Find the position of its center of mass (CM) and the moment of inertia with respect to CM.
2. Write down the Lagrange function in terms of the angle φ (see the figure)
3. Find the frequency of cylinder's oscillations in the linear regime, $\varphi \ll 1$.

Solution:
(1) Position of CM with respect to the center of the cylinder is given by the integral over cross-section

$$a = \frac{1}{M}\frac{2M}{\pi R^2}\int\limits_S ydS = \frac{2}{\pi R^2}\int\limits_0^\pi d\phi \int\limits_0^R rdr\, r\sin\phi = \frac{2}{\pi R^2}2\frac{R^3}{3} = \frac{4}{3\pi}R. \tag{4.5.22}$$

The moment of inertia of the half-cylinder with respect to the geometrical center of the cylinder is, obviously, the same as that of the cylinder:

$$I' \equiv I'_{zz} = \frac{1}{2}MR^2. \tag{4.5.23}$$

The moment of inertia with respect to CM can be obtained with the help of the Steiner theorem:

$$I = I' - Ma^2 = \left(\frac{1}{2} - \frac{16}{9\pi^2}\right)MR^2 \approx 0.32MR^2. \tag{4.5.24}$$

(2)

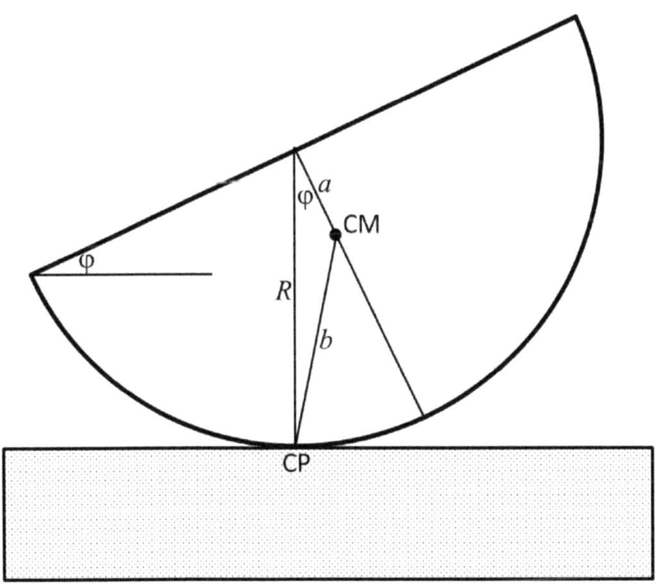

The easiest way to write down the kinetic energy is to consider rotation around the instantaneous axis going through the contact point (CP). The moment of inertia with respect to CP I'' is given by the Steiner theorem:

$$I'' = I + Mb^2 = I + M\left(R^2 + a^2 - 2Ra\cos\varphi\right), \tag{4.5.25}$$

where the cosine rule was used. Substituting a, one obtains

$$I'' = \left(\frac{3}{2} - \frac{8}{3\pi}\cos\varphi\right)MR^2. \tag{4.5.26}$$

Potential energy is similar to that of the pendulum, $U = -Mga\cos\varphi$. Finally, the Lagrangian becomes

$$\mathcal{L} = E_k - U = \frac{1}{2}I''\dot{\varphi}^2 + Mga\cos\varphi, \tag{4.5.27}$$

which in turn becomes

$$\mathcal{L} = \frac{1}{2}\left(\frac{3}{2} - \frac{8}{3\pi}\cos\varphi\right)MR^2\dot{\varphi}^2 + \frac{4}{3\pi}MgR\cos\varphi. \tag{4.5.28}$$

The system is similar to the pendulum, but the coefficient in the kinetic energy is variable.

(3) In the case of small oscillations around equilibrium, one can replace $\cos\varphi \Rightarrow 1$ in the kinetic energy, since there is already a small quantity $\dot{\varphi}^2$. After this, the problem becomes mathematically equivalent to the pendulum. One obtains

$$\mathcal{L} = \frac{1}{2}\left(\frac{3}{2} - \frac{8}{3\pi}\right)MR^2\dot{\varphi}^2 + \frac{4}{3\pi}MgR\left(1 - \frac{1}{2}\varphi^2\right), \tag{4.5.29}$$

equation of motion becomes

$$\left(\frac{3}{2} - \frac{8}{3\pi}\right)MR^2\ddot{\varphi} + \frac{4}{3\pi}MgR\varphi = 0, \tag{4.5.30}$$

and the oscillation frequency is given by

$$\omega_0^2 = \frac{\frac{4}{3\pi}}{\frac{3}{2} - \frac{8}{3\pi}}\frac{g}{R} = \frac{1}{9\pi/8 - 2}\frac{g}{R}. \tag{4.5.31}$$

4.5.4 *Cylinder rolling inside a cylinder*

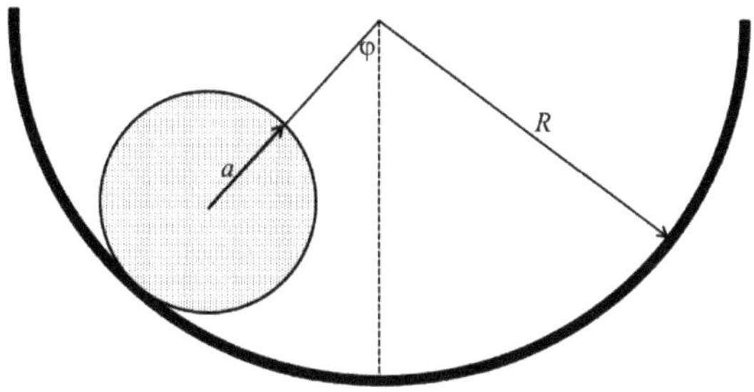

A cylinder of mass M and radius a is rolling without slipping inside a stationary cylinder of radius $R > a$. Find its kinetic energy.

Solution: The speed of CM of the cylinder is

$$V = (R - a)\,\dot{\varphi}. \tag{4.5.32}$$

From the no-slipping condition $V = a\omega$, one obtains the angular velocity of the cylinder

$$\omega = \frac{R - a}{a}\dot{\varphi}. \tag{4.5.33}$$

Now, the kinetic energy can be written as the sum of the kinetic energy of CM and the rotation around CM as

$$E_k = \frac{MV^2}{2} + \frac{I\omega^2}{2} = \frac{1}{2}\left(M + \frac{I}{a^2}\right)(R - a)^2\,\dot{\varphi}^2. \tag{4.5.34}$$

Substituting $I = MR^2/2$ for the cylinder, one obtains

$$E_k = \frac{3}{4}M\,(R - a)^2\,\dot{\varphi}^2. \tag{4.5.35}$$

4.5.5 *Cone rolling on a plane*

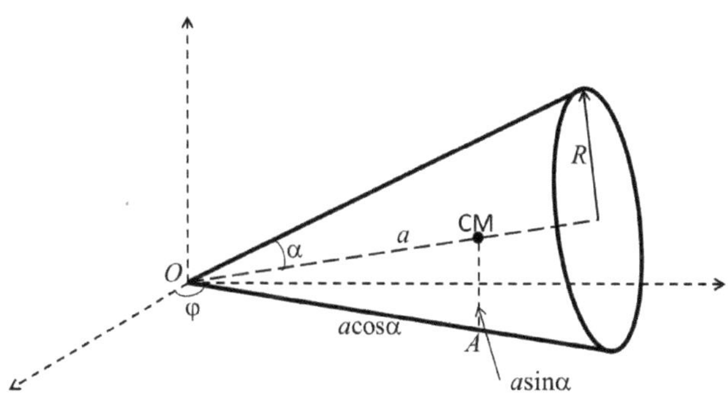

Find the kinetic energy of a cone rolling on a plane without slipping. The height of the cone is h, the apex angle is 2α. The mass is M.

Solution: The cone is rolling in such a way that its apex is stationary. We chose a frame such that the apex is at its origin O. The orientation of the cone is described by the angle φ that the contact line OA makes with one of the horizontal axes. CM of the cone is at the distance a from the apex. At any moment of time, the cone is rotating around the instantaneous axis OA. The angular velocity can be found via the speed of CM V as

$$\omega = \frac{V}{a\sin\alpha} = \frac{a\cos\alpha\dot{\varphi}}{a\sin\alpha} = \cot\alpha\dot{\varphi}. \tag{4.5.36}$$

Here, a canceled, thus one could use any point on the cone's symmetry axis instead of CM. Kinetic energy can be found by projecting $\boldsymbol{\omega}$ onto the 3 and perpendicular axes of the cone:

$$E_k = \frac{1}{2}I_3\left(\omega\cos\alpha\right)^2 + \frac{1}{2}I'_\perp\left(\omega\sin\alpha\right)^2. \tag{4.5.37}$$

Using moments of inertia with respect to the apex from the problem above,

$$\begin{aligned} I_3 &= \frac{3}{10}MR^2 = \frac{3}{10}Mh^2\tan^2\alpha, \\ I'_\perp &= \frac{3}{20}MR^2 + \frac{3}{5}Mh^2 = \frac{3}{20}Mh^2\tan^2\alpha + \frac{3}{5}Mh^2, \end{aligned} \tag{4.5.38}$$

one obtains

$$\begin{aligned} E_k &= \frac{1}{2}Mh^2\left[\frac{3}{10}\cos^2\alpha + \frac{3}{20}\sin^2\alpha + \frac{3}{5}\cos^2\alpha\right]\dot{\varphi}^2 \\ &= \frac{3}{40}Mh^2\left(1+5\cos^2\alpha\right)\dot{\varphi}^2. \end{aligned} \tag{4.5.39}$$

4.5.6 *Rolling cone with a horizontal axis*

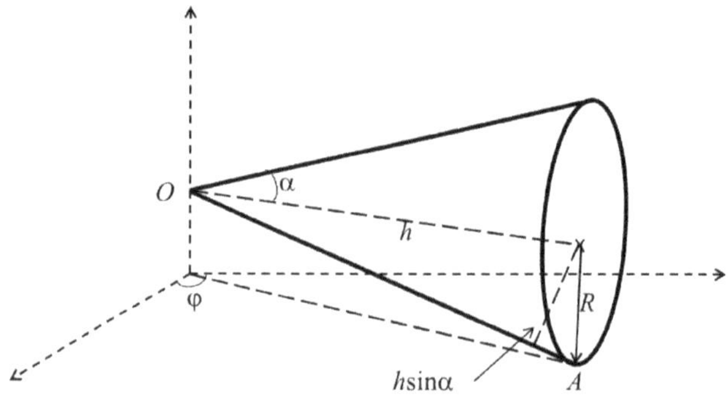

Find the kinetic energy of a cone rolling on a plane and having the symmetry axis horizontal (see the figure).

Solution: In this case, the instantaneous axis of rotation is the line OA. To find ω, one can use, say, the center of the cone's base. The speed of this point is $V = h\dot{\varphi}$, so that

$$\omega = \frac{V}{h \sin\alpha} = \frac{h\dot{\varphi}}{h \sin\alpha} = \frac{\dot{\varphi}}{\sin\alpha}. \tag{4.5.40}$$

Kinetic energy is given by

$$E_k = \frac{1}{2} I_3 (\omega \cos\alpha)^2 + \frac{1}{2} I'_\perp (\omega \sin\alpha)^2. \tag{4.5.41}$$

Using moments of inertia with respect to the apex from the problem above,

$$\begin{aligned} I_3 &= \frac{3}{10} MR^2 = \frac{3}{10} Mh^2 \tan^2\alpha, \\ I'_\perp &= \frac{3}{20} MR^2 + \frac{3}{5} Mh^2 = \frac{3}{20} Mh^2 \tan^2\alpha + \frac{3}{5} Mh^2, \end{aligned} \tag{4.5.42}$$

one obtains

$$E_k = \frac{1}{2} Mh^2 \left[\frac{3}{10} + \frac{3}{20} \tan^2\alpha + \frac{3}{5} \right] \dot{\varphi}^2 = \frac{3}{40} Mh^2 \left(\frac{1}{\cos^2\alpha} + 5 \right) \dot{\varphi}^2. \tag{4.5.43}$$

4.5.7 *Rod on the axis*

Thin rod of length l and mass M is mounted on an axis at its center.

(a) If the angle θ between the rod and the axis is fixed and the rod rotates with the angular velocity $\omega = \dot{\varphi}$ around the axis, what are the (i) kinetic energy of the rod and (ii) the breaking torque acting from the rod on the axis?

(b) Set up Lagrange equations for the rod in the case where both θ and φ can freely change. Find integrals of motion and the effective potential energy. If you have access to mathematical software, you can try to produce numerical solutions with particular initial conditions such as $\theta(0) = \theta_0$, $\dot{\theta}(0) = 0$, $\varphi(0) = 0$, $\dot{\varphi}(0) = \omega_0$.

(c) Consider the motion of this system confined to the vicinity of $\theta = \pi/2$ and try to integrate Lagrange equations analytically.

Solution:
(a) (i) Introducing the linear mass density $\gamma \equiv M/l$, for the kinetic energy, one obtains

$$E_k = \frac{\gamma}{2}\int\limits_{l/2}^{l/2} du\,[\boldsymbol{\omega}\times\mathbf{r}]^2 = \frac{\gamma\omega^2\sin^2\theta}{2}\int\limits_{l/2}^{l/2} du\,u^2 = \frac{\gamma\omega^2 l^3\sin^2\theta}{24} = \frac{M\omega^2 l^2\sin^2\theta}{24}. \tag{4.5.44}$$

This can be rewritten in the form

$$E_k = \frac{I_\theta\omega^2}{2}, \quad I_\theta = \frac{Ml^2\sin^2\theta}{12}. \tag{4.5.45}$$

(ii) To calculate the breaking torque, one has first to work out the angular momentum $\mathbf{L}$ that lies in the plane spanned by $\boldsymbol{\omega}$ (z-axis) and the axis of the rod and perpendicular to the latter:

$$\mathbf{L} = \gamma\int\limits_{l/2}^{l/2} du\,[\mathbf{r}\times[\boldsymbol{\omega}\times\mathbf{r}]] = \gamma\int\limits_{l/2}^{l/2} du\left\{\boldsymbol{\omega}r^2 - \mathbf{r}\,(\mathbf{r}\cdot\boldsymbol{\omega})\right\}$$

$$= \gamma\omega\mathbf{e}_2\sin\theta\int\limits_{l/2}^{l/2} du\,u^2 = I\omega\sin\theta\mathbf{e}_2, \tag{4.5.46}$$

where we use sliding embedded vectors, $\mathbf{e}_2 = \mathbf{e}_A$. As the rod is precessing around z-axis, $\mathbf{e}_2$ is precessing as well. Thus, the Newtonian equation for $\mathbf{L}$ has the form

$$\dot{\mathbf{L}} = I\omega\sin\theta\,[\boldsymbol{\omega}\times\mathbf{e}_2] = I\omega^2\sin\theta\cos\theta\mathbf{e}_1 = \boldsymbol{\tau}. \tag{4.5.47}$$

This yields the breaking torque $\boldsymbol{\tau}$, acting between the rod and the axis. τ reaches its maximum for $\theta = \pi/4$.

(b) In the case when both θ and φ are free to change, the kinetic energy of the rod can be written as

$$E_k = \frac{1}{2}I_1\omega_1^2 + \frac{1}{2}I_2\omega_2^2. \tag{4.5.48}$$

With the sliding embedded vectors and $I_2 = I_1 = I$, one obtains

$$\mathcal{L} = E_k = \frac{1}{2}I\left(\dot{\theta}^2 + \sin^2\theta\dot{\varphi}^2\right). \tag{4.5.49}$$

Here φ is the cyclic variable, and the corresponding integral of motion is the projection of the angular momentum on z-axis:

$$l_z \equiv l = \frac{\partial \mathcal{L}}{\partial \dot{\varphi}} = I \sin^2 \theta \dot{\varphi}. \tag{4.5.50}$$

Lagrange equation for θ reads

$$\frac{d}{dt}\frac{\partial \mathcal{L}}{\partial \dot{\theta}} - \frac{\partial \mathcal{L}}{\partial \theta} = I\ddot{\theta} - I \sin\theta \cos\theta \dot{\varphi}^2 = 0. \tag{4.5.51}$$

With the account of angular-momentum conservation, one obtains the equation

$$I\ddot{\theta} - \frac{\cos\theta}{\sin^3\theta}\frac{l}{I}^2 = 0. \tag{4.5.52}$$

This can be written in the form

$$I\ddot{\theta} = -\frac{\partial U_{\text{eff}}}{\partial \theta}, \quad U_{\text{eff}} = \frac{l}{2I}^2 \frac{1}{\sin^2\theta}. \tag{4.5.53}$$

$U_{\text{eff}}(\theta)$ has a minimum at $\theta = \pi/2$.

(c) Near the minimum of the effective potential energy, one can use

$$\theta = \frac{\pi}{2} + \delta\theta, \quad \delta\theta \ll 1. \tag{4.5.54}$$

With $\sin\theta \cong 1$ and $\cos\theta \cong -\delta\theta$, the equation of motion becomes

$$\delta\ddot{\theta} + \omega_0^2 \delta\theta = 0, \quad \omega_0 = \frac{l}{I}. \tag{4.5.55}$$

This is the equation of motion of a harmonic oscillator.

4.5.8 *Asymmetric top with a horizontal holder*

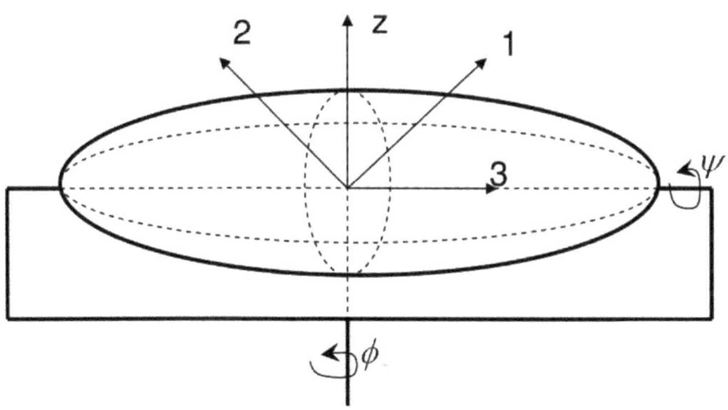

Consider an asymmetric top with moments of inertia $I_1 < I_2$ supported by a holder that allows the top to freely rotate changing its Euler angles ϕ and ψ, while preserving $\theta = \pi/2$, see the figure. The axes of the holder cross at the center of mass of the top.

(a) Set up Lagrange equations for this top, find integrals of motion.

(b) Eliminate ϕ to obtain an effective energy for ψ. What kinds of motion for ψ are possible? Analyze the behavior of ψ near the minimum of the effective potential energy.

(c) If you have access to mathematical software, you can try to produce numerical solutions with particular initial conditions.

Solution:
(a) Angular velocity is due to the two types of rotational motion:

$$\boldsymbol{\omega} = \mathbf{e}_z\dot{\phi} + \mathbf{e}_3\dot{\psi}. \tag{4.5.56}$$

Projecting $\mathbf{e}_z$ onto the principal axes $\mathbf{e}_1$ and $\mathbf{e}_2$, one obtains kinetic energy and thus the Lagrangian:

$$\begin{aligned}\mathcal{L} = E_k = E &= \frac{1}{2}I_1\omega_1^2 + \frac{1}{2}I_2\omega_2^2 + \frac{1}{2}I_3\omega_3^2 \\ &= \frac{1}{2}\left[I_1 \sin^2\psi + I_2\cos^2\psi\right]\dot{\phi}^2 + \frac{1}{2}I_3\dot{\psi}^2.\end{aligned} \tag{4.5.57}$$

At $\psi = 0$, one has $\mathbf{e}_2 = \mathbf{e}_z$. Lagrange equations have the form

$$\frac{d}{dt}\frac{\partial\mathcal{L}}{\partial\dot{\phi}} - \frac{\partial\mathcal{L}}{\partial\phi} = \frac{d}{dt}\left\{\left[I_1\sin^2\psi + I_2\cos^2\psi\right]\dot{\phi}\right\} = 0 \tag{4.5.58}$$

and

$$\frac{d}{dt}\frac{\partial\mathcal{L}}{\partial\dot{\psi}} - \frac{\partial\mathcal{L}}{\partial\psi} = I_3\ddot{\psi} - (I_2 - I_1)\sin\psi\cos\psi\dot{\phi}^2 = 0. \tag{4.5.59}$$

(b) Here, ϕ is the cyclic variable and the corresponding generalized momentum is conserved. This is nothing else than vertical component of the angular momentum:

$$l_z \equiv l = \frac{\partial\mathcal{L}}{\partial\dot{\phi}} = \left[I_1\sin^2\psi + I_2\cos^2\psi\right]\dot{\phi} = \text{const.} \tag{4.5.60}$$

Eliminating $\dot{\phi}$, one can rewrite energy as

$$E = \frac{1}{2}I_3\dot{\psi}^2 + U_{\text{eff}}(\psi), \tag{4.5.61}$$

where effective energy is given by

$$U_{\text{eff}}(\psi) \equiv \frac{l^2}{2\left[I_1 \sin^2\psi + I_2 \cos^2\psi\right]} = \frac{l^2}{2\left[I_2 - (I_2 - I_1)\sin^2\psi\right]}. \quad (4.5.62)$$

Motion of the system can now be defined as in the case of one-dimensional motion—using energy conservation, resolving for $\dot{\psi}$ and then integrating. For $I_1 < I_2$, the minimum of effective potential energy is at $\psi = 0$, which corresponds to rotating around axis 2, having the greater moment of inertia. Near $\psi = 0$, one has

$$U_{\text{eff}}(\psi) \cong \frac{l^2}{2I_2}\left[1 + \left(1 - \frac{I_1}{I_2}\right)\sin^2\psi\right] \cong \frac{l^2}{2I_2}\left(1 - \frac{I_1}{I_2}\right)\psi^2 + \text{const.} \quad (4.5.63)$$

Using the effective Newtonian equation,

$$I_3\ddot{\psi} + \frac{\partial U_{\text{eff}}(\psi)}{\partial\psi} = 0 \quad (4.5.64)$$

(that is equivalent to the second Lagrange equation), one obtains the harmonic-oscillator equation

$$\ddot{\psi} + \omega_0^2\psi = 0, \quad \omega_0 = \frac{l}{\sqrt{I_2 I_3}}\sqrt{1 - \frac{I_1}{I_2}}. \quad (4.5.65)$$

Chapter 5

Hamiltonian Mechanics

Both Newtonian and Lagrangian formalisms operate with systems of second-order differential equations for time-dependent generalized coordinates, $\ddot{q}_i = \cdots$. For a system with N degrees of freedom, N such equations can be reformulated as systems of $2N$ first-order differential equations if one considers velocities $v_i = \dot{q}_i$ as additional dynamical variables. This system of equations has the form $\dot{q}_i = v_i$, $\dot{v}_i = \cdots$, which is non-symmetric with respect to q_i and v_i.

Hamiltonian formalism uses q_i and p_i as dynamical variables, where p_i are generalized momenta defined by

$$p_i = \frac{\partial \mathcal{L}}{\partial \dot{q}_i}. \tag{5.0.1}$$

The resulting $2N$ Hamiltonian equations of motion for q_i and p_i have an elegant symmetric form that is the reason for calling them *canonical equations*. Although for most of mechanical problems Hamiltonian formalism is of no practical advantage, it is worth studying because of the similarity between its mathematical structure and that of quantum mechanics. In fact, a significant part of quantum mechanics using matrix and operator algebra grew out of Hamiltonian mechanics. The latter is invoked in constructing new field theories. Hamiltonian formalism finds application in statistical physics too.

5.1 Hamiltonian function and equations

Hamiltonian equations can be obtained from Eq. (5.0.1) and Lagrange equations that can be written in the form

$$\dot{p}_i = \frac{\partial \mathcal{L}}{\partial q_i}, \tag{5.1.1}$$

using a Legendre transformation. The differential of the Lagrange function has the form

$$d\mathcal{L} = \sum_i \left(\frac{\partial \mathcal{L}}{\partial q_i} dq_i + \frac{\partial \mathcal{L}}{\partial \dot{q}_i} d\dot{q}_i \right), \tag{5.1.2}$$

which with the help of Eqs. (5.0.1) and (5.1.1) can be rewritten as

$$d\mathcal{L} = \sum_i (\dot{p}_i dq_i + p_i d\dot{q}_i). \tag{5.1.3}$$

Let us introduce the Hamiltonian function, or the Hamiltonian, defined by

$$\mathcal{H} = \sum_i p_i \dot{q}_i - \mathcal{L}. \tag{5.1.4}$$

With the help of the above, its differential becomes

$$d\mathcal{H} = \sum_i (\dot{q}_i dp_i - \dot{p}_i dq_i). \tag{5.1.5}$$

One can see that the natural variables of $\mathcal{H}$ are q and p, and

$$\dot{q}_i = \frac{\partial \mathcal{H}}{\partial p_i}, \quad \dot{p}_i = -\frac{\partial \mathcal{H}}{\partial q_i}, \tag{5.1.6}$$

which are Hamiltonian equations of motion for q and p.

As the Lagrange function is bilinear in $\dot{q}_i$,

$$\mathcal{L} = E_k(q, \dot{q}) - U(q), \quad E_k = \frac{1}{2} \sum_{ij} m_{ij}(q) \dot{q}_i \dot{q}_j, \tag{5.1.7}$$

the first term in Eq. (5.1.4) is the double kinetic energy,

$$\sum_i p_i \dot{q}_i = \sum_i \frac{\partial \mathcal{L}}{\partial \dot{q}_i} \dot{q}_i = 2E_k. \tag{5.1.8}$$

Now, one can see that $\mathcal{H}$ is the full energy,

$$\mathcal{H} = E_k + U. \tag{5.1.9}$$

In most cases, the kinetic energy is written in a natural way in terms of $\dot{q}_i$, so that expressing it in terms of p_i may require work. For unconstrained particles, there is the linear relation $\mathbf{p}_i = m\mathbf{v}_i$ and re-expressing $\mathcal{H}$ in terms of p_i is not a problem. In particular, for a system of particles, one obtains

$$\mathcal{H} = \sum_i \frac{\mathbf{p}_i^2}{2m_i} + U. \tag{5.1.10}$$

On the other hand, for systems with constraints, such as double pendulum, different generalized velocities $\dot{q}_i$ couple since $m_{ij} \equiv (\mathbb{M})_{ij}$ in Eq. (5.1.7) is a non-diagonal matrix. Equation (5.1.7) can be written in the matrix-vector form

$$E_k = \frac{1}{2}\dot{\mathbf{q}}^T \cdot \mathbb{M} \cdot \dot{\mathbf{q}}, \tag{5.1.11}$$

where $\dot{\mathbf{q}} = (\dot{q}_1, \dot{q}_2, \ldots)$ is a column and the transposed $\dot{\mathbf{q}}^T$ is a row. Further, from Eqs. (5.0.1) and (5.1.7), one obtains

$$\mathbf{p} = \mathbb{M} \cdot \dot{\mathbf{q}}, \tag{5.1.12}$$

and thus,

$$\dot{\mathbf{q}} = \mathbb{M}^{-1} \cdot \mathbf{p}, \quad \dot{\mathbf{q}}^T = \left(\mathbb{M}^{-1} \cdot \mathbf{p}\right)^T = \mathbf{p}^T \cdot \left(\mathbb{M}^{-1}\right)^T = \mathbf{p}^T \cdot \mathbb{M}^{-1}, \tag{5.1.13}$$

where we have taken into account that $\mathbb{M}^{-1}$, as well as $\mathbb{M}$, is a symmetric matrix. Plugging this result into Eq. (5.1.11), one obtains

$$\mathcal{H} = \frac{1}{2}\mathbf{p}^T \cdot \mathbb{M}^{-1} \cdot \mathbf{p} + U. \tag{5.1.14}$$

Although inverting the mass matrix $\mathbb{M}$ is possible, at least numerically, it makes Hamiltonian formalism less appropriate. For this reason, it is never used for systems with holonomic constraints relevant in engineering, exactly where Lagrangian approach shows its strength.

Let us, as an illustration, find the Hamiltonian of a particle in spherical coordinates. Kinetic energy has the form

$$E_k = \frac{mv^2}{2} = \frac{m}{2}\left(v_r^2 + v_\theta^2 + v_\phi^2\right) = \frac{m}{2}\left[\dot{r}^2 + \left(r\dot{\theta}\right)^2 + \left(r\sin\theta\,\dot{\phi}\right)^2\right]. \tag{5.1.15}$$

Thus, the generalized momenta (5.0.1) are given by

$$\begin{aligned} p_r &= \frac{\partial E_k}{\partial \dot{r}} = m\dot{r}, \\ p_\theta &= \frac{\partial E_k}{\partial \dot{\theta}} = mr^2\dot{\theta}, \\ p_\phi &= \frac{\partial E_k}{\partial \dot{\phi}} = mr^2\sin^2\theta\,\dot{\phi}. \end{aligned} \tag{5.1.16}$$

Solving these equations for $\dot{r}$, $\dot{\theta}$, and $\dot{\phi}$ yields

$$\dot{r} = \frac{p_r}{m}, \quad \dot{\theta} = \frac{p_\theta}{mr^2}, \quad \dot{\phi} = \frac{p_\phi}{mr^2\sin^2\theta}. \tag{5.1.17}$$

Inserting the results into Eq. (5.1.15), one obtains

$$\mathcal{H} = E_k + U = \frac{1}{2m}\left(p_r^2 + \frac{p_\theta^2}{r^2} + \frac{p_\phi^2}{r^2 \sin^2\theta}\right) + U(r,\theta,\phi). \tag{5.1.18}$$

One can see that Eqs. (5.1.17) are Hamilton equations

$$\dot{r} = \frac{\partial \mathcal{H}}{\partial p_r}, \quad \dot{\theta} = \frac{\partial \mathcal{H}}{\partial p_\theta}, \quad \dot{\phi} = \frac{\partial \mathcal{H}}{\partial p_\phi}. \tag{5.1.19}$$

5.2 Hamiltonian equations from the least-action principle

As Lagrange equations follow from the least-action principle and Hamiltonian equations can be derived from Lagrange equations, one can obtain Hamiltonian equations from the least-action principle directly. The calculation starts from the action written with the help of Eq. (5.1.4) in the form

$$\mathcal{S} = \int_{t_1}^{t_2} dt\mathcal{L}(q,\dot{q},t) = \int_{t_1}^{t_2} dt\left(\sum_i p_i\dot{q}_i - \mathcal{H}\right) = \int_{t_1}^{t_2}\left(\sum_i p_i dq_i - \mathcal{H}dt\right). \tag{5.2.1}$$

Here, in contrast to Lagrange formalism, one considers both p_i and q_i as independent dynamical variables and makes a variation

$$q_i \Rightarrow q_i + \delta q_i, \quad p_i \Rightarrow p_i + \delta p_i. \tag{5.2.2}$$

The corresponding variation of the action has the form

$$\delta\mathcal{S} = \sum_i \int_{t_1}^{t_2}\left(\delta p_i dq_i + p_i d\delta q_i - \frac{\partial \mathcal{H}}{\partial q_i}\delta q_i dt - \frac{\partial \mathcal{H}}{\partial p_i}\delta p_i dt\right). \tag{5.2.3}$$

Integrating the second term by parts and using that $\delta q_i = 0$ at the beginning and at the end, one can rewrite this as

$$\delta\mathcal{S} = \sum_i \int_{t_1}^{t_2}\left[\delta p_i\left(dq_i - \frac{\partial \mathcal{H}}{\partial p_i}dt\right) + \delta q_i\left(-dp_i - \frac{\partial \mathcal{H}}{\partial q_i}dt\right)\right]. \tag{5.2.4}$$

Since for the actual trajectories $\delta\mathcal{S} = 0$ for arbitrary and independent δq_i and δp_i, one concludes that both expressions in round brackets are zero that leads to Hamilton equations (5.1.6).

5.3 Poisson brackets and conservation

Time derivative of any quantity $f = f(q, p, t)$ has the form

$$\frac{df}{dt} = \frac{\partial f}{\partial t} + \sum_i \left(\frac{\partial f}{\partial q_i} \dot{q}_i + \frac{\partial f}{\partial p_i} \dot{p}_i \right). \tag{5.3.1}$$

With the help of Eqs. (5.1.6), it can be rewritten as

$$\frac{df}{dt} = \frac{\partial f}{\partial t} + \{f, \mathcal{H}\}, \tag{5.3.2}$$

where $\{f, \mathcal{H}\}$ is a Poisson bracket defined by

$$\{f, g\} \equiv \sum_i \left(\frac{\partial f}{\partial q_i} \frac{\partial g}{\partial p_i} - \frac{\partial f}{\partial p_i} \frac{\partial g}{\partial q_i} \right). \tag{5.3.3}$$

One can see that if f does not depend on time explicitly and its Poisson bracket with the Hamiltonian is zero, f is an integral of motion. In particular, obviously $\{\mathcal{H}, \mathcal{H}\} = 0$, so that in the absence of explicit time dependence, the energy is an integral of motion, $\mathcal{H} = E = \text{const}$. Note the similarity between Eq. (5.3.2) and the quantum mechanical equation of motion for an operator in Heisenberg representation where the commutator $[f, \hat{H}]$ replaces the Poisson bracket.

Poisson brackets satisfy obvious relations such as

$$\{g, f\} = -\{f, g\}, \quad \{f, g_1 + g_2\} = \{f, g_1\} + \{f, g_2\}, \tag{5.3.4}$$

etc. There is also the non-trivial Jacobi identity

$$\{f, \{g, h\}\} + \{h, \{f, g\}\} + \{g, \{h, f\}\} = 0, \tag{5.3.5}$$

which can be proven directly. From this identity, it follows that if both f and g are integrals of motion, then $\{f, g\}$ is also an integral of motion. For the proof, one chooses $h = \mathcal{H}$, then the Jacobi identity takes the form

$$\left\{ f, \underbrace{\{g, \mathcal{H}\}}_{0} \right\} + \{\mathcal{H}, \{f, g\}\} + \left\{ g, \underbrace{\{\mathcal{H}, f\}}_{0} \right\} = \{\mathcal{H}, \{f, g\}\} = 0, \tag{5.3.6}$$

thus $\{f, g\} = \text{const}$. In principle, this could be used to obtain new integrals of motion from already known "old" integrals of motion. However, in most cases, new integrals of motion $\{f, g\}$ are trivial, such as functions of the old integrals of motion.

The straightforwardly obtainable Poisson brackets

$$\begin{aligned} \{q_i, q_j\} &= 0, \quad \{q_i, p_j\} = \delta_{ij}, \\ \{p_i, q_j\} &= -\delta_{ij}, \quad \{p_i, p_j\} = 0 \end{aligned} \tag{5.3.7}$$

are called *fundamental* Poisson brackets.

If the Hamiltonian depends on one generalized coordinate and the corresponding generalized momentum as a combination that does not contain other variables, this combination is an integral of motion. That is, for

$$\mathcal{H}(\{q_i, p_i\}) = \mathcal{H}\left(\{q_i, p_i\}_{i\neq 1}; F(q_1, p_1)\right) \tag{5.3.8}$$

(where $\{\cdots\}$ is not a Poisson bracket), $F(q_1, p_1)$ is an integral of motion. This can be proven with the help of Poisson brackets. Since differentiation with respect to the variables $i \neq 1$ gives zero, one obtains

$$\begin{aligned} \dot{F} = \{F, \mathcal{H}\} &= \frac{\partial F}{\partial q_1}\frac{\partial \mathcal{H}}{\partial p_1} - \frac{\partial F}{\partial p_1}\frac{\partial \mathcal{H}}{\partial q_1} \\ &= \frac{\partial \mathcal{H}}{\partial F}\left(\frac{\partial F}{\partial q_1}\frac{\partial F}{\partial p_1} - \frac{\partial F}{\partial p_1}\frac{\partial F}{\partial q_1}\right) = \frac{\partial \mathcal{H}}{\partial F}\{F, F\} = 0. \end{aligned} \tag{5.3.9}$$

This result can be generalized for *nested* dependences of the type

$$\mathcal{H}(\{q_i, p_i\}) = \mathcal{H}\left(\{q_i, p_i\}_{i\neq 1,2}; F_2(q_2, p_2; F_1(q_1, p_1))\right). \tag{5.3.10}$$

Here,

$$F_1(q_1, p_1) = \alpha_1, \quad F_2(q_2, p_2; \alpha_1) = \alpha_2 \tag{5.3.11}$$

are integrals of motion. For F_2, one obtains

$$\begin{aligned} \dot{F}_2 = \{F_2, \mathcal{H}\} &= \frac{\partial F_2}{\partial q_1}\frac{\partial \mathcal{H}}{\partial p_1} - \frac{\partial F_2}{\partial p_1}\frac{\partial \mathcal{H}}{\partial q_1} + \frac{\partial F_2}{\partial q_2}\frac{\partial \mathcal{H}}{\partial p_2} - \frac{\partial F_2}{\partial p_2}\frac{\partial \mathcal{H}}{\partial q_2} \\ &= \left(\frac{\partial F_2}{\partial F_1}\right)^2 \frac{\partial \mathcal{H}}{\partial F_2}\{F_1, F_1\} + \frac{\partial \mathcal{H}}{\partial F_2}\{F_2, F_2\} = 0. \end{aligned} \tag{5.3.12}$$

5.3.1 *Example: Particle's motion in spherical coordinates*

As an example of non-trivial integrals of motion detectable with the help of Poisson brackets, consider a particle moving in the potential

$$U(r,\theta,\phi)=a(r)+\frac{b(\theta)}{r^2}+\frac{c(\phi)}{r^2\sin^2\theta} \tag{5.3.13}$$

in the spherical coordinates. In this case, with the help of Eq. (5.1.18), one obtains the Hamiltonian of the *nested* form

$$\mathcal{H}=\frac{1}{2m}\left[p_r^2+2ma(r)+\frac{1}{r^2}\left(p_\theta^2+2mb(\theta)+\frac{p_\phi^2+2mc(\phi)}{\sin^2\theta}\right)\right]. \tag{5.3.14}$$

In accordance with the above, one has integrals of motion

$$p_\phi^2+2mc(\phi)=\alpha_\phi \tag{5.3.15}$$

and

$$p_\theta^2+2mb(\theta)+\frac{\alpha_\phi}{\sin^2\theta}=\alpha_\theta. \tag{5.3.16}$$

If $c(\phi)=0$, ϕ is the cyclic variable and the first integral of motion simplifies to $p_\phi=\text{const}$. With these two integrals of motion, the problem becomes effectively one-dimensional with respect to r,

$$\mathcal{H}=\frac{1}{2m}\left[p_r^2+2ma(r)+\frac{\alpha_\theta}{r^2}\right], \tag{5.3.17}$$

and can be integrated directly using energy conservation, $\mathcal{H}=E=\text{const}$.

Integration of the whole problem can be done in three steps. From Eq. (5.3.14), one obtains

$$p_r=m\dot{r}=\sqrt{2mE-2ma(r)-\alpha_\theta/r^2}, \tag{5.3.18}$$

which can be integrated to implicitly find $r(t)$ via

$$t=m\int\frac{dr}{\sqrt{2mE-2ma(r)-\alpha_\theta/r^2}}. \tag{5.3.19}$$

Second, $\theta(t)$ can be defined using the integral of motion α_θ, Eq. (5.3.16):

$$p_\theta=mr^2\dot{\theta}=\sqrt{\alpha_\theta-2mb(\theta)-\frac{\alpha_\phi}{\sin^2\theta}}. \tag{5.3.20}$$

This equation can be integrated to define $\theta(t)$ implicitly via

$$\int\frac{d\theta}{\sqrt{\alpha_\theta-2mb(\theta)-\alpha_\phi/\sin^2\theta}}=\int\frac{dt}{mr^2(t)}. \tag{5.3.21}$$

Note that to work out the integral in the rhs, one at first has to find $r(t)$ from Eq. (5.3.19).

Finally, $\phi(t)$ can be defined using the integral of motion α_ϕ:

$$p_\phi = mr^2 \sin^2\theta\, \dot{\phi} = \sqrt{\alpha_\phi - 2mc(\phi)}. \tag{5.3.22}$$

This equation can be integrated to define $\phi(t)$ implicitly via

$$\int \frac{d\phi}{\sqrt{\alpha_\phi - 2mc(\phi)}} = \int \frac{dt}{mr^2(t)\sin^2\theta(t)}. \tag{5.3.23}$$

Note that the previous two steps are needed to work out the integral on the right.

The situation allowing us to obtain the solution of this and similar problems in quadratures is called *separation of variables.* Note that building integrals of motion goes in the direction inside→outside, whereas integrating the resulting separated equations goes in the direction outside→inside.

5.3.2 *Poisson brackets and commutators*

Let us finally work out the correspondence between Poisson brackets and commutators. In quantum mechanics, the fundamental commutators

$$\hat{q}_i = q_i, \quad \hat{p}_i = -i\hbar\frac{d}{dq_i}, \quad [\hat{q}_i, \hat{p}_j] = i\hbar\delta_{ij} \tag{5.3.24}$$

are similar to the fundamental Poisson brackets, Eq. (5.3.7). Operator functions like $f(\hat{q}, \hat{p})$ that correspond to classical functions $f(q, p)$ are interpreted as Taylor series with full symmetrization over permutations of terms. The commutator $[f, g]$ can be calculated, for instance, by expanding both f and g in Taylor series in $\hat{q}$ and $\hat{p}$, commuting term by term, and then summing the series back. The final result of this procedure can be obtained much easier by formally substituting

$$[f, g] \Rightarrow \left[\sum_i \left(\frac{\partial f}{\partial \hat{q}_i}\hat{q}_i + \frac{\partial f}{\partial \hat{p}_i}\hat{p}_i\right), \sum_j \left(\frac{\partial g}{\partial \hat{q}_j}\hat{q}_j + \frac{\partial g}{\partial \hat{p}_j}\hat{p}_j\right)\right] \tag{5.3.25}$$

and then commuting the momentum and coordinate operators considering partial derivatives as numbers. With the help of Eq. (5.3.24), this yields the relation

$$[f, g] = i\hbar\{f, g\}. \tag{5.3.26}$$

In this formula, the Poisson bracket should be symmetrized over permutations, $\partial_{\hat{q}} f\, \partial_{\hat{p}} g \Rightarrow (\partial_{\hat{q}} f\, \partial_{\hat{p}} g + \partial_{\hat{p}} g\, \partial_{\hat{q}} f)/2$, etc., since $\hat{q}$ and $\hat{p}$ are operators. If $\hbar$ is formally considered as small, as is sometimes done in the analysis of the semiclassical case, the symmetrization is irrelevant. This is because each commutation introduces a factor $\hbar$, so that changing the order of terms changes the result by terms starting with $\hbar^2$.

5.4 Canonical transformations

As any system of differential equations, Hamiltonian equations (5.1.6) allow the change of variables

$$Q_i = Q_i\,(q,p,t), \quad P_i = P_i\,(q,p,t), \tag{5.4.1}$$

where q, p are "old" variables and Q, P are "new" variables. The transformation of variables above is called *canonical* if the transformed Hamiltonian equations also have a Hamiltonian (canonical) form

$$\dot{Q}_i = \frac{\partial \mathcal{H}'}{\partial P_i}, \quad \dot{P}_i = -\frac{\partial \mathcal{H}'}{\partial Q_i}, \tag{5.4.2}$$

where, in principle, $\mathcal{H}'$ can differ from $\mathcal{H}$. Sometimes, one can find a canonical transformation that results in a simple Hamiltonian function, allowing for an easy solution. In particular, if $\mathcal{H}'$ does not depend on Q_i, the variable Q_i is called *cyclic*. The second of the above equations then yields $\dot{P}_i = 0$ and $P_i = \text{const}$. Now, in the case of one degree of freedom, the rhs of the first equation above is an integral of motion, $\dot{Q} = \alpha = \text{const}$. This equation can be easily integrated: $Q = \alpha t + \text{const}$.

5.4.1 *Least-action principle and generating functions*

Possible relations between old and new variables can be obtained in the elegant and powerful form from the least-action principle. The validity of both Eqs. (5.1.6) and (5.4.2) implies that action $\mathcal{S}$ in terms of the old variables, Eq. (5.2.1), is equivalent to the action in terms of the new variables. It is sufficient to require that the integrands of the two actions differ by a full differential of some function F. Then variations of both actions coincide, so that $\delta\mathcal{S} = 0$ for the old variables entails $\delta\mathcal{S} = 0$ for the new variables. Then the validity of Eq. (5.1.6) entails the validity of Eq. (5.4.2). The difference between the two action integrands having the form

$$\sum_i p_i dq_i - \sum_i P_i dQ_i + (\mathcal{H}' - \mathcal{H})\, dt = dF \tag{5.4.3}$$

suggests that $F = F(q, Q, t)$ and

$$p_i = \frac{\partial F}{\partial q_i}, \quad P_i = -\frac{\partial F}{\partial Q_i}, \quad \mathcal{H}' - \mathcal{H} = \frac{\partial F}{\partial t}. \tag{5.4.4}$$

These formulas establish relations between new and old variables. Function $F(q, Q, t)$ is the *generating function* of the canonical transformation. In the absence of explicit time dependence of the canonical transformation, one has $\mathcal{H}' = \mathcal{H}$.

Canonical transformation specified by Eq. (5.4.4) can be put in other equivalent forms with the help of different Legendre transformations. For instance, differential of the function

$$\Phi = F + \sum_i Q_i P_i \tag{5.4.5}$$

with the help of Eq. (5.4.3) takes the form

$$d\Phi = \sum_i p_i dq_i + \sum_i Q_i dP_i + (\mathcal{H}' - \mathcal{H})\, dt, \tag{5.4.6}$$

thus $\Phi = \Phi\,(q, P, t)$ and

$$p_i = \frac{\partial \Phi}{\partial q_i}, \quad Q_i = \frac{\partial \Phi}{\partial P_i}, \quad \mathcal{H}' - \mathcal{H} = \frac{\partial \Phi}{\partial t}. \tag{5.4.7}$$

One can introduce other types of generating functions such as $F - \sum_i q_i p_i$, $F - \sum_i q_i p_i + \sum_i Q_i P_i$, and many functions in which Legendre transformation is done with respect to some select canonical pairs $q_i p_i$ or $Q_i P_i$ with particular values of i. In all cases, generating functions contain old and new variables for each degree of freedom i. It should be stressed that Legendre transformations lead to other expressions for the same canonical transformation specified by the generating function F rather than to new canonical transformations.

5.4.2 *Trivial examples of canonical transformations*

Transformation defined by the generating function

$$\Phi(q, P) = \sum_i q_i P_i \tag{5.4.8}$$

has the form

$$p_i = \frac{\partial \Phi}{\partial q_i} = P_i, \quad Q_i = \frac{\partial \Phi}{\partial P_i} = q_i \tag{5.4.9}$$

or

$$Q_i = q_i, \quad P_i = p_i. \tag{5.4.10}$$

This is the identity transformation.

Transformation with

$$F(q,Q)=\sum_i q_iQ_i \tag{5.4.11}$$

has the form

$$p_i=\frac{\partial F}{\partial q_i}=Q_i,\quad P_i=-\frac{\partial F}{\partial Q_i}=-q_i \tag{5.4.12}$$

or

$$Q_i=p_i,\quad P_i=-q_i. \tag{5.4.13}$$

This transformation interchanges generalized coordinates and momenta. The above example shows that there is no essential difference between generalized coordinates and momenta in the Hamiltonian formalism. One cannot say that generalized momenta are related to velocities while generalized coordinates not.

It should be noted that for the above two transformations that are obviously canonical, an attempt to make Legendre transformation (5.4.5) does not work. For instance, one cannot find the primary form $F(q,Q)$ of the transformation, the secondary form $\Phi(q,P)$ of which is given by Eq. (5.4.8). Since only $F(q,Q)$ follows from the least-action principle and it does not exist here, $\Phi(q,P)$ loses its relation to the general formalism. It looks like there are more canonical transformations than those following from the least-action principle.

5.4.3 *Poisson-brackets criterion of canonicity*

Canonicity of transformations that do not explicitly depend on time can be checked with the help of the Poisson-brackets criterion. This criterion can be obtained directly by changing the variables in Hamilton equations. The key point here is that, according to Eq. (5.4.7), the Hamiltonian is the same for both sets of dynamic variables, $\mathcal{H}'=\mathcal{H}$. For instance, equations for Q_i follow as

$$\begin{aligned}\dot{Q}_i&=\frac{\partial Q_i}{\partial q_j}\dot{q}_j+\frac{\partial Q_i}{\partial p_j}\dot{p}_j=\frac{\partial Q_i}{\partial q_j}\frac{\partial \mathcal{H}}{\partial p_j}-\frac{\partial Q_i}{\partial p_j}\frac{\partial \mathcal{H}}{\partial q_j}\\&=\frac{\partial Q_i}{\partial q_j}\left(\frac{\partial \mathcal{H}}{\partial Q_l}\frac{\partial Q_l}{\partial p_j}+\frac{\partial \mathcal{H}}{\partial P_l}\frac{\partial P_l}{\partial p_j}\right)-\frac{\partial Q_i}{\partial p_j}\left(\frac{\partial \mathcal{H}}{\partial Q_l}\frac{\partial Q_l}{\partial q_j}+\frac{\partial \mathcal{H}}{\partial P_l}\frac{\partial P_l}{\partial q_j}\right)\end{aligned}$$

$$= \left(\frac{\partial Q_i}{\partial q_j} \frac{\partial Q_l}{\partial p_j} - \frac{\partial Q_i}{\partial p_j} \frac{\partial Q_l}{\partial q_j} \right) \frac{\partial \mathcal{H}}{\partial Q_l} + \left(\frac{\partial Q_i}{\partial q_j} \frac{\partial P_l}{\partial p_j} - \frac{\partial Q_i}{\partial p_j} \frac{\partial P_l}{\partial q_j} \right) \frac{\partial \mathcal{H}}{\partial P_l}$$

$$= \{Q_i, Q_l\} \frac{\partial \mathcal{H}}{\partial Q_l} + \{Q_i, P_l\} \frac{\partial \mathcal{H}}{\partial P_l}. \tag{5.4.14}$$

Here, summation over repeated indices is assumed. Equations for Q_i are canonical if

$$\{Q_i, Q_l\}_{qp} = 0, \quad \{Q_i, P_l\}_{qp} = \delta_{il}. \tag{5.4.15}$$

Similarly, one can prove that equations for P_i are canonical if

$$\{P_i, Q_l\}_{qp} = -\delta_{il}, \quad \{P_i, P_l\}_{qp} = 0. \tag{5.4.16}$$

Index qp shows that Poisson brackets are calculated with respect to the old (original) variables. The above means that fundamental Poisson brackets have the same form as those for the old variables, Eq. (5.3.7). However, Eq. (5.3.7) is trivial, whereas Eqs. (5.4.15) and (5.4.16) are not. As transformation can also be done in the opposite direction, $(Q, P) \Rightarrow (q, p)$, by the same method, one obtains

$$\begin{aligned} &\{q_i, q_j\}_{QP} = 0, \quad \{q_i, p_j\}_{QP} = \delta_{ij}, \\ &\{p_i, q_j\}_{QP} = -\delta_{ij}, \quad \{p_i, p_j\}_{QP} = 0 \end{aligned} \tag{5.4.17}$$

as canonicity criterion.

5.4.4 *Harmonic oscillator via canonical transformation*

Another example of canonical transformations is the harmonic oscillator having the Hamiltonian

$$\mathcal{H} = \frac{p^2}{2m} + \frac{kq^2}{2} = \frac{1}{2m} \left(p^2 + m^2\omega^2 q^2 \right), \tag{5.4.18}$$

where $\omega = \sqrt{k/m}$ is the oscillator's eigenfrequency. Quadratic dependence on q and p suggests the use of a transformation of the type

$$q = \frac{f(P)}{m\omega} \sin Q, \quad p = f(P) \cos Q, \tag{5.4.19}$$

which leads to the transformed Hamiltonian

$$\mathcal{H} = \frac{f^2(P)}{2m} \left(\cos^2 Q + \sin^2 Q \right) = \frac{f^2(P)}{2m}, \tag{5.4.20}$$

which is independent of Q. Here, function $f(P)$ can be obtained from the Poisson-brackets canonicity criterion, Eq. (5.4.17). In particular, one should have

$$\begin{aligned}
1 &= \frac{\partial q}{\partial Q}\frac{\partial p}{\partial P} - \frac{\partial q}{\partial P}\frac{\partial p}{\partial Q} \\
&= \frac{f(P)}{m\omega}\cos Q \times \frac{df(P)}{dP}\cos Q + \frac{1}{m\omega}\frac{df(P)}{dP}\sin Q \times f(P)\sin Q \\
&= \frac{f(P)}{m\omega}\frac{df(P)}{dP} = \frac{1}{2m\omega}\frac{d}{dP}f^2(P).
\end{aligned} \tag{5.4.21}$$

Integrating this equation, one obtains

$$f(P) = \sqrt{2Pm\omega}. \tag{5.4.22}$$

This yields

$$q = \sqrt{\frac{2P}{m\omega}}\sin Q, \quad p = \sqrt{2Pm\omega}\cos Q. \tag{5.4.23}$$

Now, the new Hamiltonian has the form

$$\mathcal{H} = \omega P. \tag{5.4.24}$$

The generalized momentum

$$P = \frac{\mathcal{H}}{\omega} = \frac{E}{\omega} \tag{5.4.25}$$

is conserved. In Sec. 5.6, P will be identified with the action variable I. The interpretation of P is action over the period of motion I divided by 2π, see Eq. (5.5.37). The equation of motion for the cyclic variable Q is

$$\dot{Q} = \frac{\partial \mathcal{H}}{\partial P} = \omega. \tag{5.4.26}$$

Its solution reads

$$Q = \omega t + \varphi_0, \tag{5.4.27}$$

where $\varphi_0 = \text{const}$. One can see that Q is the oscillator's phase angle. Inserting the above results into Eq. (5.4.23), one finally obtains the solution

$$q = \sqrt{\frac{2E}{m\omega^2}}\sin(\omega t + \varphi_0), \quad p = \sqrt{2mE}\cos(\omega t + \varphi_0). \tag{5.4.28}$$

Generating function $F(q, Q)$ of the above canonical transformation can be found by integrating the equations

$$p = \frac{\partial F}{\partial q}, \quad P = -\frac{\partial F}{\partial Q}. \tag{5.4.29}$$

Before this, one has to express p or P via q and Q with the help of Eq. (5.4.23). For instance, from the first of Eqs. (5.4.23), one obtains

$$P = \frac{m\omega q^2}{2}\frac{1}{\sin^2 Q}. \tag{5.4.30}$$

Integrating the rhs of equation over Q and discarding the integration constant, one obtains

$$F(q,Q) = \frac{m\omega q^2}{2}\cot Q. \tag{5.4.31}$$

From this follows $p = \partial F/\partial q = m\omega q \cot Q$, and after elimination of Q with the help of the first of Eqs. (5.4.23), one obtains the second of Eq. (5.4.23).

One can also find the second-kind generating function $\Phi(q,P)$ by Legendre transformation, Eq. (5.4.5). Expressing Q with the help of the first of Eqs. (5.4.23) as

$$Q = \arcsin\tilde{q}, \quad \tilde{q} \equiv \sqrt{\frac{m\omega}{2P}}q, \tag{5.4.32}$$

one obtains

$$\Phi(q,P) = \frac{m\omega q^2}{2}\cot Q + QP = P\left(\tilde{q}\sqrt{1-\tilde{q}^2} + \arcsin\tilde{q}\right). \tag{5.4.33}$$

5.4.5 *Symplectic formalism*

Hamiltonian equations can be put into a compact and elegant symplectic form. Introducing the dynamical vector

$$\mathbf{x} = \{x_1, x_2, \ldots, x_{2N}\} = \{q_1, q_2, \ldots, q_N, p_1, p_2, \ldots, p_N\}, \tag{5.4.34}$$

one can write Hamiltonian equations (5.1.6) in the form

$$\dot{\mathbf{x}} = \mathbb{J}\cdot\frac{\partial\mathcal{H}}{\partial\mathbf{x}}, \quad \dot{x}_i = J_{ij}\frac{\partial\mathcal{H}}{\partial x_j}, \tag{5.4.35}$$

where summation over repeated indices is assumed and matrix $\mathbb{J}$ is given by

$$J_{ij} = \delta_{i+N,j} - \delta_{i-N,j}. \tag{5.4.36}$$

Let us now perform a canonical transformation (5.4.1) without time dependence and introduce the new dynamical vector

$$\mathbf{y} = \{y_1, y_2, \ldots, y_{2N}\} = \{Q_1, Q_2, \ldots, Q_N, P_1, P_2, \ldots, P_N\}. \tag{5.4.37}$$

The equation of motion for $\mathbf{y}$ follows from Eq. (5.4.35):

$$\dot{y}_i = \frac{\partial y_i}{\partial x_j}\dot{x}_j = \frac{\partial y_i}{\partial x_j}J_{jk}\frac{\partial \mathcal{H}}{\partial x_k} = \frac{\partial y_i}{\partial x_j}J_{jk}\frac{\partial y_l}{\partial x_k}\frac{\partial \mathcal{H}}{\partial y_l}. \tag{5.4.38}$$

The condition that the resulting equation is Hamiltonian and thus the transformation is canonical in the vector and component forms reads

$$\mathbb{A}\cdot\mathbb{J}\cdot\mathbb{A}^T = \mathbb{J}, \quad \frac{\partial y_i}{\partial x_j}J_{jk}\frac{\partial y_l}{\partial x_k} = J_{il}, \tag{5.4.39}$$

where $A_{ij} \equiv \partial y_i/\partial x_j$ is the Jacobian matrix of the transformation. The condition above contains all the four conditions of the standard formalism, Eqs. (5.4.15) and (5.4.16). One can prove the inverse and more general statement: If the transformation is canonical, Poisson brackets of *any* two variables A and B are invariant with respect to the transformation. The proof uses Eq. (5.4.39):

$$\{A,B\}_x = \frac{\partial A}{\partial x_i}J_{ij}\frac{\partial B}{\partial x_j} = \frac{\partial A}{\partial y_k}\frac{\partial y_k}{\partial x_i}J_{ij}\frac{\partial y_l}{\partial x_j}\frac{\partial B}{\partial y_l} = \frac{\partial A}{\partial y_l}J_{kl}\frac{\partial B}{\partial y_l} = \{A,B\}_y\,. \tag{5.4.40}$$

5.4.6 *Relation between generating-function and Poisson-brackets canonicity criteria*

It is of interest to perform the direct canonicity check provided by Eqs. (5.4.15) and (5.4.16) for canonical transformations via generating functions obtained from the least-action principle in Sec. 5.4.1. The resulting equivalence between the two seemingly unrelated approaches is referred to as the Carathéodory theorem. For simplicity, the proof for one degree of freedom will be shown. Using the second of relation of Eq. (5.4.4), and using the generating function in the form $F = F(q, Q(q,p))$, one can write

$$\begin{aligned}
\{Q,P\}_{qp} &= \frac{\partial Q}{\partial q}\frac{\partial P}{\partial p} - \frac{\partial Q}{\partial p}\frac{\partial P}{\partial q} \\
&= -\frac{\partial Q}{\partial q}\frac{\partial}{\partial p}\frac{\partial F(q,Q(q,p))}{\partial Q} + \frac{\partial Q}{\partial p}\frac{\partial}{\partial q}\frac{\partial F(q,Q(q,p))}{\partial Q} \\
&= -\frac{\partial Q}{\partial q}\frac{\partial^2 F(q,Q)}{\partial^2 Q}\frac{\partial Q}{\partial p} + \frac{\partial Q}{\partial p}\frac{\partial^2 F(q,Q)}{\partial Q\partial q} + \frac{\partial Q}{\partial p}\frac{\partial^2 F(q,Q)}{\partial Q^2}\frac{\partial Q}{\partial q} \\
&= \frac{\partial Q}{\partial p}\frac{\partial p}{\partial Q} = \frac{\partial Q(q,p(q,Q))}{\partial p(q,Q)}\frac{\partial p(q,Q)}{\partial Q} = \frac{\partial Q}{\partial Q} = 1
\end{aligned} \tag{5.4.41}$$

which proves canonicity. In the term remaining after canceling, the first relation of Eq. (5.4.4) was used. For more than one degree of freedom, the proof seems to be much more involved.

5.5 Action as function of coordinates and Hamilton–Jacobi equation

5.5.1 *General formulation with a simple example*

Action $\mathcal{S}$ given by Eq. (5.2.1) was used to derive Lagrangian and Hamiltonian equations of motion from the least-action principle, $\delta\mathcal{S} = 0$. This condition singles out the real physical trajectory from all other competing trajectories. Here, the action for the real trajectory as a function of the upper-limit variables $t_2 \Rightarrow t$ will be considered. The expression for $\mathcal{S}$ in the Hamiltonian form, the last term in Eq. (5.2.1), shows that as q_i and t change in the course of motion, action acquires corresponding increments, so that infinitesimally one has

$$d\mathcal{S} = \sum_i p_i dq_i - \mathcal{H}dt. \tag{5.5.1}$$

This implies that $\mathcal{S} = \mathcal{S}(q, t)$ and

$$\frac{\partial \mathcal{S}}{\partial q_i} = p_i, \quad \frac{\partial \mathcal{S}}{\partial t} = -\mathcal{H}. \tag{5.5.2}$$

Let us illustrate this on a free particle in one dimension. The Lagrangian has the form $\mathcal{L} = T = mv^2/2$, and the trajectory is described by $q = vt$. Thus, the action reads

$$\mathcal{S} = \int^t dt\, \mathcal{L} = \frac{mv^2}{2}t + \text{const.} = \frac{mq^2}{2t} + \text{const.} \tag{5.5.3}$$

Now, one can check that

$$\begin{aligned} \frac{\partial \mathcal{S}}{\partial q} &= \frac{mq}{t} = mv = p, \\ \frac{\partial \mathcal{S}}{\partial t} &= -\frac{mq^2}{2t^2} = -\frac{mv^2}{2} = -\frac{p^2}{2m} = -\mathcal{H}, \end{aligned} \tag{5.5.4}$$

as it should be.

If the problem does not explicitly depend on time, energy is conserved, $\mathcal{H} = \text{const.} = E$. Then the second of Eq. (5.5.2) can be integrated, which yields

$$\mathcal{S}(q,t) = \mathcal{S}_0(q) - Et. \tag{5.5.5}$$

Here, $\mathcal{S}_0(q)$ is the so-called short or abbreviated action that can be obtained by integrating the first of equation (5.5.2). Alternatively, using Eq. (5.2.1) with $\int_{t_1}^{t_2} \Rightarrow \int^{t}$and $\mathcal{H} \Rightarrow E$, one obtains Eq. (5.5.5) with

$$\mathcal{S}_0(q) = \sum_i \int^{q_i} p_i dq_i. \tag{5.5.6}$$

Equations (5.5.2) can be used to set up the famous Hamilton–Jacobi equation that, together with canonical transformations, is an efficient tool for finding the analytical solutions of mechanical problems. The Hamilton–Jacobi equation

$$\frac{\partial \mathcal{S}}{\partial t} + \mathcal{H}\left(q, \frac{\partial \mathcal{S}}{\partial q}, t\right) = 0 \tag{5.5.7}$$

is a nonlinear first-order partial differential equation (PDE) for the function $\mathcal{S}(q,t)$. As usual, q and $\partial\mathcal{S}/\partial q$ in the arguments stand for the whole sets of q_i and $\partial\mathcal{S}/\partial q_i = p_i$. For practical purposes, it is sufficient to find just some solution of Hamilton–Jacobi equation rather than its most general solution.

In particular, for the free particle considered above, Eq. (5.5.7) becomes

$$\frac{\partial \mathcal{S}}{\partial t} + \frac{1}{2m}\left(\frac{\partial \mathcal{S}}{\partial q}\right)^2 = 0. \tag{5.5.8}$$

The solution of this equation can be searched for in the form

$$\mathcal{S}(q,t) = \mathcal{S}_0(q) + \mathcal{S}'(t). \tag{5.5.9}$$

Then Eq. (5.5.8) becomes

$$\frac{\partial \mathcal{S}'}{\partial t} + \frac{1}{2m}\left(\frac{\partial \mathcal{S}_0}{\partial q}\right)^2 = 0. \tag{5.5.10}$$

To satisfy this equation, both terms should be constants compensating each other, so that

$$\frac{\partial \mathcal{S}'}{\partial t} = \text{const} = -E, \quad \frac{\partial \mathcal{S}_0}{\partial q} = \text{const} = p, \tag{5.5.11}$$

and the constants satisfy

$$E = \frac{p^2}{2m}, \quad p = \sqrt{2mE}. \tag{5.5.12}$$

Integrating Eq. (5.5.11), one obtains $\mathcal{S}(q,t)$ in the two alternative forms

$$\mathcal{S}(q,t) = pq - \frac{p^2}{2m}t = \sqrt{2mE}q - Et \tag{5.5.13}$$

up to an irrelevant constant. One can check that this result coincides with Eq. (5.5.3):

$$\mathcal{S}(q,t) = pq - \frac{p^2}{2m}t = \frac{mq}{t}q - \left(\frac{mq}{t}\right)^2 \frac{t}{2m} = \frac{mq^2}{2t}. \tag{5.5.14}$$

Whereas the solution of the Hamilton–Jacobi equation for one degree of freedom such as Eq. (5.5.13) depends on one constant, the so-called *complete integral* of Hamilton–Jacobi equation for N degrees of freedom depends on N constants that we call P_i. On the top of it, one can always add an irrelevant constant to $\mathcal{S}(q,t)$ that has been suppressed in Eq. (5.5.13). The complete integral yields the solution for the system's dynamics if one uses it as the generating function of a canonical transformation in terms of the old coordinates and new momenta, $\Phi(q,P,t)$ of Eq. (5.4.5). The new momenta are constants P_i in the complete integral. The new Hamiltonian $\mathcal{H}'$ given by Eq. (5.4.7) vanishes according to the Hamilton–Jacobi equation:

$$\mathcal{H}' = \mathcal{H} + \frac{\partial \Phi}{\partial t} = \mathcal{H} + \frac{\partial \mathcal{S}}{\partial t} = \mathcal{H} - \mathcal{H} = 0. \tag{5.5.15}$$

Thus, Hamiltonian equations for the new dynamic variables Q_i and P_i become trivial:

$$\begin{aligned} \dot{Q}_i = 0 \quad &\Longrightarrow \quad Q_i = \text{const.}, \\ \dot{P}_i = 0 \quad &\Longrightarrow \quad P_i = \text{const.} \end{aligned} \tag{5.5.16}$$

Time dependences of the old variables q_i and p_i can be obtained from the first two Eqs. (5.4.7). At first, q_i are found by resolving the equations

$$Q_i = \frac{\partial \mathcal{S}}{\partial P_i}. \tag{5.5.17}$$

Then p_i are given by the formulas

$$p_i = \frac{\partial \mathcal{S}}{\partial q_i}. \tag{5.5.18}$$

Some literature uses α_i and β_i as new momenta and coordinates within Hamilton–Jacobi formalism, $\alpha_i \equiv P_i$ and $\beta_i \equiv Q_i$.

Let us illustrate finding the dynamical solution for the free particle. One can choose constant p in the first expression in Eq. (5.5.13) as the new momentum P:

$$\mathcal{S}(q, P, t) = Pq - \frac{P^2}{2m}t. \tag{5.5.19}$$

Equation (5.5.17) takes the form

$$Q = \frac{\partial \mathcal{S}}{\partial P} = q - \frac{P}{m}t, \tag{5.5.20}$$

which yields the solution

$$q = \frac{P}{m}t + Q = \frac{P}{m}t + \text{const}. \tag{5.5.21}$$

Then momentum is defined by Eq. (5.5.18):

$$p = \frac{\partial \mathcal{S}}{\partial q} = P = \text{const}. \tag{5.5.22}$$

Thus, Eq. (5.5.21) reproduces the well-known solution $q = (p/m)\,t + \text{const.}$ for a free particle.

Alternatively, one can use the second expression in Eq. (5.5.13) and choose energy E as conserved new momentum P. Now, instead of Eq. (5.5.20), one obtains

$$Q = \frac{\partial \mathcal{S}}{\partial E} = \sqrt{\frac{m}{2E}}q - t, \tag{5.5.23}$$

which yields the familiar solution

$$q = \sqrt{\frac{2E}{m}}t + \sqrt{\frac{2E}{m}}Q = \sqrt{\frac{2E}{m}}t + \text{const}. \tag{5.5.24}$$

The old momentum is given by the familiar formula

$$p = \frac{\partial \mathcal{S}}{\partial q} = \sqrt{2mE}. \tag{5.5.25}$$

This alternative solution using energy rather than momentum as conserved new momentum is preferred because it survives for systems with non-trivial potential energy where momentum is not conserved. As an example, we will consider the harmonic oscillator in the following section.

5.5.2 *Harmonic oscillator by Hamilton–Jacobi method*

With the harmonic-oscillator Hamiltonian of Eq. (5.4.18), the Hamilton–Jacobi equation (5.5.7) becomes

$$\frac{\partial \mathcal{S}}{\partial t} + \frac{1}{2m}\left(\frac{\partial \mathcal{S}}{\partial q}\right)^2 + U(q) = 0, \quad U(q) = \frac{m\omega^2 q^2}{2}. \tag{5.5.26}$$

Action has the form of Eq. (5.5.5) with $\mathcal{S}_0$ satisfying the equation

$$\frac{1}{2m}\left(\frac{\partial \mathcal{S}_0}{\partial q}\right)^2 + U(q) = E. \tag{5.5.27}$$

Resolving this equation and integrating, one obtains

$$\mathcal{S}(q, E, t) = \mathcal{S}_0(q, E) - Et = \int dq \sqrt{2m\left[E - U(q)\right]} - Et. \tag{5.5.28}$$

Using this as generating function $\Phi(q, P, t)$ with $P = E$, one obtains the implicit formula for $q(t)$

$$Q = \frac{\partial \mathcal{S}}{\partial E} = \int dq \sqrt{\frac{m}{2\left[E - U(q)\right]}} - t. \tag{5.5.29}$$

For the harmonic oscillator, one can calculate the integral analytically as follows:

$$Q = \int dq \sqrt{\frac{m}{2E - m\omega^2 q^2}} - t = \frac{1}{\omega}\int \frac{d\tilde{q}}{\sqrt{1 - \tilde{q}^2}} - t = \frac{1}{\omega}\arcsin\tilde{q} - t, \tag{5.5.30}$$

where

$$\tilde{q} \equiv \sqrt{\frac{m\omega^2}{2E}}\, q. \tag{5.5.31}$$

Inverting Eq. (5.5.30), one obtains the well-known solution

$$\tilde{q} = \sin(\omega t + \omega Q) = \sin(\omega t + \varphi_0) \tag{5.5.32}$$

or

$$q = \sqrt{\frac{2E}{m\omega^2}} \sin(\omega t + \varphi_0). \tag{5.5.33}$$

After that, one finds p as

$$p = \frac{\partial \mathcal{S}}{\partial q} = \sqrt{2m\left[E - U(q)\right]}, \tag{5.5.34}$$

which for the harmonic oscillator yields the well-known expression

$$p = \sqrt{2mE}\sqrt{1-\tilde{q}^2} = \sqrt{2mE}\cos(\omega t + \varphi_0). \tag{5.5.35}$$

Finally, we calculate the short action $\mathcal{S}_0$, the integral term in Eq. (5.5.28). The result has the form

$$\mathcal{S}_0(q) = I\left(\tilde{q}\sqrt{1-\tilde{q}^2} + \arcsin\tilde{q}\right), \quad I = \frac{E}{\omega}, \tag{5.5.36}$$

which coincides with the generating function Φ given by Eq. (5.4.33) with $P = I$. The latter is the harmonic-oscillator form of the so-called action variable that will be introduced in Sec. (5.6). $\mathcal{S}_0(q)$ is a multivalued function of q, taking into account different branches of the $\sqrt{\dots}$ and $\arcsin(\cdots)$ functions, see Fig. 5.5.1. As $\tilde{q}$ changes from 0 to 1, which is the quarter of the oscillation period, the expression in brackets changes from 0 to $\pi/2$. The same happens each quarter period, so that the change over the period makes up 2π. Thus, short action over the period is given by

$$\Delta\mathcal{S}_0^{(\text{Period})} = 2\pi I. \tag{5.5.37}$$

Unlike the short action $\mathcal{S}_0$, the full action $\mathcal{S}$ does not increase with time on average. Since $\mathcal{L} = E_k - U$ and both kinetic and potential energies oscillate having the same average values, $\mathcal{S} = \int dt\mathcal{L}$ oscillates without growing.

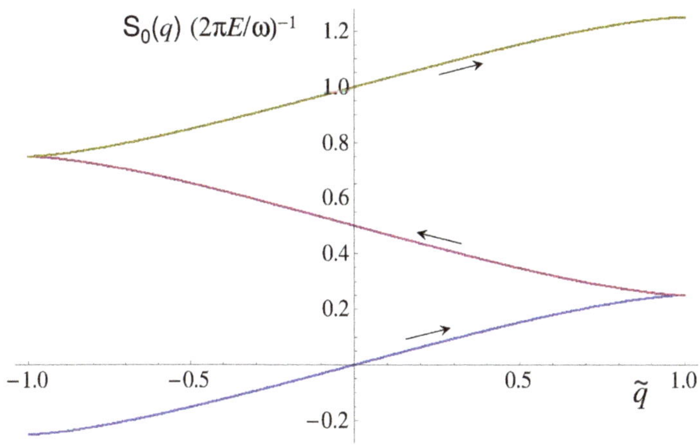

Figure 5.5.1. Short action of the harmonic oscillator.

5.5.3 *Separation of variables*

The method of solving Hamilton–Jacobi equation applied to the harmonic oscillator in the preceding section works for any system with one degree of freedom, described by a potential energy $U(q)$. This method can be generalized for systems with N degrees of freedom that allow the separation of variables. In such systems, one or more canonical pairs (q_i, p_i) enter the Hamiltonian $\mathcal{H}$ as combinations that do not contain other dynamical variables. In the case of one such pair, the Hamilton–Jacobi equation has the form

$$\frac{\partial \mathcal{S}}{\partial t} + \mathcal{H}\left(\left\{q_i, \frac{\partial \mathcal{S}}{\partial q_i}\right\}_{i\neq 1}, F_1\left(q_1, \frac{\partial \mathcal{S}}{\partial q_1}\right), t\right) = 0. \tag{5.5.38}$$

The solution can be searched for in the form of the sum

$$\mathcal{S} = \mathcal{S}^{(N-1)}\left(\{q_i\}_{i\neq 1}, t\right) + \mathcal{S}_1(q_1), \tag{5.5.39}$$

where $\mathcal{S}_1(q_1)$ is the short action corresponding to a single degree of freedom, q_1. With this *ansatz,* Eq. (5.5.38) becomes

$$\frac{\partial \mathcal{S}^{(N-1)}}{\partial t} + \mathcal{H}\left(\left\{q_i, \frac{\partial \mathcal{S}^{(N-1)}}{\partial q_i}\right\}_{i\neq 1}, F_1\left(q_1, \frac{d\mathcal{S}_1}{dq_1}\right), t\right) = 0. \tag{5.5.40}$$

Since this equation has to be valid for any value of q_1, the condition $F_1 = \text{const.} = \alpha_1$ should be fulfilled. Thus, Eq. (5.5.40) splits up into two equations:

$$\begin{aligned} &F_1\left(q_1, \frac{d\mathcal{S}_1}{dq_1}\right) = \alpha_1, \\ &\frac{\partial \mathcal{S}^{(N-1)}}{\partial t} + \mathcal{H}\left(\left\{q_i, \frac{\partial \mathcal{S}^{(N-1)}}{\partial q_i}\right\}_{i\neq 1}, \alpha_1, t\right) = 0. \end{aligned} \tag{5.5.41}$$

This is why this situation is called the separation of variables. The first equation here is an ordinary differential equation that allows for a solution in quadratures. The second equation is a Hamilton–Jacobi equation with $N-1$ degrees of freedom.

If the problem is time-independent, one can search for the solution in the form

$$\mathcal{S} = -Et + \mathcal{S}_0^{(N-1)}\left(\{q_i\}_{i\neq 1}\right) + \mathcal{S}_1(q_1). \tag{5.5.42}$$

This results in simpler equations

$$F_1\left(q_1, \frac{\partial \mathcal{S}_1}{\partial q_1}\right) = \alpha_1,$$

$$\mathcal{H}\left(\left\{q_i, \frac{\partial \mathcal{S}_0^{(N-1)}}{\partial q_i}\right\}_{i\neq 1}, \alpha_1\right) = E. \tag{5.5.43}$$

A particular case of separation of variables is the case of a cyclic variable q_1 that does not enter the Hamiltonian. Then $F_1(q_1, \partial\mathcal{S}_1/\partial q_1)$ reduces to $\partial\mathcal{S}_1/\partial q_1$, so that the first equation (5.5.41) can be easily integrated and Eq. (5.5.39) becomes

$$\mathcal{S} = \mathcal{S}^{(N-1)}\left(\{q_i\}_{i\neq 1}, t\right) + \alpha_1 q_1. \tag{5.5.44}$$

Time is also a cyclic variable, and $-Et$ in Eq. (5.5.42) is similar to $\alpha_1 q_1$ in the above equation.

If there is another separating variable q_2, the second equation (5.5.41) can be further simplified in terms of α_2, $\mathcal{S}_2$, and the remainder action $\mathcal{S}^{(N-2)}$. If all N variables separate, the procedure results in the complete integral of Hamilton–Jacobi equation of the completely additive form

$$\mathcal{S} = \mathcal{S}^{(0)}(t) + \sum_i^N \mathcal{S}_i(q_i, \{\alpha_i\}), \tag{5.5.45}$$

whereas $\mathcal{S}^{(0)}$ satisfies the equation

$$\frac{\partial \mathcal{S}^{(0)}}{\partial t} + \mathcal{H}(\alpha_1, \alpha_2, \ldots, \alpha_N, t) = 0, \tag{5.5.46}$$

which can also be solved in quadratures. Each term of Eq. (5.5.45) depends on one or more constants α_i. For time-independent problems, one obtains

$$\mathcal{S}^{(0)}(t) = -Et, \quad E = \mathcal{H}(\alpha_1, \alpha_2, \ldots, \alpha_N). \tag{5.5.47}$$

The complete integral above can be used now to define the system's dynamics as was explained in Sec. 5.5.1, see Eqs. (5.5.17) and (5.5.18). Let us write these equations again:

$$\beta_i = \frac{\partial \mathcal{S}}{\partial \alpha_i}, \quad \beta_i = \text{const.} \tag{5.5.48}$$

Note the equivalence $\alpha_i \Leftrightarrow P_i$ and $\beta_i \Leftrightarrow Q_i$. As, in general, $\mathcal{S}_i(q_i, \{\alpha_i\})$ depends not only on its own constant α_i but also on other constants as well, these equations for different i are coupled. For simpler problems such as three-dimensional harmonic oscillator, one has $\mathcal{S}_i(q_i, \{\alpha_i\}) = \mathcal{S}_i(q_i, \alpha_i)$, and different equations above are uncoupled.

5.5.4 *Example: Particle's motion in spherical coordinates*

As an example, consider a generalization of the Kepler problem in spherical coordinates having the Hamiltonian (5.1.18). As we have seen in Sec. 5.3.1, here the variables separate for

$$U(r, \theta, \phi) = a(r) + \frac{b(\theta)}{r^2} + \frac{c(\phi)}{r^2 \sin^2\theta}, \tag{5.5.49}$$

and integrals of motion can be found with the help of Poisson brackets, after which the problem can be integrated. Now, we solve the same problem by the Hamilton–Jacobi formalism. Hamilton–Jacobi equation has the form

$$\frac{\partial \mathcal{S}}{\partial t} + \frac{1}{2m}\left(\frac{\partial \mathcal{S}}{\partial r}\right)^2 + a(r) + \frac{1}{2mr^2}\left\{\left(\frac{\partial \mathcal{S}}{\partial \theta}\right)^2 + 2mb(\theta)\right.$$
$$\left. + \frac{1}{\sin^2\theta}\left[\left(\frac{\partial \mathcal{S}}{\partial \phi}\right)^2 + 2mc(\phi)\right]\right\} = 0. \tag{5.5.50}$$

Searching for action $\mathcal{S}$ in the additive form

$$\mathcal{S} = -Et + \mathcal{S}_r(r) + \mathcal{S}_\theta(\theta) + \mathcal{S}_\phi(\phi), \tag{5.5.51}$$

for the parts of action, one obtains equations

$$\left(\frac{\partial \mathcal{S}_\phi}{\partial \phi}\right)^2 + 2mc(\phi) = \alpha_\phi = \text{const.},$$

$$\left(\frac{\partial \mathcal{S}_\theta}{\partial \theta}\right)^2 + 2mb(\theta) + \frac{\alpha_\phi}{\sin^2\theta} = \alpha_\theta = \text{const.}, \tag{5.5.52}$$

$$\left(\frac{\partial \mathcal{S}_r}{\partial r}\right)^2 + 2ma(r) + \frac{\alpha_\theta}{r^2} = 2mE = \text{const.} \tag{5.5.53}$$

Note that $\mathcal{S}_\theta$ depends on constants α_ϕ and α_θ, whereas $\mathcal{S}_r$ depends on α_θ and E. Integration of these equations yields

$$\mathcal{S} = -Et + \int dr\sqrt{2m\left[E - a(r)\right] - \frac{\alpha_\theta}{r^2}} + \int d\theta\sqrt{\alpha_\theta - 2mb(\theta) - \frac{\alpha_\phi}{\sin^2\theta}}$$
$$+ \int d\phi\sqrt{\alpha_\phi - 2mc(\phi)}, \tag{5.5.54}$$

which depends on three constants, E, α_ϕ, and α_θ. Differentiating $\mathcal{S}$ with respect to these constants and equating the results to other constants, one obtains equations of motion in the implicit form that include six constants, as it should be. These equations are as follows:

$$\begin{aligned}
\beta_r &= \frac{\partial \mathcal{S}}{\partial E} = -t + \int \frac{m dr}{\sqrt{2m\left[E - a(r)\right] - \alpha_\theta / r^2}}, \\
\beta_\theta &= \frac{\partial \mathcal{S}}{\partial \alpha_\theta} = \int \frac{d\theta}{2\sqrt{\alpha_\theta - 2mb(\theta) - \alpha_\phi / \sin^2\theta}} \\
&\quad - \int \frac{dr}{2r^2\sqrt{2m\left[E - a(r)\right] - \alpha_\theta / r^2}}, \\
\beta_\phi &= \frac{\partial \mathcal{S}}{\partial \alpha_\phi} = \int \frac{d\phi}{2\sqrt{\alpha_\phi - 2mc(\phi)}} - \int \frac{d\theta}{2\sin^2\theta\sqrt{\alpha_\theta - 2mb(\theta) - \alpha_\phi / \sin^2\theta}}.
\end{aligned} \tag{5.5.55}$$

The first of these equations after integration and solving for r yields $r(t)$. Then the second equation after integration and solving for θ yields $\theta(t)$, expressed via $r(t)$. Finally, the third equation yields the solution $\phi(t)$ expressed via $\theta(t)$, which is in turn expressed via $r(t)$. The solution above is equivalent to the solution using Poisson brackets in Sec. 5.3.1 that is much less involved.

Practical usefulness of these solutions is questionable since, in general, the integrals cannot be done analytically. In this case, direct numerical solution of the original problem is much simpler than the numerical treatment of these integrals.

5.6 Angle-action variables and slow evolution of mechanical systems

5.6.1 *Definition of angle-action variables*

Consider a system with one degree of freedom with a time-independent Hamiltonian

$$\mathcal{H} = \frac{p^2}{2m} + U(q). \tag{5.6.1}$$

The Hamiltonian is conserved, $\mathcal{H} = E = \text{const}$. Assume that potential energy $U(q)$ is such that the system performs a periodic motion between turning points $q_{\pm}$. The short action

$$\mathcal{S}_0(q) = \int p dq = \int \sqrt{2m\,[E - U(q)]} dq \tag{5.6.2}$$

depends on energy E as a parameter. Let us introduce the so-called action variable

$$I = \frac{1}{2\pi} \oint p dq, \tag{5.6.3}$$

which is proportional to the action over the period of motion, cf. Eq. (5.5.37). One can see that I depends on E only. Thus, one can consider $\mathcal{S}_0$ of Eq. (5.6.2) as parametrically dependent on I and use the function $\mathcal{S}_0(q, I)$ as a generating function of a time-independent canonical transformation, $I \equiv P$ being the new momentum and its canonically conjugate coordinate $\varphi \equiv Q$ being the angle (see the following). The relation between the old and new dynamic variables is given by Eq. (5.4.7) in the form

$$\varphi = \frac{\partial \mathcal{S}_0(q, I)}{\partial I}, \quad p = \frac{\partial \mathcal{S}_0(q, I)}{\partial q}. \tag{5.6.4}$$

Since the Hamiltonian $\mathcal{H} = E(I)$ does not depend on φ, the latter is a cyclic variable. The Hamilton equations have the form

$$\begin{aligned} \dot{I} &= -\frac{\partial \mathcal{H}}{\partial \varphi} = 0, \\ \dot{\varphi} &= \frac{\partial \mathcal{H}}{\partial I} = \frac{dE}{dI} = \omega, \end{aligned} \tag{5.6.5}$$

where $\omega = 2\pi/T$ is the frequency of motion, $T = T(E)$ being the period of motion. The proof of the last relation is as follows:

$$\begin{aligned} \frac{dI}{dE} &= \frac{d}{dE}\frac{1}{2\pi} \oint \sqrt{2m\,[E - U(q)]} dq = \frac{1}{2\pi} \oint \sqrt{\frac{m}{2\,[E - U(q)]}} dq \\ &= \frac{1}{2\pi} \oint \frac{dq}{\dot{q}} = \frac{1}{2\pi} \oint dt = \frac{T}{2\pi} = \frac{1}{\omega}. \end{aligned} \tag{5.6.6}$$

Note the difference between the relations $\omega = dE/dI$ and $\omega = E/I$ in Eq. (5.5.36). The latter is valid only for a harmonic oscillator and it can be obtained by integration of $dE/dI = \omega$ in the second equation (5.6.5) with $\omega =$ const. The choice of the action variable I above anticipates the nice result $\dot{\varphi} = \omega$ that allows us to interpret φ as the phase or the *angle* of the periodic motion that, according to the second of Eqs. (5.6.5), linearly grows with time, $\varphi(t) = \omega t + \varphi_0$. Time dependence of q can be found from Eq. (5.6.4).

The relation between the (q, p) and (φ, I) variables that follows from Eq. (5.6.4) reads

$$q = \sqrt{\frac{2I}{m\omega}} \sin\varphi, \quad p = \sqrt{2mI\omega} \cos\varphi. \tag{5.6.7}$$

This is obviously a form of Eq. (5.4.28) or Eqs. (5.5.33) and (5.5.35). Using this, one can obtain a convenient presentation of $\mathcal{S}_0$ in terms of new variables (φ, I) as

$$\mathcal{S}_0 = \int p dq = \int p(\varphi) \frac{dq(\varphi)}{d\varphi} d\varphi = 2I \int \cos^2 \varphi d\varphi = I \left[\varphi + \frac{1}{2} \sin(2\varphi) \right]. \tag{5.6.8}$$

Together with $\tilde{q} = \sin\varphi$, this formula gives a parametric presentation of $\mathcal{S}_0(q)$ shown in Fig. 5.5.1.

Equation (5.6.3) can be written in the form

$$I = \frac{1}{2\pi} \int\int dp dq, \tag{5.6.9}$$

where the double integral is the area circumscribed by the closed orbit in the phase plane (q, p). To the contrary, the trajectory of the system in the phase plane (φ, I) is just a straight line. One can say that transformation to the angle-action variables straightens the trajectory.

The action-angle formalism is merely a variant of the Hamilton–Jacobi method. In the latter, the full action $\mathcal{S}$ is used as the generating function of a canonical transformation rather than the short action $\mathcal{S}_0$ here. The Hamilton–Jacobi method is even more radical because after the canonical transformation, trajectories reduce to points. In the case of separation of variables, Eq. (5.5.45), one can define the action variable I_i, as well as the conjugate angle φ_i, for each part of the short action, $\mathcal{S}_i$. Note that one can introduce the angle-action formalism only in the case of finite motion that excludes scattering problems.

From the above, it is unclear, what the advantage of using the action variable I instead of the energy E is. In the following, it will be shown that I, unlike E, is the so-called *adiabatic invariant* that is nearly constant in the case of a slowly changing parameter.

5.6.2 *Integrable and non-integrable systems*

If the motion of a system with N degrees of freedom can be determined by the Hamilton–Jacobi method, e.g., in the case of separation of variables, one can also find a transformation to the angle-action variables $\{\varphi_i, I_i\}$, $i = 1, 2, \ldots, N$. Then transformed Hamiltonian of the system depends only on N constants I_i as $\mathcal{H} = \mathcal{H}(I_1, I_2, \ldots, I_N)$, whereas all angles linearly increase with time, $\varphi_i(t) = \omega_i t + \varphi_{0i}$. Thus, again, the trajectory of the system is a straight line in a $2N$-dimensional space. The corresponding motion in the real phase space $\{q_i, p_i\}$ is a multiperiodic motion within general incommensurate periods. Topologically, this is a motion on multi-dimensional *tori* that are known as *invariant tori*.

The systems that allow a complete transformation to angle-action variables are called *integrable systems*. Unfortunately, this definition does not provide an explicit integrability check for a particular system. Obviously, systems with separating variables are integrable. However, one can transform a system with separating variables to a non-separating form by some wild canonical transformation. The resulting system will be integrable too, although non-separable. It will be very difficult to integrate this system without knowing the canonical transformation. One may give up and erroneously conclude that this system is non-integrable.

Trajectories of integrable systems corresponding to different initial conditions are straight lines that do not cross and depend smoothly on the initial conditions. On the contrary, for non-integrable systems (that are the majority of mechanical systems), angle-action variables cannot be found and trajectories cannot be straightened. This usually leads to an apparently irregular behavior known as *dynamical chaos*.

One apparent difference between integrable and non-integrable systems is that the former have many integrals of motion I_i, one for each separable degree of freedom i. These integrals of motion, expressed through the natural variables $\{q_i, p_i\}$, impose limitations on the regions in the phase space accessible to the system. This makes the motion of the system regular. On the contrary, non-integrable systems do not have integrals of motion, depending on a small subset of dynamical variables. Thus, much more phase space becomes accessible to them.

5.6.3 *Time-dependent systems*

Let us now consider a system with one degree of freedom and the Hamiltonian depending on time via a parameter $\lambda(t)$, i.e., $\mathcal{H} = \mathcal{H}(q, p, \lambda(t))$.

If the time dependence of λ is weak (slow or fast but with a small amplitude), the energy E is a slow function of time. The formalism of angle-action variables allows one to separate this slow dynamics from the fast (or regular) orbiting dynamics of the system. One can define the function $\mathcal{S}_0$ by the same Eq. (5.6.2), formally considering λ as a parameter depending on the current time. That is, $\mathcal{S}_0$ is calculated as before with $\lambda =$ const. and then replacement $\lambda \Rightarrow \lambda(t)$ is done. The function $\mathcal{S}_0(q, I, \lambda(t))$ that can be called the *adiabatic action* is not the short action since the latter would take into account the time dependence of λ in the integration over the time-dependent trajectory in Eq. (5.6.2). Similar to Sec. (5.6.1), we use $\mathcal{S}_0(q, I, \lambda(t))$ as the generating function of a canonical transformation. Since this canonical transformation is *time-dependent*, it changes the Hamiltonian:

$$\mathcal{H}' = \mathcal{H} + \frac{\partial \mathcal{S}_0}{\partial t} = E\left(I, \lambda\right) + \Lambda \dot{\lambda}, \tag{5.6.10}$$

where

$$\Lambda \equiv \left. \frac{\partial \mathcal{S}_0(q, I, \lambda)}{\partial \lambda} \right|_{q,I}. \tag{5.6.11}$$

In the function Λ, one has to eliminate q with the help of transformation formulas

$$\varphi = \frac{\partial \mathcal{S}_0(q, I, \lambda)}{\partial I}, \quad p = \frac{\partial \mathcal{S}_0(q, I, \lambda)}{\partial q}, \tag{5.6.12}$$

after which $\Lambda = \Lambda\left(\varphi, I, \lambda\right)$. Hamiltonian equations are modified by the term with $\dot{\lambda}$:

$$\begin{aligned} \dot{I} &= -\frac{\partial \mathcal{H}'}{\partial \varphi} = -\frac{\partial \Lambda}{\partial \varphi}\dot{\lambda}, \\ \dot{\varphi} &= \frac{\partial \mathcal{H}'}{\partial I} = \omega\left(I, \lambda\right) + \frac{\partial \Lambda}{\partial I}\dot{\lambda}, \end{aligned} \tag{5.6.13}$$

where $\omega\left(I, \lambda\right) = \partial E\left(I, \lambda\right)/\partial I$. One can see that I and hence E are no longer integrals of motion because of the time dependence of λ.

Let us consider, as an illustration, a harmonic oscillator with time-dependent frequency, the Hamiltonian being given by Eq. (5.4.18) with $\omega = \omega(t)$. Short action is given by Eq. (5.5.36), where the dependence on

$\lambda = \omega$ enters via $\tilde{q}$ defined by

$$\tilde{q} \equiv \sqrt{\frac{m\omega}{2I}}q, \tag{5.6.14}$$

which follows from Eqs. (5.5.31) and (5.5.36). Thus, for Λ in Eq. (5.6.11) with the help of Eq. (5.5.32), one obtains

$$\Lambda = \left.\frac{\partial \mathcal{S}_0}{\partial \tilde{q}}\right|_I \left.\frac{\partial \tilde{q}}{\partial \omega}\right|_{q,I} = 2I\sqrt{1-\tilde{q}^2}\frac{\tilde{q}}{2\omega} = \frac{I}{\omega}\sin\varphi\cos\varphi = \frac{I}{2\omega}\sin(2\varphi). \tag{5.6.15}$$

Now, Eqs. (5.6.13) take the form

$$\begin{aligned} \dot{I} &= -I\cos(2\varphi)\,\frac{\dot{\omega}}{\omega}, \\ \dot{\varphi} &= \omega + \frac{1}{2}\sin(2\varphi)\,\frac{\dot{\omega}}{\omega}. \end{aligned} \tag{5.6.16}$$

Note that the second of these equations is autonomous.

5.6.4 *Adiabatic invariants*

Function Λ of Eq. (5.6.11) is a periodic function of φ with zero average. Although short action $\mathcal{S}_0$ increases acquiring the increment $2\pi I$ every period of motion, Eq. (5.5.37), this increment does not depend on λ and makes no contribution into Λ. An example is Λ for the harmonic oscillator given by Eq. (5.6.15). Periodicity of Λ can be proven rigorously if one uses Legendre-transformed generating function

$$\mathcal{S}_0^*(q,\varphi,\lambda(t)) = \mathcal{S}_0(q,I,\lambda(t)) - I\varphi \tag{5.6.17}$$

instead of $\mathcal{S}_0(q,I,\lambda(t))$ for the same canonical transformation $(q,p) \rightarrow (\varphi, I)$. Then one obtains the same equations (5.6.10) and (5.6.13), whereas Λ is now given by

$$\Lambda \equiv \left.\frac{\partial \mathcal{S}_0^*(q,\varphi,\lambda)}{\partial \lambda}\right|_{q,\varphi} = \left.\frac{\partial \mathcal{S}_0(q,I,\lambda)}{\partial \lambda}\right|_{q,I}. \tag{5.6.18}$$

One can see that this Λ is the same as the above. However, $\mathcal{S}_0^*(q,\varphi,\lambda)$ does not grow on average with φ because the increment $2\pi I$ of $\mathcal{S}_0(q,I,\lambda)$ over the period of motion is exactly compensated for by the term $-I\times 2\pi$ in Eq. (5.6.17). Thus, it is obvious that Λ is periodic with zero average.

In the first of Eq. (5.6.13), the coefficient $\partial\Lambda/\partial\varphi$ is also periodic with zero average. If the change of λ over the period of oscillations T is small,

$$\left|\frac{\dot{\lambda}T}{\lambda}\right| \ll 1, \tag{5.6.19}$$

the change of I becomes very small upon integration on time as oscillations in the rhs average out. Thus, I is the so-called *adiabatic invariant* of motion. On the contrary, the energy E is not an adiabatic invariant. For instance, if the frequency ω of a harmonic oscillator is slowly changing, one has

$$I = \frac{E}{\omega} = \text{const}, \quad E \propto \omega. \tag{5.6.20}$$

Adiabatic invariants also emerge in the motion of a charged particle in a weakly non-uniform magnetic field and in quantum mechanics. Bohr–Sommerfeld quasiclassical quantization condition has the form

$$I = \frac{1}{2\pi}\oint p dq = n\hbar, \tag{5.6.21}$$

where n is an integer. If parameters of the system change slowly, I practically does not change, and neither does n. Were I not an adiabatic invariant, it would change continuously with the parameters of the system. However, n cannot change continuously. Thus, we conclude that only adiabatic invariants are suitable to impose quantization, while going from classical to quantum mechanics.

Change of I becomes especially small if λ slowly changes over the time interval $(-\infty, \infty)$. In this case, the integral

$$\Delta I = -\int_{-\infty}^{\infty} \frac{\partial \Lambda}{\partial \varphi}\dot{\lambda}dt \tag{5.6.22}$$

can be usually transformed by shifting the integration contour into the complex plane to suppress the oscillations of the integrand. Then the dominant contribution to ΔI comes from the singularity of the integrand closest to the real axis and ΔI becomes *exponentially* small. The smaller the $\dot{\lambda}$, the greater the negative exponential in ΔI.

As a numerical example, one can consider a harmonic oscillator with the energy function

$$E = \frac{m}{2}\left[\dot{x}^2 + \omega^2(t)x^2\right], \tag{5.6.23}$$

$\omega(t)$ changing from $\omega(-\infty)$ to $\omega(\infty)$ according to

$$\omega(t) = \omega_0 + \Delta\omega\frac{1 + \tanh(\alpha t)}{2}, \tag{5.6.24}$$

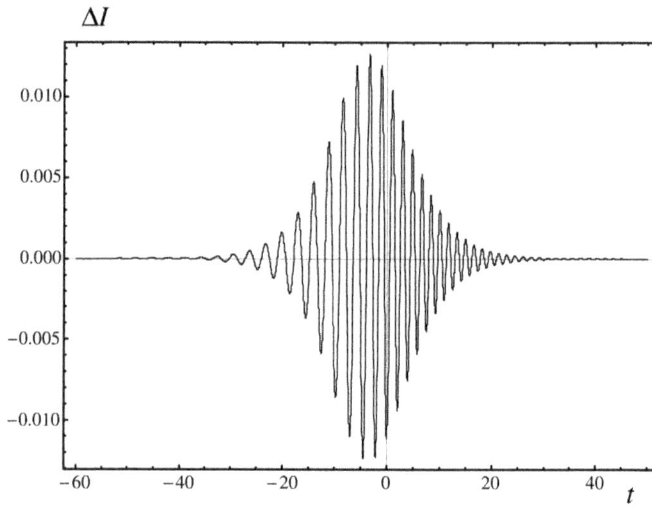

Figure 5.6.1. Time dependence of the change of the action variable I of a harmonic oscillator with slowly changing frequency. Since I is an adiabatic invariant, ΔI is small at all times. (See parameters for the numerical calculation in the text.)

with $\alpha > 0$. The change of $\omega(t)$ is slow if $\alpha \ll \omega_0$. Let us set $\alpha = \omega_0/10$. Also, we set the oscillator mass $m = 2$. For $\omega_0 = \Delta\omega = 1$, one has $\omega(-\infty) = 1$ and $\omega(\infty) = 2$. For the initial state $x(-\infty) = 1$ and $\dot{x}(-\infty) = 0$, the initial oscillator energy is $E(-\infty) = 1$. Initial value of the action variable is $I(-\infty) = E(-\infty)/\omega(-\infty) = 1$. Numerical integration of the equation of motion $\ddot{x} + \omega^2(t)x = 0$ yields the results for $I(t)$ shown in Fig. 5.6.1. One can see that the change $\Delta I(t) \equiv I(t) - I(-\infty)$ is small at all times, i.e., I is indeed an adiabatic invariant. Whereas the asymptotic change $\Delta I(\infty)$ is so small that it cannot be seen in the plot, $\Delta I(t)$ at t around zero, where the change of $\omega(t)$ occurs, is noticeable. Although $\Delta I(\infty)$ is exponentially small, $\Delta I(t)$ for a general t is not.

5.6.5 *Charged particles in a slowly varying magnetic field*

Charged particles in a uniform magnetic field B are cycling with the frequency $\omega_c = qB/m$ in the direction perpendicular to the magnetic field and moving with a constant velocity in the direction parallel to the magnetic field, see Sec. 1.1.4. The transverse part of the particle's dynamics described by Eq. (1.1.49) can be put into the Hamiltonian form by choosing generalized coordinate and momentum as

$$\mathcal{Q} \equiv \alpha v_x, \quad \mathcal{P} \equiv \alpha v_y, \tag{5.6.25}$$

where α is to be determined. Then the first two equations of Eq. (1.1.49) become Hamiltonian equations

$$\dot{\mathcal{Q}} = \frac{\partial \mathcal{H}}{\partial \mathcal{P}} = \omega_c \mathcal{P}, \quad \dot{\mathcal{P}} = -\frac{\partial \mathcal{H}}{\partial \mathcal{Q}} = -\omega_c \mathcal{Q}, \tag{5.6.26}$$

where

$$\mathcal{H} = \frac{\omega_c}{2}\left(\mathcal{Q}^2 + \mathcal{P}^2\right). \tag{5.6.27}$$

The constant α is determined from the condition

$$\mathcal{H} = E = \frac{m}{2}\left(v_x^2 + v_y^2\right) = \frac{p_\perp^2}{2m}. \tag{5.6.28}$$

One obtains $\alpha = \sqrt{m/\omega_c}$.

Now, the action variable I is defined in a standard way:

$$I = \frac{1}{2\pi}\oint \mathcal{P} d\mathcal{Q} = \frac{4}{2\pi}\int\limits_0^{2E/\omega_c} \sqrt{\frac{2E}{\omega_c} - \mathcal{Q}^2 d\mathcal{Q}} = \frac{E}{\omega_c} = \frac{p_\perp^2}{2qB}. \tag{5.6.29}$$

As for the harmonic oscillator, the action variable I is an adiabatic invariant. If the magnetic field is slowly changing with time, kinetic energy is changing proportional to the cyclotron frequency.

If the magnetic field is slowly changing in space along its lines, then one can consider the time dependence of the magnetic field via the longitudinal coordinate $z(t)$, i.e., $B(t) = B(z(t))$ and use the adiabatic invariant together with the total energy conservation:

$$E_{\text{tot}} = \frac{mv^2}{2} = \frac{mv_z^2}{2} + E = \frac{m\dot{z}^2}{2} + I\omega_c(z) = \text{const.} \tag{5.6.30}$$

This is a differential equation for $z(t)$ that can be integrated. One can see that there can be a turning point where $v_z = 0$. The particle cannot enter the region with a too strong magnetic field beyond the turning point, where $E > E_{\text{tot}}$.

5.6.6 *Parametric resonance via angle-action variables*

The formalism of angle-action variables can be used for a more insightful solution of the parametric-resonance problem considered in Sec. 3.2. The time-dependent frequency of the oscillator can be written as

$$\omega_0(t) = \omega_0\,(1 + \alpha\cos\omega t), \quad \alpha \ll 1. \tag{5.6.31}$$

The second autonomous equation (5.6.16) with $\omega \Rightarrow \omega_0$ has the explicit form

$$\dot{\varphi} = \omega_0\,(1 + \alpha\cos\omega t) - \frac{1}{2}\sin(2\varphi)\frac{\alpha\omega\sin\omega t}{1 + \alpha\cos\omega t}. \tag{5.6.32}$$

As we shall see, parametric resonance occurs if ω is close to $2\omega_0$. Thus, we use

$$\omega = 2\omega_0 + \epsilon, \tag{5.6.33}$$

with a small resonance detuning ϵ. Solution of the equation for φ can be searched for in the form

$$\varphi(t) = \frac{1}{2}\left[\omega t + f(t)\right], \tag{5.6.34}$$

where $f(t)$ is a slow phase. The latter satisfies the equation

$$\dot{f}(t) = -\epsilon + 2\omega_0\alpha\cos\omega t - \sin\left[\omega t + f(t)\right]\frac{\alpha\omega\sin\omega t}{1 + \alpha\cos\omega t}, \tag{5.6.35}$$

which is still exact. Using the formula

$$2\sin\left[\omega t + f(t)\right]\sin\omega t = \cos f(t) - \cos\left[2\omega t + f(t)\right], \tag{5.6.36}$$

dropping the term with α in the denominator, approximating $\alpha\omega = \alpha\,(2\omega_0 + \epsilon) \cong 2\alpha\omega_0$, and discarding the fast oscillating terms that average to zero, one obtains the slow equation

$$\dot{f}(t) = -\epsilon - \alpha\omega_0\cos f(t). \tag{5.6.37}$$

This equation can be written in the "potential" form

$$\dot{f} = -\frac{dU_{\text{eff}}}{df}, \quad U_{\text{eff}}(f) = \epsilon f + \alpha\omega_0\sin f, \tag{5.6.38}$$

a tilted washboard potential. For small resonance detuning $|\epsilon| < \alpha\omega_0$, the potential U_{eff} has local maxima and minima, so that the phase $f(t)$ relaxes down to a constant value that satisfies

$$\cos f = -\frac{\epsilon}{\alpha\omega_0}, \tag{5.6.39}$$

thus

$$\sin f = -\sqrt{1 - \left(\frac{\epsilon}{\alpha\omega_0}\right)^2}. \tag{5.6.40}$$

This result for $\sin f$ corresponds to the minima of $U_{\text{eff}}(f)$, whereas the solution with the sign $(+)$ in front of the square root corresponds to the maxima of $U_{\text{eff}}(f)$. The latter is unstable and it should be discarded. The result above means that the oscillator locks into the frequency $\omega/2$, as also follows from the solution in natural variables. In the case $|\epsilon| > \alpha\omega_0$, the potential $U_{\text{eff}}(f)$ is monotonic, and the $f(t)$ performs slow nonlinear motion with a *variable* rate without stopping anywhere.

To see the parametric instability that develops for $|\epsilon| < \alpha\omega_0$, let us now consider the first of the angle-action equations (5.6.16). This equation can be written as

$$\frac{d\ln I}{dt} \cong 2\omega_0\alpha\cos\left[\omega t + f(t)\right]\sin\omega t. \tag{5.6.41}$$

Discarding the fast oscillating term similar to the above, one obtains the slow equation

$$\frac{d\ln I}{dt} \cong -\omega_0\alpha\sin f(t). \tag{5.6.42}$$

After $f(t)$ approaches the constant given by Eq. (5.6.40), this equation becomes

$$\frac{d\ln I}{dt} = 2\lambda, \quad \lambda = \frac{1}{2}\sqrt{(\alpha\omega_0)^2 - \epsilon^2}, \tag{5.6.43}$$

where λ is the parametric resonance exponent. Solution of Eq. (5.6.43) has the form

$$I = I_0 e^{2\lambda t}, \tag{5.6.44}$$

where I_0 is the initial value of I. One can see the exponential divergence of I and thus of the oscillator's energy E in the region of parametric resonance. On the other hand, for large detuning, $|\epsilon| > \alpha\omega_0$, exponent λ is imaginary and I oscillates without growing.

As we have seen, the analytical solution using angle-action variables is more elegant than the straightforward Newtonian solution and it provides more insight. Numerical solution can be done with both formalisms to the same effect. However, including damping and nonlinearities in the Newtonian formalism is straightforward, whereas in the angle-action formalism, it requires a significant work.

5.7 Problems on Hamiltonian dynamics

5.7.1 *Double pendulum*

Consider a double pendulum that consists of two massless rods of lengths $l_1 = l_2 = l$ with masses $m_1 = m_2 = m$ attached to their ends. The first pendulum is attached to a fixed point and can freely swing about it. The second pendulum is attached to the end of the first one and can freely swing too. The motion of both pendulums is confined to a plane, so that it can be described in terms of their angles with respect to the vertical, θ_1 and θ_2.

(a) Write down the Lagrange function for this system.
(b) Introduce generalized momenta p_1 and p_2 and change to Hamiltonian description. Find the transformation matrix that yields the velocities $\theta_1$and θ_2 in terms of the momenta p_1 and p_2. Write down the Hamiltonian function $\mathcal{H}(\theta_1, p_1, \theta_2, p_2)$ using the transformation matrix.
(c) Obtain Hamiltonian equations.
(d) Linearize the Hamiltonian equations near equilibrium.
(e) Eliminate the momenta to obtain second-order differential equations for the angles.
(f) Evaluate the efficiency of the Hamiltonian approach to this problem, comparing it to the Lagrangian approach.

Solution:

(a) Describing the pendula by the angles θ_1 and θ_2, one writes

$$\begin{aligned} x_1 &= l_1 \sin\theta_1, \quad x_2 = l\sin\theta_1 + l\sin\theta_2, \\ y_1 &= -l\cos\theta_1, \quad y_2 = -l\cos\theta_1 - l\cos\theta_2. \end{aligned} \tag{5.7.1}$$

Potential energy reads

$$U = mg\,(y_1 + y_2) = -mgl(2\cos\theta_2 + \cos\theta_1), \tag{5.7.2}$$

and the kinetic energy is given by

$$\begin{aligned} E_k &= \frac{m}{2}[\dot{x}_1^2 + \dot{y}_1^2 + \dot{x}_2^2 + \dot{y}_2^2] \\ &= \frac{ml^2}{2}[\dot{\theta}_1^2 + (\cos\theta_1\dot{\theta}_1 + \cos\theta_2\dot{\theta}_2)^2 + (\sin\theta_1\dot{\theta}_1 + \sin\theta_2\dot{\theta}_2)^2] \\ &= \frac{ml^2}{2}[2\dot{\theta}_1^2 + \dot{\theta}_2^2 + 2(\cos\theta_1\cos\theta_2 + \sin\theta_1\sin\theta_2)\dot{\theta}_1\dot{\theta}_2] \\ &= \frac{ml^2}{2}[2\dot{\theta}_1^2 + \dot{\theta}_2^2 + 2\cos(\theta_1 - \theta_2)\,\dot{\theta}_1\dot{\theta}_2]. \end{aligned} \tag{5.7.3}$$

This kinetic energy can be written in the matrix form

$$E_k = \frac{1}{2}\boldsymbol{\theta}^T \cdot \mathbb{A} \cdot \dot{\boldsymbol{\theta}}, \tag{5.7.4}$$

where $\boldsymbol{\theta}^T = (\theta_1, \theta_2)$ and

$$\mathbb{A} = ml^2 \begin{pmatrix} 2 & \cos(\theta_1 - \theta_2) \\ \cos(\theta_1 - \theta_2) & 1 \end{pmatrix}. \tag{5.7.5}$$

Lagrange function reads

$$\mathcal{L} = E_k - U = \frac{ml^2}{2}\left[2\dot{\theta}_1^2 + \dot{\theta}_2^2 + 2\cos(\theta_1 - \theta_2)\,\dot{\theta}_1\dot{\theta}_2\right] + mgl\,(2\cos\theta_2 + \cos\theta_1). \tag{5.7.6}$$

(b) Generalized momenta has the form

$$\begin{aligned} p_1 &= \frac{\partial \mathcal{L}}{\partial \dot{\theta}_1} = ml^2\left[2\dot{\theta}_1 + \cos(\theta_1 - \theta_2)\,\dot{\theta}_2\right] \\ p_2 &= \frac{\partial \mathcal{L}}{\partial \dot{\theta}_2} = ml^2\left[\dot{\theta}_2 + \cos(\theta_1 - \theta_2)\,\dot{\theta}_1\right]. \end{aligned} \tag{5.7.7}$$

This can be written in the matrix form as $\mathbf{p} = \mathbb{A} \cdot \dot{\boldsymbol{\theta}}$. Resolving these equations for $\dot{\theta}_1$ and $\dot{\theta}_2$, one obtains

$$\begin{aligned} \dot{\theta}_1 &= \frac{1}{ml^2}\frac{p_1 - \cos(\theta_1 - \theta_2)\,p_2}{2 - \cos^2(\theta_1 - \theta_2)}, \\ \dot{\theta}_2 &= \frac{1}{ml^2}\frac{2p_2 - \cos(\theta_1 - \theta_2)\,p_1}{2 - \cos^2(\theta_1 - \theta_2)}. \end{aligned} \tag{5.7.8}$$

This can be written in the matrix form $\dot{\boldsymbol{\theta}} = \mathbb{A}^{-1} \cdot \mathbf{p}$, where

$$\mathbb{A}^{-1} = \frac{1}{ml^2}\frac{1}{2 - \cos^2(\theta_1 - \theta_2)} \begin{pmatrix} 1 & -\cos(\theta_1 - \theta_2) \\ -\cos(\theta_1 - \theta_2) & 2 \end{pmatrix}. \tag{5.7.9}$$

Now, Hamilton function becomes

$$\mathcal{H} = \frac{1}{2}\mathbf{p}^T \cdot \mathbb{A}^{-1} \cdot \mathbf{p} + U = \frac{1}{2}\mathbf{p}^T \cdot \dot{\boldsymbol{\theta}} + U \tag{5.7.10}$$

or, in components,

$$\mathcal{H} = \frac{1}{2ml^2}\frac{p_1^2 - 2\cos(\theta_1 - \theta_2)\,p_1p_2 + 2p_2^2}{2 - \cos^2(\theta_1 - \theta_2)} - mgl\,(2\cos\theta_2 + \cos\theta_1). \tag{5.7.11}$$

(c) Hamiltonian equations are as follows:

$$\dot{\theta}_1 = \frac{\partial \mathcal{H}}{\partial p_1} = \frac{1}{ml^2}\frac{p_1 - \cos(\theta_1 - \theta_2)\,p_2}{2 - \cos^2(\theta_1 - \theta_2)}, \tag{5.7.12}$$

$$\dot{\theta}_2 = \frac{\partial \mathcal{H}}{\partial p_2} = \frac{1}{ml^2}\frac{2p_2 - \cos(\theta_1 - \theta_2)\,p_1}{2 - \cos^2(\theta_1 - \theta_2)}. \tag{5.7.13}$$

These two equations have already been written above. Other Hamiltonian equations are

$$\begin{aligned}
\dot{p}_1 &= -\frac{\partial \mathcal{H}}{\partial \theta_1} \\
&= -\frac{1}{ml^2}\frac{\sin(\theta_1 - \theta_2)\left\{\left[2 + \cos^2(\theta_1 - \theta_2)\right]p_2 - 2p_1\cos(\theta_1 - \theta_2)\right\}}{\left[2 - \cos^2(\theta_1 - \theta_2)\right]^2} \\
&\quad - mgl\sin\theta_1, \\
\dot{p}_2 &= -\frac{\partial \mathcal{H}}{\partial \theta_2} \\
&= -\frac{1}{ml^2}\frac{\sin(\theta_2 - \theta_1)\left\{\left[2 + \cos^2(\theta_1 - \theta_2)\right]p_1 - 2p_2\cos(\theta_1 - \theta_2)\right\}}{\left[2 - \cos^2(\theta_1 - \theta_2)\right]^2} \\
&\quad - 2mgl\sin\theta_2.
\end{aligned} \tag{5.7.14}$$

(d) Near equilibrium ($|\theta_{1,2}| \ll 1$, $|p_{1,2}| \ll 1$), linearization of these equations results in

$$\begin{aligned}
\dot{\theta}_1 &= \frac{p_1 - p_2}{ml^2}, \quad \dot{\theta}_2 = \frac{2p_2 - p_1}{ml^2}, \\
\dot{p}_1 &= -mgl\theta_1, \quad \dot{p}_2 = -2mgl\theta_2.
\end{aligned} \tag{5.7.15}$$

(e) Differentiating the θ-equations and substituting the p-derivatives, one obtains

$$\begin{aligned}
\ddot{\theta}_1 &= \omega_0^2(-\theta_1 + 2\theta_2), \\
\ddot{\theta}_2 &= \omega_0^2(\theta_1 - 4\theta_2).
\end{aligned} \tag{5.7.16}$$

These equations coincide with the similar equations obtained within the Lagrangian approach, Eqs. (2.4.58).

(f) For this problem (as well as for all problems with holonomic constraints), the Hamiltonian approach is inferior to the Lagrangian approach. Derivation of the general system of Hamiltonian equations above requires more work, and these equations are more complicated than Lagrangian equations because of the functional denominators. When linearized near equilibrium, the Hamiltonian equations practically coincide with Lagrangian equations.

5.7.2 *Vortex dynamics via Hamiltonian formalism*

Consider the equations of motions describing vortices of strength γ_i at positions $r_i = (x_i, y_i)$ in the plane:

$$\dot{x}_i = -\sum_{j\neq i}\gamma_j \frac{y_i - y_j}{|\mathbf{r}_i - \mathbf{r}_j|^2}, \quad \dot{y}_i = \sum_{j\neq i}\gamma_j \frac{x_i - x_j}{|\mathbf{r}_i - \mathbf{r}_j|^2}. \tag{5.7.17}$$

Consider the Hamiltonian $\mathcal{H}$ and the following modified Poisson brackets:

$$\mathcal{H} = -\frac{1}{2}\sum_{j\neq i}\gamma_i\gamma_j \ln|\mathbf{r}_i - \mathbf{r}_j|, \quad \{f, g\} \equiv \sum_i \frac{1}{\gamma_i}\left(\frac{\partial f}{\partial x_i}\frac{\partial g}{\partial y_i} - \frac{\partial f}{\partial y_i}\frac{\partial g}{\partial x_i}\right). \tag{5.7.18}$$

(a) Check that Hamiltonian equations

$$\dot{x}_i = \{x_i, \mathcal{H}\}, \quad \dot{y}_i = \{y_i, \mathcal{H}\} \tag{5.7.19}$$

reproduce the equations of motions.

(b) Show that the following quantities are conserved:

$$P_x \equiv \sum_i \gamma_i y_i, \quad P_y \equiv -\sum_i \gamma_i x_i. \tag{5.7.20}$$

Solution:

(a) Work out Poisson brackets:

$$\begin{aligned} \dot{x}_i &= \{x_i, \mathcal{H}\} = \sum_j \frac{1}{\gamma_j}\left(\frac{\partial x_i}{\partial x_j}\frac{\partial \mathcal{H}}{\partial y_j} - \frac{\partial x_i}{\partial y_j}\frac{\partial \mathcal{H}}{\partial x_j}\right) = \frac{1}{\gamma_i}\frac{\partial \mathcal{H}}{\partial y_i}, \\ \dot{y}_i &= \{y_i, \mathcal{H}\} = \sum_j \frac{1}{\gamma_j}\left(\frac{\partial y_i}{\partial x_j}\frac{\partial \mathcal{H}}{\partial y_j} - \frac{\partial y_i}{\partial y_j}\frac{\partial \mathcal{H}}{\partial x_j}\right) = -\frac{1}{\gamma_i}\frac{\partial \mathcal{H}}{\partial x_i}. \end{aligned} \tag{5.7.21}$$

This is the standard form of Hamiltonian equations, except for the factors $1/\gamma_i$. Next, one differentiates the Hamiltonian, taking into account that there are two occurrences of a summation index coinciding with i: $j' = i$ and $j = i$. One proceeds as

$$\begin{aligned} \dot{x}_i &= \frac{1}{\gamma_i}\frac{\partial \mathcal{H}}{\partial y_i} = \frac{1}{\gamma_i}\frac{\partial}{\partial y_i}\left(-\frac{1}{2}\sum_{j\neq j'}\gamma_{j'}\gamma_j \ln|\mathbf{r}_{j'} - \mathbf{r}_j|\right) \\ &= -\sum_{j\neq i}\gamma_j \frac{\partial}{\partial y_i}\ln|\mathbf{r}_i - \mathbf{r}_j| = -\sum_{j\neq i}\gamma_j \frac{1}{|\mathbf{r}_i - \mathbf{r}_j|}\frac{\partial|\mathbf{r}_i - \mathbf{r}_j|}{\partial y_i} \\ &= -\sum_{j\neq i}\gamma_j \frac{y_i - y_j}{|\mathbf{r}_i - \mathbf{r}_j|^2}, \end{aligned} \tag{5.7.22}$$

which is the first equation of motion. The second equation of motion can be obtained in a similar way.

(b) Check conservation directly:

$$\dot{P}_x = \sum_i \gamma_i \dot{y}_i = \sum_i \gamma_i \sum_{j\neq i} \gamma_j \frac{x_i - x_j}{|\mathbf{r}_i - \mathbf{r}_j|^2} = \sum_{i\neq j} \gamma_i\gamma_j \frac{x_i - x_j}{|\mathbf{r}_i - \mathbf{r}_j|^2}. \tag{5.7.23}$$

Interchanging summation indices $i \rightleftharpoons j$ changes the sign of this expression:

$$\sum_{i\neq j} \gamma_i\gamma_j \frac{x_i - x_j}{|\mathbf{r}_i - \mathbf{r}_j|^2} = \sum_{j\neq i} \gamma_j\gamma_i \frac{x_j - x_i}{|\mathbf{r}_j - \mathbf{r}_i|^2} = -\sum_{i\neq j} \gamma_i\gamma_j \frac{x_i - x_j}{|\mathbf{r}_i - \mathbf{r}_j|^2}. \tag{5.7.24}$$

Thus, it is zero. In a similar way, one obtains $\dot{P}_y = 0$.

5.7.3 *Canonical transformations, A*

Canonical transformation between the two sets of variables is

$$Q = \ln(1 + \sqrt{q}\cos p), \tag{5.7.25a}$$

$$P = 2(1 + \sqrt{q}\cos p)\sqrt{q}\sin p. \tag{5.7.25b}$$

Show directly that this transformation is canonical. Show that

$$F_{pQ}(p, Q) = -\left(e^Q - 1\right)^2 \tan p \tag{5.7.26}$$

is generating the function of this transformation.

Solution: Since the transformation does not explicitly depend on time, its canonicity can be checked by the Poisson-brackets criterion. One calculates

$$\begin{aligned}
\{Q, P\} &= \frac{\partial Q}{\partial q}\frac{\partial P}{\partial p} - \frac{\partial Q}{\partial p}\frac{\partial P}{\partial q} \\
&= \frac{1}{1 + \sqrt{q}\cos p}\frac{2\cos p}{2\sqrt{q}}\left[(1 + \sqrt{q}\cos p)\sqrt{q}\cos p - q\sin^2 p\right] \\
&\quad - \frac{-2\sqrt{q}\sin p}{1 + \sqrt{q}\cos p}\left[\frac{\cos p}{2\sqrt{q}}\sqrt{q}\sin p + (1 + \sqrt{q}\cos p)\frac{\sin p}{2\sqrt{q}}\right] \\
&= \frac{1}{1 + \sqrt{q}\cos p}\frac{1}{\sqrt{q}}[\sqrt{q}\cos^2 p + q\cos p\left(\cos^2 p - \sin^2 p\right) \\
&\quad + 2q\sin^2 p\cos p + \sqrt{q}\sin^2 p] \\
&= \frac{1}{1 + \sqrt{q}\cos p}\frac{1}{\sqrt{q}}\left[\sqrt{q} + q\cos p\right] = 1.
\end{aligned} \tag{5.7.27}$$

Thus, the transformation is canonical. Next, the proposed generating function is obtained from the primary generating function $F_{qQ}(q, Q)$ by Legendre transformation

$$F_{pQ} = F_{qQ} - pq. \tag{5.7.28}$$

Its differential reads

$$dF_{pQ} = dF_{qQ} - pdq - qdp = pdq - PdQ - pdq - qdp = -qdp - PdQ. \tag{5.7.29}$$

Thus, $F_{pQ} = F_{pQ}(p, Q)$ and our canonical transformation is defined by

$$q = -\frac{\partial F_{pQ}}{\partial p}, \quad P = -\frac{\partial F_{pQ}}{\partial Q}. \tag{5.7.30}$$

Let us work out these formulas. One has

$$q = -\frac{\partial F_{pQ}}{\partial p} = \left(e^Q - 1\right)^2 \frac{1}{\cos^2 p}. \tag{5.7.31}$$

Resolving this for Q, one obtains Eq. (5.7.25a). Next, one obtains

$$P = -\frac{\partial F_{pQ}}{\partial Q} = \frac{\partial}{\partial Q}\left(e^Q - 1\right)^2 \tan p = 2\left(e^Q - 1\right) e^Q \tan p. \tag{5.7.32}$$

Inserting Eq. (5.7.25a) here, one obtains Eq. (5.7.25b).

5.7.4 *Canonical transformations, B*

For what values of α and β do the equations

$$Q = q^\alpha \cos(\beta p), \quad P = q^\alpha \sin(\beta p) \tag{5.7.33}$$

represent a canonical transformation? What is the form of the generating function $F_{pQ}(p, Q)$ in this case?

Solution: Let us calculate the Poisson bracket,

$$\begin{aligned}\{Q, P\} &= \frac{\partial Q}{\partial q}\frac{\partial P}{\partial p} - \frac{\partial Q}{\partial p}\frac{\partial P}{\partial q} \\ &= \alpha q^{\alpha-1}\cos(\beta p) q^\alpha \beta \cos(\beta p) + q^\alpha \beta \sin(\beta p) \alpha q^{\alpha-1} \sin(\beta p) \\ &= \alpha\beta q^{2\alpha-1}. \end{aligned} \tag{5.7.34}$$

The result is 1 and thus the transformation is canonical for $\alpha = 1/2$ and $\beta = 2$.

Let us now find a generating function for the canonical transformation

$$Q = \sqrt{q}\cos(2p), \quad P = \sqrt{q}\sin(2p). \tag{5.7.35}$$

To use Eq. (5.7.30), first one has to express the transformation in terms of p and Q. One obtains

$$Q^2 + P^2 = q, \tag{5.7.36}$$

thus

$$P = \sqrt{Q^2 + P^2}\sin(2p). \tag{5.7.37}$$

Resolving for P, one obtains

$$P = Q\tan(2p). \tag{5.7.38}$$

Now, integrating the second equation of (5.7.30), one obtains

$$F_{pQ} = -\frac{1}{2}Q^2\tan(2p) + f(p), \tag{5.7.39}$$

where $f(p)$ is an integration constant. Now, the first equation of (5.7.30) becomes

$$q = -\frac{\partial F_{pQ}}{\partial p} = Q^2\frac{1}{\cos^2(2p)} + f'(p). \tag{5.7.40}$$

Setting $f(p) = 0$ and resolving for Q, one obtains the first equation in (5.7.35). Thus, Eq. (5.7.39) with $f(p) = 0$ is the generating function of our transformation.

5.7.5 *Coulomb + uniform field via parabolic coordinates*

Find the complete integral of Hamilton–Jacobi equations for the problem with potential energy

$$U = \frac{\alpha}{r} + Fz. \tag{5.7.41}$$

(a) Write Hamilton–Jacobi equation in spherical coordinates and check whether variables separate.
(b) Write Hamilton–Jacobi equation in cylindrical coordinates and check whether variables separate.
(c) Use parabolic coordinates (ξ, η, φ) that are related to cylindrical coordinates (ρ, z, φ) as

$$z = \frac{1}{2}(\xi - \eta), \quad \rho = \sqrt{\xi\eta}. \tag{5.7.42}$$

Show that variables separate and find the complete integral of Hamilton–Jacobi equations.

Solution:

(a) In spherical coordinates, the Hamiltonian has the form

$$\mathcal{H} = \frac{1}{2m}\left(p_r^2 + \frac{p_\theta^2}{r^2} + \frac{p_\phi^2}{r^2 \sin^2\theta}\right) + \frac{\alpha}{r} + Fr\cos\theta. \tag{5.7.43}$$

Solution of Hamilton–Jacobi (HJ) equation

$$\frac{\partial \mathcal{S}}{\partial t} + \mathcal{H}\left(q, \frac{\partial \mathcal{S}}{\partial q}\right) = 0 \tag{5.7.44}$$

for this time-independent problem can be searched for in the form

$$\mathcal{S} = -Et + \mathcal{S}_0(q), \tag{5.7.45}$$

which yields

$$-E + \frac{1}{2m}\left[\left(\frac{\partial \mathcal{S}_0}{\partial r}\right)^2 + \frac{1}{r^2}\left(\frac{\partial \mathcal{S}_0}{\partial \theta}\right)^2 + \frac{1}{r^2\sin^2\theta}\left(\frac{\partial \mathcal{S}_0}{\partial \varphi}\right)^2\right] + \frac{\alpha}{r} + Fr\cos\theta = 0. \tag{5.7.46}$$

Here, φ is a cyclic variable, thus the solution has the form

$$\mathcal{S}_0 = \mathcal{S}_0^{(r,\theta)} + l\varphi, \tag{5.7.47}$$

where l is the constant z component of the angular momentum. HJ equation for the remaining function $\mathcal{S}_0^{(r,\theta)}$ takes the form

$$-E + \frac{1}{2m}\left[\left(\frac{\partial \mathcal{S}_0^{(r,\theta)}}{\partial r}\right)^2 + \frac{1}{r^2}\left(\frac{\partial \mathcal{S}_0^{(r,\theta)}}{\partial \theta}\right)^2 + \frac{l^2}{r^2\sin^2\theta}\right] + \frac{\alpha}{r} + Fr\cos\theta = 0. \tag{5.7.48}$$

Here, because of the last term, one cannot split a fragment of this equation that depends only on θ (also only on r). Thus, variables do not separate.

(b) Cylindrical coordinates are defined as

$$x = \rho\cos\varphi, \quad y = \rho\sin\varphi, \quad z = z. \tag{5.7.49}$$

Kinetic energy has the form

$$E_k = \frac{m}{2}\left(\dot{\rho}^2 + \rho^2\dot{\varphi}^2 + \dot{z}^2\right). \tag{5.7.50}$$

Momenta are defined by

$$p_\rho = \frac{\partial E_k}{\partial \dot{\rho}} = m\dot{\rho}, \quad p_\varphi = \frac{\partial E_k}{\partial \dot{\varphi}} = m\rho^2\dot{\varphi}, \quad p_z = m\dot{z}. \tag{5.7.51}$$

Thus, the Hamiltonian has the form

$$\mathcal{H} = \frac{1}{2m}\left(p_\rho^2 + \frac{p_\varphi^2}{\rho^2} + p_z^2\right) + \frac{\alpha}{\sqrt{\rho^2+z^2}} + Fz. \tag{5.7.52}$$

Here, again, φ is the cyclic variable and the solution of HJ equation can be found in the form

$$\mathcal{S} = -Et + l\varphi + \mathcal{S}_0^{(\rho,z)}, \tag{5.7.53}$$

which yields the short HJ equation

$$-E + \frac{1}{2m}\left[\left(\frac{\partial \mathcal{S}_0^{(\rho,z)}}{\partial \rho}\right)^2 + \left(\frac{\partial \mathcal{S}_0^{(\rho,z)}}{\partial z}\right)^2 + \frac{l^2}{\rho^2}\right] + \frac{\alpha}{\sqrt{\rho^2+z^2}} + Fz = 0. \tag{5.7.54}$$

Here, variables do not separate because of the Coulomb-energy term.

(c) Now, use Eq. (5.7.42) and additional relations following from it:

$$r = \sqrt{\rho^2+z^2} = \frac{1}{2}(\xi+\eta) \tag{5.7.55}$$

and

$$\xi = r + z, \quad \eta = r - z. \tag{5.7.56}$$

First, calculate $\dot{\rho}^2 + \dot{z}^2$ in Eq. (5.7.50):

$$\begin{aligned}\dot{\rho}^2 + \dot{z}^2 &= \left(\frac{1}{2}\sqrt{\frac{\eta}{\xi}}\dot{\xi} + \frac{1}{2}\sqrt{\frac{\xi}{\eta}}\dot{\eta}\right)^2 + \left(\frac{1}{2}\dot{\xi} - \frac{1}{2}\dot{\eta}\right)^2 \\ &= \frac{1}{4}\left(\frac{\eta}{\xi}\dot{\xi}^2 + 2\dot{\xi}\dot{\eta} + \frac{\xi}{\eta}\dot{\eta}^2 + \dot{\xi}^2 - 2\dot{\xi}\dot{\eta} + \dot{\eta}^2\right) = \frac{(\xi+\eta)}{4}\left(\frac{\dot{\xi}^2}{\xi} + \frac{\dot{\eta}^2}{\eta}\right).\end{aligned} \tag{5.7.57}$$

Now, kinetic energy of Eq. (5.7.50) becomes

$$E_k = \frac{m}{2}\left[\frac{(\xi+\eta)}{4}\left(\frac{\dot{\xi}^2}{\xi} + \frac{\dot{\eta}^2}{\eta}\right) + \xi\eta\dot{\varphi}^2\right]. \tag{5.7.58}$$

Generalized momenta are given by

$$p_\xi = \frac{\partial E_k}{\partial \xi} = \frac{m}{4}(\xi+\eta)\frac{\dot{\xi}}{\xi}, \quad p_\eta = \frac{\partial E_k}{\partial \eta} = \frac{m}{4}(\xi+\eta)\frac{\dot{\eta}}{\eta}, \quad p_\varphi = m\xi\eta\dot{\varphi}. \tag{5.7.59}$$

Expressing velocities through momenta, inserting them into kinetic energy and adding potential energy yield the Hamiltonian

$$\mathcal{H} = \frac{2}{m}\frac{\xi p_\xi^2 + \eta p_\eta^2}{\xi+\eta} + \frac{p_\varphi^2}{2m\xi\eta} + U, \tag{5.7.60}$$

where

$$U = \frac{2\alpha}{\xi+\eta} + \frac{F}{2}(\xi-\eta) = \frac{4\alpha + F\left(\xi^2-\eta^2\right)}{2(\xi+\eta)}. \tag{5.7.61}$$

Now, searching for the action in the form

$$\mathcal{S} = -Et + l\varphi + \mathcal{S}_0^{(\xi,\eta)}, \tag{5.7.62}$$

one obtains HJ equation in the form

$$\begin{aligned}-E &+ \frac{2}{m(\xi+\eta)}\left[\xi\left(\frac{\partial \mathcal{S}_0^{(\xi,\eta)}}{\partial \xi}\right)^2 + \eta\left(\frac{\partial \mathcal{S}_0^{(\xi,\eta)}}{\partial \eta}\right)^2\right] \\ &+ \frac{l^2}{2m\xi\eta} + \frac{4\alpha + F\left(\xi^2-\eta^2\right)}{2(\xi+\eta)} = 0.\end{aligned} \tag{5.7.63}$$

After multiplying by $\xi+\eta$, variables separate:

$$\begin{aligned}-E(\xi+\eta) &+ \frac{2}{m}\left[\xi\left(\frac{\partial \mathcal{S}_0^{(\xi,\eta)}}{\partial \xi}\right)^2 + \eta\left(\frac{\partial \mathcal{S}_0^{(\xi,\eta)}}{\partial \eta}\right)^2\right] \\ &+ \frac{l^2}{2m}\left(\frac{1}{\xi}+\frac{1}{\eta}\right) + \frac{4\alpha + F\left(\xi^2-\eta^2\right)}{2} = 0.\end{aligned} \tag{5.7.64}$$

Solution of this equation can be searched for in the form

$$\mathcal{S}_0^{(\xi,\eta)}(\xi,\eta) = \mathcal{S}_0^{(\xi)}(\xi) + \mathcal{S}_0^{(\eta)}(\eta). \tag{5.7.65}$$

Equating the two parts of HJ equation to constants P and $-P$, one obtains two separate ODEs:

$$-E\xi + \frac{2}{m}\xi\left(\frac{\partial \mathcal{S}_0^{(\xi)}}{\partial \xi}\right)^2 + \frac{l^2}{2m\xi} + \alpha + \frac{F}{2}\xi^2 = P \tag{5.7.66}$$

and

$$-E\eta + \frac{2}{m}\eta\left(\frac{\partial \mathcal{S}_0^{(\eta)}}{\partial \eta}\right)^2 + \frac{l^2}{2m\eta} + \alpha + \frac{F}{2}\eta^2 = -P. \tag{5.7.67}$$

Integrating these equations, one obtains

$$\mathcal{S}_0^{(\xi)}(\xi) = \sqrt{\frac{m}{2}}\int d\xi\sqrt{E - U_\xi(\xi)}, \quad U_\xi(\xi) \equiv \frac{\alpha - P}{\xi} + \frac{l^2}{2m\xi^2} + \frac{F}{2}\xi \tag{5.7.68}$$

and

$$\mathcal{S}_0^{(\eta)}(\eta) = \sqrt{\frac{m}{2}}\int d\eta\sqrt{E - U_\eta(\eta)}, \quad U_\eta(\eta) \equiv \frac{\alpha + P}{\eta} + \frac{l^2}{2m\eta^2} + \frac{F}{2}\eta. \tag{5.7.69}$$

As the result, we have obtained the complete integral of Hamilton–Jacobi equations having the form

$$\mathcal{S} = \mathcal{S}(\xi, \eta, \varphi; E, l, P; t) = -Et + l\varphi + \mathcal{S}_0^{(\xi)}(\xi; E, l, P) + \mathcal{S}_0^{(\eta)}(\eta; E, l, P). \tag{5.7.70}$$

It depends on three generalized coordinates ξ, η, φ and three integration constants E, l, P. Considering $\mathcal{S}$ as the generating function of a canonic transformation with E, l, P being new momenta, one defines the dynamics of the system implicitly from the three equations

$$Q_E = \frac{\partial \mathcal{S}}{\partial E}, \quad Q_l = \frac{\partial \mathcal{S}}{\partial l}, \quad Q_P = \frac{\partial \mathcal{S}}{\partial P}, \tag{5.7.71}$$

where Q_E, Q_l, and Q_P are another three constants. As the system has three degrees of freedom, its solution is specified by the total six integration constants, as it should be. Constants Q_E and Q_l can be set to zero because reference points for the time and angle φ are arbitrary. The third equation yields the relation between ξ and η. Together with the second equation, this yields the trajectory without time dependence. The latter is obtained using the first equation. Doing this analytically is hardly possible because of complicated integrals.

5.7.6 *Generalized harmonic oscillator*

The motion of a particle in one dimension is described by the Hamiltonian

$$\mathcal{H} = \frac{1}{2}\left(p^2 + \omega_0^2 q^2\right) + \frac{\lambda}{4}\left(p^2 + \omega_0^2 q^2\right)^2. \tag{5.7.72}$$

(a) Using the generating function $F(q, Q) = \left(\omega_0 q^2/2\right)\cot Q$, define new canonical variables Q and P and find the transformed Hamiltonian $\mathcal{H}(Q, P)$.
(b) Set up Hamilton–Jacobi equation for the Hamiltonian $\mathcal{H}(Q, P)$ and the action $\mathcal{S}(Q, \alpha, t)$, where α is an integration constant.
(c) Find $q(t)$ and $p(t)$ using Hamilton–Jacobi method and the function $\mathcal{S}(Q, \alpha, t)$. Use initial conditions $q(0) = q_0$ and $p = 0$.

Solution:

(a) Transformation formulas have the form

$$p = \frac{\partial F}{\partial q}, \quad P = -\frac{\partial F}{\partial Q}. \tag{5.7.73}$$

From this, one obtains

$$p = \omega_0 q \cot Q, \quad P = \frac{\omega_0 q^2}{2}\frac{1}{\sin^2 Q}. \tag{5.7.74}$$

Resolving the second equation for q and substituting the result into the first equation, one obtains

$$q = \sqrt{\frac{2P}{\omega_0}}\sin Q, \quad p = \sqrt{2\omega_0 P}\cos Q. \tag{5.7.75}$$

Inserting this into the Hamiltonian, one obtains

$$\mathcal{H} = \omega_0 P + \lambda\left(\omega_0 P\right)^2. \tag{5.7.76}$$

(b) Using the formula

$$d\mathcal{S} = PdQ - \mathcal{H}dt \tag{5.7.77}$$

along the trajectory, thus

$$P = \frac{\partial \mathcal{S}}{\partial Q}, \quad \mathcal{H} = -\frac{\partial \mathcal{S}}{\partial t}, \tag{5.7.78}$$

one obtains the Hamilton–Jacobi equation

$$\frac{\partial \mathcal{S}}{\partial t} + \mathcal{H}\left(Q, \frac{\partial \mathcal{S}}{\partial Q}\right) = 0. \tag{5.7.79}$$

With the transformed Hamiltonian above, this becomes

$$\frac{\partial \mathcal{S}}{\partial t} + \omega_0\frac{\partial \mathcal{S}}{\partial Q} + \lambda\left(\omega_0\frac{\partial \mathcal{S}}{\partial Q}\right)^2 = 0. \tag{5.7.80}$$

The complete integral of this equation can be searched for in the form

$$\mathcal{S} = -Et + \mathcal{S}_0(Q). \tag{5.7.81}$$

For $\mathcal{S}_0(Q)$, one obtains the equation

$$\omega_0 \frac{\partial \mathcal{S}_0}{\partial Q} + \lambda \left(\omega_0 \frac{\partial \mathcal{S}}{\partial Q} \right)^2 = E. \tag{5.7.82}$$

Its solution can be searched for in the form

$$\mathcal{S}_0(Q) = \alpha Q, \tag{5.7.83}$$

where the irrelevant integration constant has been discarded and constant α can be expressed via the conserved energy E using

$$\omega_0 \alpha + \lambda \left(\omega_0 \alpha \right)^2 = E. \tag{5.7.84}$$

However, it is more convenient to work with α than with E. The full action reads

$$\mathcal{S}\left(Q, \alpha, t\right) = - \left[\omega_0 \alpha + \lambda \left(\omega_0 \alpha \right)^2 \right] t + \alpha Q. \tag{5.7.85}$$

(c) Using $\mathcal{S}\left(Q, \alpha, t\right)$ as the generating function of another canonical transformation that nullifies the transformed Hamiltonian, one obtains the solution for the dynamics in the form

$$\beta = \frac{\partial \mathcal{S}}{\partial \alpha}, \tag{5.7.86}$$

where β is another integration constant. Thus, one obtains

$$\beta = - \left[\omega_0 + 2\lambda \omega_0^2 \alpha \right] t + Q \tag{5.7.87}$$

and

$$Q = \tilde{\omega}_0 t + \beta, \quad \tilde{\omega}_0 \equiv \omega_0 + 2\lambda \omega_0^2 \alpha, \tag{5.7.88}$$

$\tilde{\omega}_0$ being renormalized frequency. Generalized momentum P is obtained from the transformation formula

$$P = \frac{\partial \mathcal{S}}{\partial Q} = \alpha, \tag{5.7.89}$$

and it is constant. Now the dynamics of original variables q, p can be obtained as

$$q = \sqrt{\frac{2P}{\omega_0}} \sin Q = \sqrt{\frac{2\alpha}{\omega_0}} \sin\left(\tilde{\omega}_0 t + \beta \right) \tag{5.7.90}$$

and

$$p = \sqrt{2\omega_0 P} \cos Q = \sqrt{2\omega_0 \alpha} \cos\left(\tilde{\omega}_0 t + \beta \right). \tag{5.7.91}$$

Chapter 6

Dynamical Chaos

Progress in computing led to the discovery of apparently chaotic behavior of systems with only a few degrees of freedom in 70th. The systems in question can be either mechanical or non-mechanical, including examples from chemistry, biology, etc. Mathematically, the models involved are described by systems of differential equations describing their dynamics, so that one calls them *dynamical systems* and the chaos found in them *dynamical chaos.*

Two different kinds of dynamical chaos are observed in *conservative* and *dissipative* systems, i.e., in systems without and with relaxation.

Observation of chaos requires sufficient time. For this reason, dynamical chaos can take place in systems with variables evolving within a finite range, but there is usually no chaos in scattering problems.

The signature of dynamical chaos is apparently irregular trajectories that differ drastically from regular periodic or multiperiodic trajectories. Irregularity means that the form of trajectories is very complicated and seemingly unpredictable. The latter implies great sensitivity of trajectories to initial conditions. Although trajectories are deterministic and repeated numerical solutions yield the same result, the slightest change in initial conditions or any parameter of the problem leads to a completely different trajectory. Two trajectories with very close initial conditions diverge exponentially with time. This is the mathematical criterion of the dynamical chaos. Practically, great sensitivity to initial conditions or problem parameters makes trajectories not well defined. This stimulates ideas of statistical description of such systems.

Dynamical chaos in systems with few degrees of freedom differs from the *molecular chaos* in macroscopically large systems. The latter is not necessarily a real chaos but rather our way of thinking of large systems. Since it is difficult to follow the motion of an Avogadro number $N_A \sim 10^{23}$

of different particles, coarse-grained variables such as the particle's density are introduced. To justify reduced description, statistical concepts such as molecular chaos are used. In fact, even regular motion of many-particle systems may be apparently chaotic. An example is the model of a harmonic oscillator coupled to a bath of harmonic oscillators that obeys linear equations of motion. Statistical properties of this model can be described with perfectly regular solutions of the dynamical equations. In general, in large systems, the effect of dynamical chaos (defined in the preceding paragraph) and that of a large number of degrees of freedom coexist. It is the latter and not the former that is responsible for macroscopic behavior of large systems.

In the following sections, different aspects of the dynamical chaos will be discussed. At first, we consider conservative systems and then discuss a different behavior of dissipative systems.

6.1 Dynamical systems and degrees of freedom

The common feature of the systems showing dynamical chaos is that mathematically they are described by systems of *nonlinear* differential equations

$$\dot{x}_i = f_i\left(\{x_i\}, t\right), \quad i = 1, 2, \ldots, N. \tag{6.1.1}$$

Mechanical problems can be put into the above form if one considers coordinates and velocities (or momenta) of particles as independent variables. In particular, Newtonian equations of motion for one particle with one degree of freedom can be written as

$$\begin{aligned} \dot{x} &= v, \\ \dot{v} &= \frac{1}{m} F(x, v, t). \end{aligned} \tag{6.1.2}$$

Explicit time dependence in the rhs of Eq. (6.1.1) is equivalent to an additional dynamical variable t that obeys the equation

$$\dot{t} = 1. \tag{6.1.3}$$

Thus, a system of N differential equations of the type of Eq. (6.1.1) with explicit time dependence is mathematically equivalent to an extended system of $N + 1$ differential equations without explicit time dependence. One can speak of the effective number of degrees of freedom $N/2$ and $(N+1)/2$ for the systems without and with the explicit time dependence.

For instance, Eq. (6.1.2) without explicit time dependence has one degree of freedom, whereas the same with explicit time dependence has 3/2 degrees of freedom.

Dynamical chaos can exist in systems starting with 3/2 degrees of freedom. For one degree of freedom, the phase space of the system is two-dimensional and trajectories cannot cross each other. Note that the crossing of two different trajectories would contradict the uniqueness of the solution of the dynamical problem with the initial condition set at the would-be crossing point. This imposes a severe restriction on the form of trajectories, making them streamlined and regular. For 3/2 degrees of freedom, trajectories fill a three-dimensional space, so that the non-crossing condition becomes irrelevant and trajectories can meander unrestrictedly and independently of each other.

Integrals of motion restrict the solution and make dynamical chaos less likely. To count the effective number of the degrees of freedom N_F relevant to the dynamical chaos, one has to subtract the number of integrals of motion N_I from the number of equations and then divide by 2:

$$N_F = \frac{1}{2}(N - N_I). \tag{6.1.4}$$

If Eq. (6.1.2) does not have an explicit time dependence and the energy is conserved, $N_F = 1/2$. For a motion in a central field, there are three degrees of freedom and thus $N = 6$ equations. However, there are also $N_I = 4$ integrals of motion, the energy and three components of the angular momentum $\mathbf{L}$. Thus, $N_F = 1$, and there is no chaos. Adding a non-central field breaks conservation of at least one of the $\mathbf{L}$ components and N_F increases above 1. This can lead to dynamical chaos.

6.2 Conservative systems

6.2.1 *Integrable and non-integrable systems: KAM theorem*

Clearly, $N_F \geq 3/2$ does not automatically imply that dynamic chaos is developing in a system. A counter-example is a *separable* system of a large numbers of uncoupled particles. Although for the whole system N_F is large, the problem splits up into many subproblems for individual particles with $N_F \leq 1$, and hence, there is no chaos.

Another example is small oscillations in systems with many degrees of freedom. Approximating potentials by quadratic functions near the minima, one obtains *linear* equations of motions that can be decoupled by changing to the normal modes. Again, there is no dynamical chaos in such systems, and the example explains why dynamical chaos requires nonlinearity of the equations.

The two examples above are the so-called *integrable systems* that allow the separation of variables immediately or after some manipulations. There are also nonlinear dynamical systems in which the separation of variables can be done and solution can be obtained in a closed form. All integrable systems do not show dynamical chaos. Each uncoupled subsystem of a large system performs a periodic motion with its own period. Thus, the motion of the whole system is multiperiodic and geometrically it can be represented as regular motion on a multidimensional torus (circle for one degree of freedom and a simple torus for two degrees of freedom).

The simplest way to obtain an integrable nonlinear dynamical system is to start with an uncoupled dynamical system that can be easily integrated and make a transformation of variables that results in a coupled dynamical system. The latter may look complicated, but it is integrable since the transformation to original variables is known.

If the starting point is a coupled nonlinear dynamical system, it is usually difficult to prove that a transformation leading to the separation of variables exists. Systems that cannot be integrated by the separation of variables are called *non-integrable systems* and they can show chaos. However, there is no straightforward integrability criterion. The deficiency of this definition is thus that it relies on the skills of an individual researcher.

An important question is what happens if a small perturbation is added to an integrable system so that the latter becomes non-integrable. Does it lead to dynamical chaos? Is the regular motion on a torus destroyed? The answer is given by the Kolmogorov–Arnold–Moser (KAM) theorem:

If

- *the perturbation is small, and*
- *the periods of the unperturbed problem are incommensurate,*

then the motion remains confined to a vicinity of the torus, except for a small subset of initial conditions.

That is, all or almost all trajectories remain regular if a small non-integrable perturbation is added. The proof of this theorem is very sophisticated. Further, one can ask what are the possible scenarios of the onset of dynamical chaos as the strength of the perturbation is increased. In particular, if all trajectories remain regular for a small perturbation, there should be a threshold for the chaos onset on the perturbation strength. What are the trajectories just above the chaos threshold: strongly chaotic or regular with a small chaotic component?

6.2.2 *Divergence of trajectories and Lyapunov exponents*

As stated above, the mathematical criterion of the dynamical chaos is a strong divergence of initially close trajectories with time. Usually, divergence of trajectories, as instability, is described in the literature by exponential dependences of the type

$$s(t) = s(0)e^{\lambda t}, \tag{6.2.1}$$

where $s(t)$ is the appropriately defined distance between two initially close trajectories. In Eq. (6.2.1), λ is the Lyapunov exponent, the same as in the theory of stability/instability. If $\mathrm{Re}\lambda > 0$, the two trajectories exponentially diverge. Equation (6.2.1) is idealized, however, and it is difficult to obtain λ for particular models.

To study Lyapunov exponents, one can consider two close trajectories, $x_i(t)$ and $x_i(t) + \delta x_i(t)$, and linearize Eq. (6.1.1) in $\delta x_i(t)$. This yields the system of linear differential equations for δx_i:

$$\delta \dot{x}_i = \sum_{k=1}^{N} \frac{\partial f_i}{\partial x_k} \delta x_k, \quad i, k = 1, 2, \ldots, N, \tag{6.2.2}$$

with coefficients $\partial f_i / \partial x_k$, depending on time at least through $x_i(t)$. Since in the case of dynamical chaos, the problem is non-integrable, and thus, it cannot be solved analytically for $x_i(t)$, analysis of Eq. (6.2.2) is problematic. On top of it, even if $\partial f_i / \partial x_k$ were known, systems of linear differential equations with variable cofficients cannot usually be solved analytically. Anyway, their solution does not have a simple exponential form of Eq. (6.2.1).

A toy model for which Lyapunov exponents can be calculated is the so-called Sinai billiard. It is a two-dimensional region bounded by concave walls. Particles freely fly between the walls and elastically rebound from them like the light reflects from a convex mirror. If two trajectories (straight in the region between the walls) make a small angle α with each other, the angle after reflection will be the same multiplied by the factor $n > 1$, depending on the wall curvature and the angle of incidence. The number $n - 1$ plays the role of the Lyapunov exponent.

A simple practical way to test dynamical chaos is to solve Eq. (6.1.1) numerically for two close trajectories and see if they diverge. Usually, the behavior of trajectories depends on where they originate. Some trajectories do not diverge and thus are regular, other trajectories diverge and thus are chaotic. Even better is to solve Eq. (6.1.1) for one trajectory $x_i(t)$ and then to solve Eq. (6.2.2) for the linear deviation $\delta x_i(t)$. This does not require the high numerical accuracy needed to subtract two close trajectories.

6.2.3 *Fourier analysis*

Another tool to analyze dynamical chaos is the Fourier analysis of the trajectories, taking particular components, $x_i(t)$. In the case of regular motion, $x_i(t)$ is multiperiodic with periods T_p and each of the periodic motions is, in general, anharmonic. Thus, the Fourier spectrum is discrete and consists of spikes at fundamental frequencies $\omega_p = 2\pi/T_p$ and their harmonics $\omega_{pn} = 2\pi n/T_p$. In the case of dynamical chaos, the motion is not strictly periodic and there is a continuous component in the Fourier spectrum, although discrete components are present as well.

6.2.4 *Poincaré maps*

One more method to analyze regular and chaotic behavior of trajectories is the method of Poincaré maps. This method was initially proposed for conservative systems with two degrees of freedom, thus $N_F = 3/2$. Motion of such systems is confined to the three-dimensional hyperspace of constant energy within the four-dimensional phase space. Visualizing such trajectories is difficult, and the method of Poincaré maps simplifies the presentation by looking only at, say, variables x and $\dot{x}$ for, say, $y = 0$, whereas $\dot{y}$ is fixed by the energy conservation law and thus is not independent. Although the information in Poincaré maps is reduced, they give a very good idea of the behavior of the system. In the case of regular motion, there are usually two incommensurate periods. As a result, after a sufficiently large computation time, successive points $(x, \dot{x})$ fill curves in the plane that is a signature of a regular motion. On the contrary, in the case of chaotic motion, points $(x, \dot{x})$ look randomly placed. One can plot different trajectories, including regular and chaotic ones, on the same plot. The method of Poincaré maps is more sensitive to the weak chaos (predominantly regular motion with a random chaotic component) than the Fourier analysis.

A simplified version of the method of Poincaré maps is plotting points $(x, \dot{x})$ after equal time intervals. This version of the method does not require us to solve equations of motion for $y(t) = 0$ and it is especially convenient for systems with an explicit periodic time dependence (periodically driven systems). In this case, one just plots $(x, \dot{x})$ with the period of the external perturbation. For a system with one degree of freedom ($N_F = 3/2$), regular trajectories have a period incommensurate with that of the external perturbation, thus the points $(x, \dot{x})$ fill closed orbits. For more than one degree of freedom, the chosen pair of variables $(x, \dot{x})$, or any other pair of variables, is multiperiodic in the case of regular motion, so that the Poincaré maps become more and more complicated. As the number of degrees of freedom increases, the Poincaré map resembles more and more that of a chaotic motion, even if the motion is regular. This is the *molecular chaos* mentioned above, as opposed to the dynamical chaos in systems with few degrees of freedom.

6.2.5 *Example 1: Separatrix chaos in a driven double-well system*

An important example of dynamical chaos in conservative systems is the separatrix chaos in a driven one-dimensional mechanical system with mass m and a double-well potential,

$$U(x,t) = \frac{kx_0^2}{8}\left[1 - \left(\frac{x}{x_0}\right)^2\right]^2 - F(t)x, \quad F(t) = F_0 \sin(\omega t) \tag{6.2.3}$$

(see Fig. 6.2.1) described by the equation of motion

$$m\ddot{x} = -\frac{\partial U(x,t)}{\partial x}. \tag{6.2.4}$$

The coefficient in Eq. (6.2.3) is chosen so that the frequency of free harmonic oscillations near the potential minima at $x = \pm x_0$ is $\omega_0 = \sqrt{k/m}$. Linearization of Eq. (6.2.4) in the deviation δx from the trajectory $x(t)$ yields the linear equation

$$m\delta\ddot{x} = -\frac{\partial^2 U(x,t)}{\partial x^2}\delta x, \tag{6.2.5}$$

which is a form of Eq. (6.2.2). Explicitly, in our case, one has

$$\delta\ddot{x} + \omega_0^2\left[3\left(\frac{x}{x_0}\right)^2 - 1\right]\delta x = 0, \quad \omega_0^2 = \frac{k}{m}. \tag{6.2.6}$$

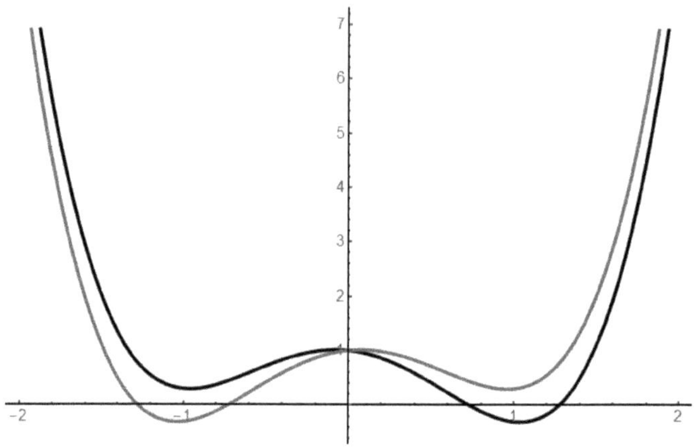

Figure 6.2.1. The double-well potential biased by the external force at different moments of time, $k_0 = 8$, $x_0 = 1$.

The concrete form of $U(x,t)$ can differ from that of Eq. (6.2.3), but an important thing is the saddle that in our example is located at $x = x_{\text{sad}} = 0$ and corresponds to the energy

$$E = E_{\text{sad}} = kx_0^2/8 \tag{6.2.7}$$

for $F = 0$. Trajectories with $E < E_{\text{sad}}$ are bounded to the left or right potential wells, whereas trajectories with $E > E_{\text{sad}}$ go over both wells. The trajectory with $E = E_{\text{sad}}$ separates both kinds of trajectories and it is called separatrix. In the presence of a time-dependent force $F(t)$, the saddle and the separatrix change with time. The instantaneous position of the saddle is defined by

$$0 = \frac{\partial U(x,t)}{\partial x} = -\frac{kx_0^2}{8} 2 \left[1 - \left(\frac{x}{x_0}\right)^2\right] \frac{2x}{x_0^2} - F(t). \tag{6.2.8}$$

This cubic equation can be solved perturbatively for small force by setting $x \to 0$ in the square brackets. This yields

$$x_{\text{sad}}(t) \cong -\frac{2F(t)}{k}, \quad E_{\text{sad}}(t) \cong \frac{kx_0^2}{8} + \frac{F^2(t)}{k}. \tag{6.2.9}$$

This means that particles with the energy close to the saddle energy can go over the barrier between the two wells or not, depending on the instantaneous value of the force $F(t)$. Of the two close trajectories in the vicinity of the separatrix, one can go over the barrier and the other not, which implies divergence of trajectories and dynamical chaos.

Indeed, numerical calculations show that for small force amplitude F_0, the trajectories in the vicinity of the separatrix become chaotic, while trajectories far from the separatrix remain regular. This is in accordance with the KAM theory, stating that for a small non-integrable perturbation, most of the trajectories remain regular and only a very small subset of trajectories become chaotic. The separatrix is exactly the chaotic trajectory referred to in the KAM theorem. With increasing F_0, the region of chaos around the separatrix becomes broader.

The results of numerical solution with $m = 1$, $k = 8$, $x_0 = 1$, and $\omega = 1$ are presented in Fig. 6.2.2 for different amplitudes of the force F_0 and different initial conditions $(x(0), v(0))$. First, Poincaré maps taken with the period of the external force $2\pi/\omega$ show the chaotization of trajectories with increasing F_0, starting from the trajectories near the separatrix. When the separate regular trajectories merge into chaos, they become the same chaotic trajectory, no matter what the initial conditions were. For $F_0 = 0.7$ (the last panel), all trajectories shown are chaotic. Since the Poincaré maps

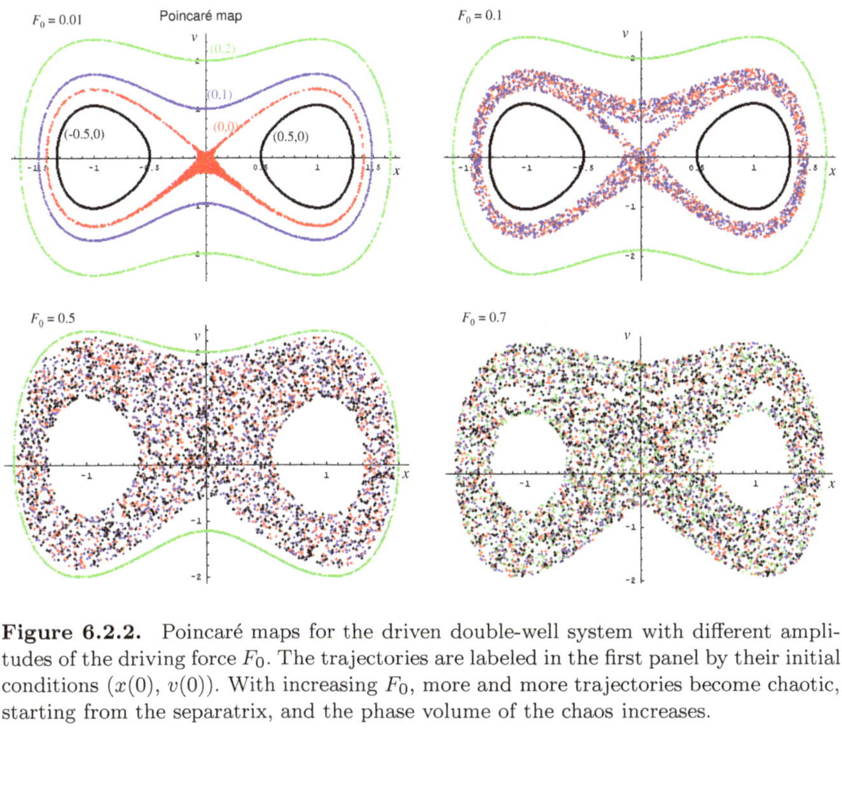

Figure 6.2.2. Poincaré maps for the driven double-well system with different amplitudes of the driving force F_0. The trajectories are labeled in the first panel by their initial conditions $(x(0), v(0))$. With increasing F_0, more and more trajectories become chaotic, starting from the separatrix, and the phase volume of the chaos increases.

contain a lot of points, the chaotic region is well defined by its boundaries. The regions with no points in the last panel correspond to the regions of regular motion. With increasing F_0 up to 2, the regular regions near the two potential minima disappear in chaos.

Let us now investigate the divergence of trajectories. We at first solve Eq. (6.2.4) with some initial conditions and then Eq. (6.2.5) with the initial condition $\delta x(0) = \delta v(0) = \delta$ with $\delta = 10^{-50}$. The distance between the trajectories (the divergence) is defined as

$$s(t) \equiv \sqrt{[\delta x(t)]^2 + [\delta v(t)]^2}. \tag{6.2.10}$$

Other initial conditions for δx and δv and other definitions of s would serve well too. In Fig. 6.2.3, one can see that near the separatrix (left panel, red) there is a strong divergence of trajectories (i.e., the dynamical chaos) even for very small F_0. For trajectories away from the separatrix (right panel, blue) chaos is a threshold phenomenon. In the particular case shown, it occurs for $F_0 \gtrsim 0.05$. Note that the exponential law of Eq. (6.2.1) is obeyed only on average, as argued above. The large deviations $|\delta x| \gg 1$ seen in Fig. 6.2.3 are, of course, unphysical, as the linearization of equations requires small $\delta x(t)$. In reality, $|\delta x|$ defined as the distance between two trajectories will saturate at $|\delta x| \sim 1$. Still, this form of presentation gives a good idea of the chaos, and the formal problem with large δx can be solved by taking δ appropriately small.

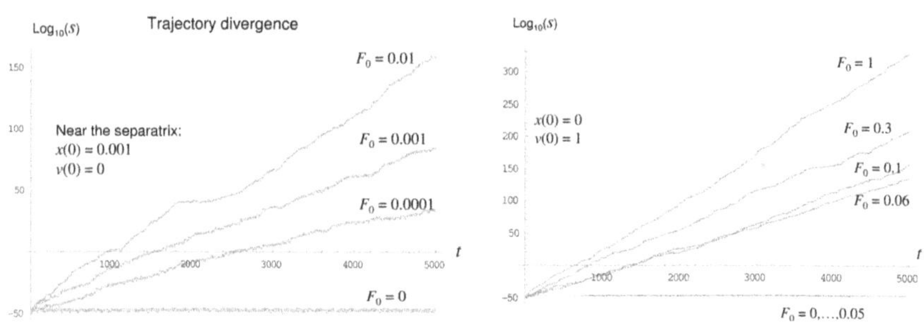

Figure 6.2.3. Divergence of trajectories: *Left panel*: Near the separatrix, trajectories become chaotic already for very small F_0. *Right panel*: Away from the separatrix, trajectories become chaotic only for sufficiently strong perturbation, according to the KAM theorem. The colors match those in the Poincaré maps above.

6.2.6 *Example 2: Chaos in a strongly driven one-well system*

Another example of the dynamical chaos in driven systems with one degree of freedom is provided by the model with the quartic potential

$$U(x,t) = \frac{kx_0^2}{4}\left(\frac{x}{x_0}\right)^4 - F(t)x, \quad F(t) = F_0 \sin(\omega t). \tag{6.2.11}$$

There is no separatrix in this model, thus there is no reason for any particular trajectory to become chaotic already for a very small F_0. This model is a pure example of the KAM theorem with the dynamical chaos emerging above a threshold on F_0, so that below the threshold, all trajectories remain regular. The Poincaré maps calculated with $m = 1$, $k = 4$, and $\omega = 1$ are shown in Fig. 6.2.4. The threshold of chaos in this case is comparable to that for the trajectories away from the separatrix in the double-well model. Note that one can see different chaotic trajectories separated from each other in the Poincaré maps.

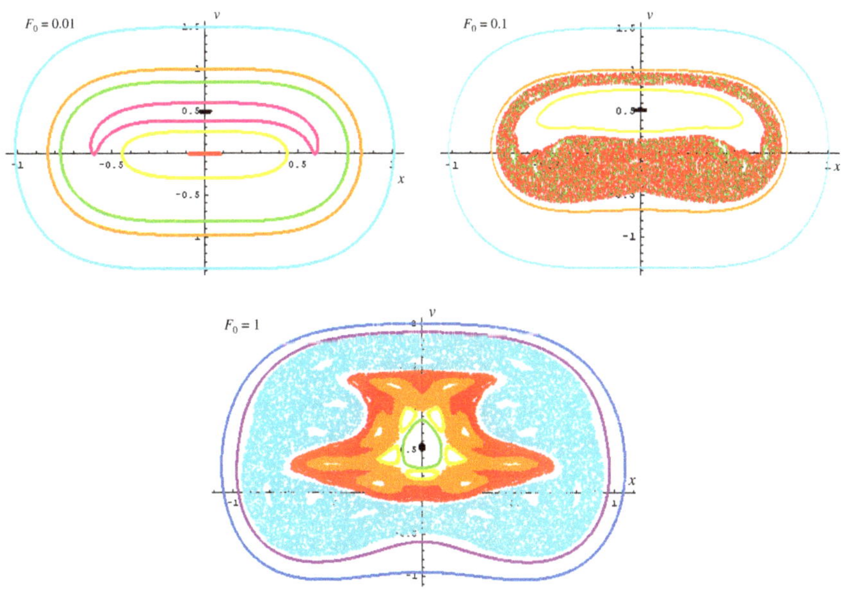

Figure 6.2.4. Poincaré maps for the driven system with a quartic potential with different amplitudes of the driving force F_0. In this model, dynamical chaos emerges above a threshold on F_0. Below the threshold, all trajectories remain regular. With increasing F_0 above the threshold, more and more trajectories become chaotic.

One can ask why the Poincaré maps, especially the last panel in Fig. 6.2.4, are non-symmetric with respect to the transformation $v \to -v$, although the problem is symmetric. The answer is that the Poincaré maps are only a tool to visualize chaos, but they do not provide a full description of trajectories since only a small subset of points is used. The Poincaré points in Fig. 6.2.4 are taken with the period of $F(t)$ at $t = 0, T, 2T$, etc. At these times, $F(t) = 0$, but $\dot{F}(t) > 0$ that breaks the symmetry of the problem with respect to velocities. Choosing another periodic set of Poincaré points that is shifted by a constant amount results in a different Poincaré map that is not symmetric with respect to both x and v in general. Thus, the exact form of the Poincaré map is not very significant.

6.3 Dissipative systems

In the presence of dissipation, a mechanical system relaxes down to one of its local energy minima. Trajectories in the phase space asymptotically approach the points ($x = x_{\min}$, $v = 0$) that are called *point attractors*. Point attractors have dimension zero. In non-physical systems such as chemical and biological systems, there may also be point attractors, which do not correspond to any energy minimum. A more complicated system is a generator that asymptotically approaches a stationary time-dependent state being a cycle in the phase space. This state is called a limiting cycle or a *cycle attractor*. A simpler example of a cycle attractor is the stationary motion of a dissipative system (such as a damped oscillator) under the influence of a periodic force. In the simplest case, the system follows the force, so that the period of the cycle attractor coincides with that of the force. This is called the *simple cycle attractor*. In the case of one degree of freedom, the cycle attractor is a line and its dimension is one. The Poincaré map corresponding to a simple cycle attractor, taken with the period of the external force, consists of a single point.

Generally, dissipation tends to make chaotic motion regular. For instance, in the case of a driven double-well system, trajectory beginning in the chaotic region near the separatrix will leave this region with time and approach a simple cycle attractor, which is a regular motion. Chaos in dissipative systems is a threshold phenomenon on the amplitude of the driving force F_0, friction constant γ, or other parameters. The onset of chaos in dissipative systems manifests itself in the system's response going out of sync with the applied periodic force. This happens via a sequence of *bifurcations* that change the system's behavior discontinuously. A typical first bifurcation is a period doubling after which the system does not return completely to its initial state after one period of the force, the mismatch being small close to the bifurcation threshold. However, after two periods of the external force, the system returns to its initial state. Further increase in the control parameter leads to the increase in the mismatch after one external period, followed by another bifurcation. This can be another period doubling or, in general, period multiplying. This makes the motion of the system more complicated, the attractor becoming a *multiple cycle attractor*. The number of external periods in the multiple cycle attractor is the number of points on the Poincaré map. The dimension of this kind of attractors is still the same as that of the corresponding simple cycle attractor. The Fourier spectrum is discrete and consists of the fundamental frequency corresponding to the period of motion of the system and its harmonics.

After a few bifurcations, the motion of the system can become *incommensurate* with the external force. This increases the dimension of the attractor. For instance, in the case of one degree of freedom, the simple and multiple cycle attractors are closed contours and they have dimension one. An incommensurate cycle attractor is a torus and it has dimension two. The Poincaré map corresponding to this kind of attractor consists of many points that fill a line after a sufficient amount of data is generated.

With further increase in the control parameter, chaos breaks in as the final bifurcation. The chaotic state of a driven dissipative system is also a kind of attractor that is reached after some transient time. The structure of chaotic attractors is complicated. Looking at the Poincaré map, it is difficult to judge about the exact dimension of the attractor. It seems to be between two and three for one degree of freedom since the points on the Poincaré map fill two-dimensional regions but not in a dense way. Magnification of regions in the Poincaré map (that costs a lot of computation time to generate enough data) shows a self-similar, i.e., *fractal*, structure. The same patterns infinitely repeat themselves upon magnification. One can extend the definition of dimensionality upon fractals and it turns out to be a non-integer indeed. As the motion on a chaotic attractor is non-periodic, the Fourier spectrum is continuous. Still there are spikes as discrete frequencies, the same as for the regular motion.

Lyapunov exponents for the motion on attractors can be obtained by linearizing the equation of motion in the deviation δx as above. Of course, this should be done by skipping the transient time needed to approach the attractor. For regular attractors, the Lyapunov exponents are negative, i.e., the initially created trajectory deviation disappears with time because of dissipation. For chaotic attractors, on the contrary, the Lyapunov exponents are positive, so that initially close trajectories diverge without leaving the attractor. It should be stressed that Eq. (6.2.1) that defines Lyapunov exponents is obeyed only on average since chaotic solutions cannot have an exact simple exponential form.

6.3.1 *Example: Driven double-well system with dissipation*

As an example, consider the driven double-well system described by Eq. (6.2.3) with added damping γ, so that the equation of motion has the form

$$m\ddot{x} + \gamma m\dot{x} = -\frac{\partial U(x,t)}{\partial x}. \tag{6.3.1}$$

First, fix $m = 1$, $k = 8$, $\omega = 1$, and $\gamma = 0.01$ and study the attractor for different values of the force amplitude F_0. For $F_0 < F_{0c} = \left(2/\sqrt{3}\right)^3 \simeq 1.54$, the potential $U(x,t)$ has two minima at all times. Thus, there should be two simple cycle attractors around each of the minima. The system ends up in one of them, depending on the initial condition. This behavior is seen in Fig. 6.3.1.

For larger force amplitudes such as $F_0 = 1.5$ and 2 in Fig. 6.3.2, attractors extend over both wells. Note that only one of the two possible attractors is shown for $F_0 = 1.5$. Another attractor can be obtained by the reflection $x \to -x$. For $F_0 = 2$, on the contrary, the attractor is symmetric and unique. More importantly, it is a triple cycle attractor. One can see the peak at the frequency $\omega/3$ in the Fourier spectrum. The period-tripling bifurcation occurs between $F_0 = 1.9$ and $F_0 = 1.95$. There is no period doubling for the chosen values of parameters in this model. Further, around $F_0 = 2.05$, the inverse bifurcation occurs and the system returns to a simple cycle attractor. At $F_0 = 2.05$, both asymmetric simple cycle attractors and the symmetric triple-cycle attractor are realized for different initial conditions.

At $F_0 = 3.0$ (not shown), there are two asymmetric simple cycle attractors similar to $F_0 = 2$. At $F_0 = 3.3$, one obtains either one of the simple asymmetric attractors as for $F_0 = 3.0$ or a five-fold cycle attractor, depending on initial conditions. The five-fold cycle attractor is shown in Fig. 6.3.3. At $F_0 = 3.5$, the attractor is chaotic. The Poincaré maps in both cases are shown in Fig. 6.3.4. With further increase in F_0, the chaos disappears again. In particular, for $F_0 = 4.5$, there are two asymmetric

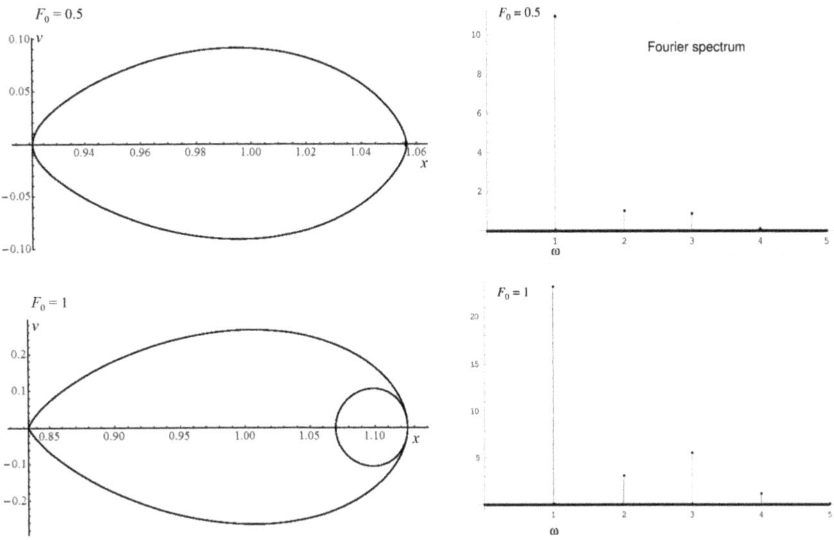

Figure 6.3.1. Attractors and Fourier spectra for the driven dissipative double-well model with small sinusoidal force, $F_0 = 0.5$ and 1. In both cases, the attractor is located in one well. Intensity of harmonics of the fundamental frequency ω increase with F_0.

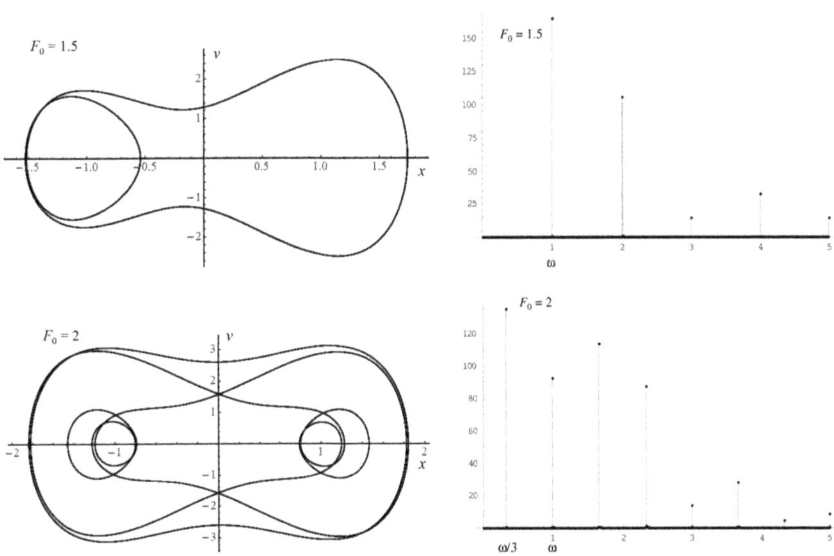

Figure 6.3.2. Attractors and Fourier spectra for the driven dissipative double-well model with larger sinusoidal force, $F_0 = 1.5$ and 2. In both cases, the attractor extends over both wells.

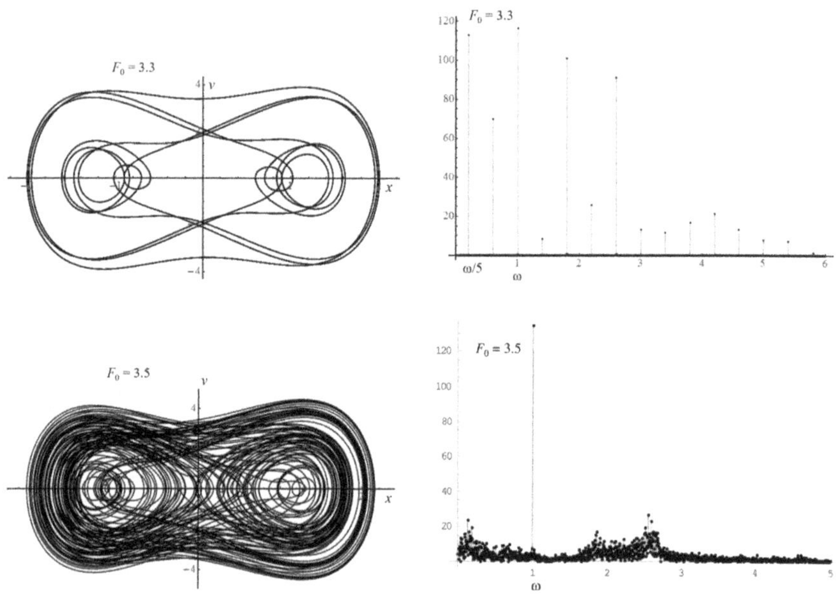

Figure 6.3.3. Attractors and Fourier spectra for the driven dissipative double-well model with $F_0 = 3.3$ (five-fold cycle attractor) and 3.5 (chaotic attractor). The Fourier spectrum for chaotic attractor is continuous with a spike at the fundamental frequency ω.

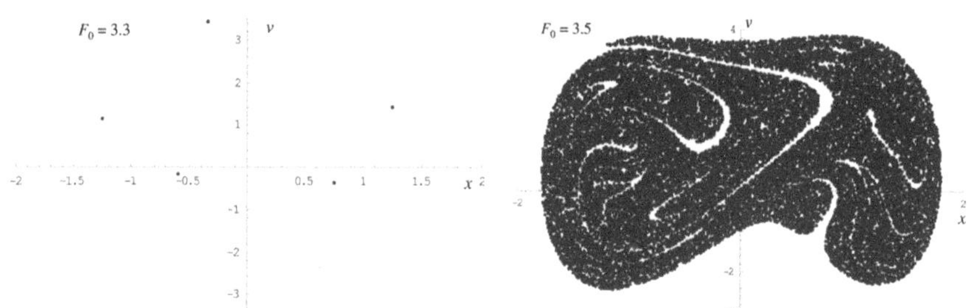

Figure 6.3.4. Poincaré maps for $F_0 = 3.3$ (five-fold cycle attractor) and $F_0 = 3.5$ (chaotic attractor).

eight-fold cycle attractors, for $F_0 = 5$, there are two asymmetric simple attractors, and for $F_0 = 10$, there are two asymmetric double cycle attractors. In all cases, attractors with an odd number of cycles are symmetric and attractors with even number of cycles occur in asymmetric pairs.

Let us now consider the case of slow external force, $\omega = 0.1$. The force amplitude will be kept constant, $F_0 = 1.6 > F_{0c}$, so that the particle slides out of a disappearing metastable energy minimum into the stable one, approaches the stable energy minimum and tends to stay there until this minimum becomes metastable and eventually disappears, etc. The control parameter will be the damping γ. For large γ, the attractor is a simple cycle, whereas for small γ, it is chaotic. The attractor and its Fourier spectrum for $\gamma = 1$ are shown in Fig. 6.3.5. Decrease of γ leads to a period-doubling bifurcation, see Fig. 6.3.6 for $\gamma = 0.3$ and 0.25. For $\gamma = 0.24$ (see Fig. 6.3.7), there is already a multicycle attractor with about 14 cycles, as one can see on the Poincaré map. For $\gamma = 0.23$ (see Fig. 6.3.8), there is an incommensurate cycle attractor of dimension 2, since the points on the Poincaré map fill a line. Further decrease in γ lead to the increase in the length of the line in the Poincaré map, i.e., the size of the incommensurate attractor, see Fig. 6.3.9. For even smaller γ, the form of the Poincaré map changes and the motion becomes apparently chaotic, see Fig. 6.3.10 for

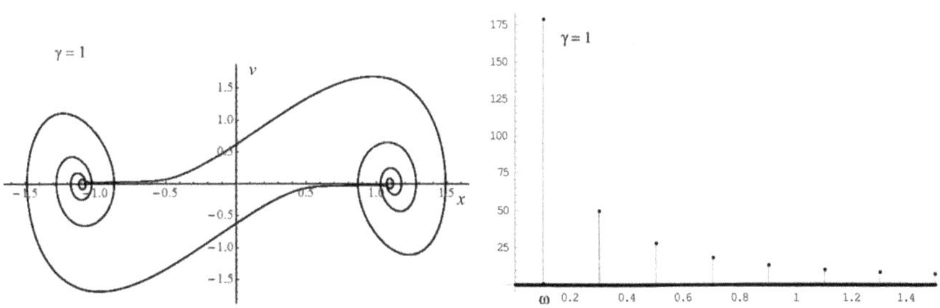

Figure 6.3.5. Simple attractor and its Fourier spectrum for $\omega = 0.1$, $F_0 = 1.6$, and $\gamma = 1$.

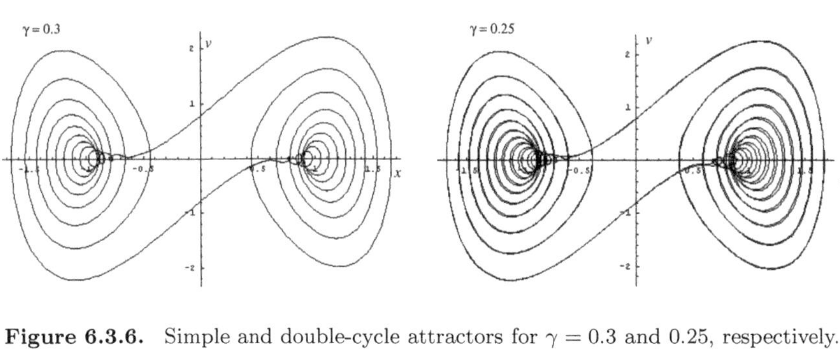

Figure 6.3.6. Simple and double-cycle attractors for $\gamma = 0.3$ and 0.25, respectively.

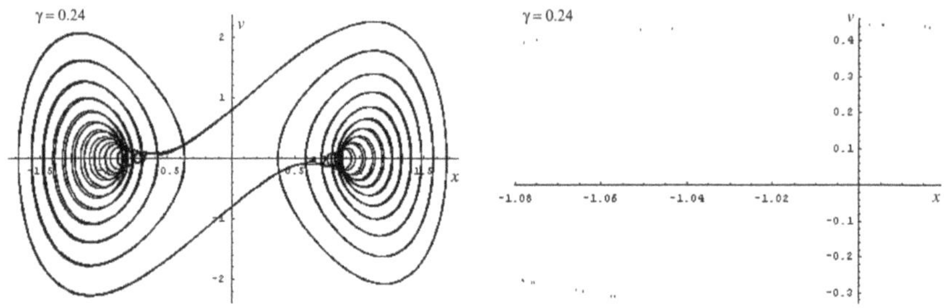

Figure 6.3.7. Multicycle attractor with about 14 cycles for $\gamma = 0.24$ and its Poincaré map.

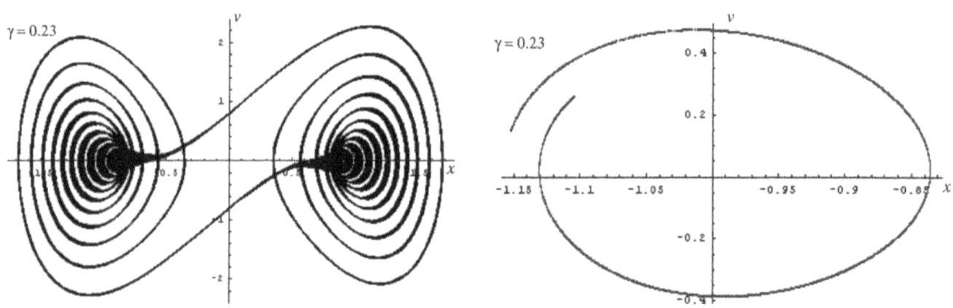

Figure 6.3.8. Incommensurate cycle attractor for $\gamma = 0.23$.

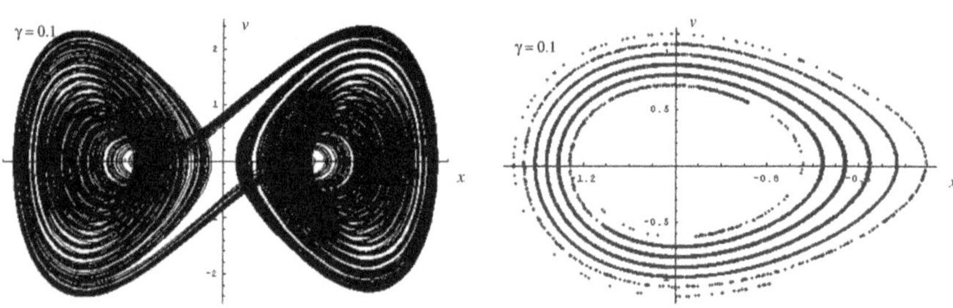

Figure 6.3.9. Incommensurate cycle attractor for $\gamma = 0.1$.

$\gamma = 0.03$. With further decrease of γ, the chaotic attractor grows and its Poincaré map approaches that of the conservative model considered above, see Fig. 6.3.11 for $\gamma = 0.003$.

Finally, the time dependence of the trajectory deviations $s(t)$ for $\omega = 0.1$, $F_0 = 1.6$, and different values of γ are shown in Fig. 6.3.12. One can see

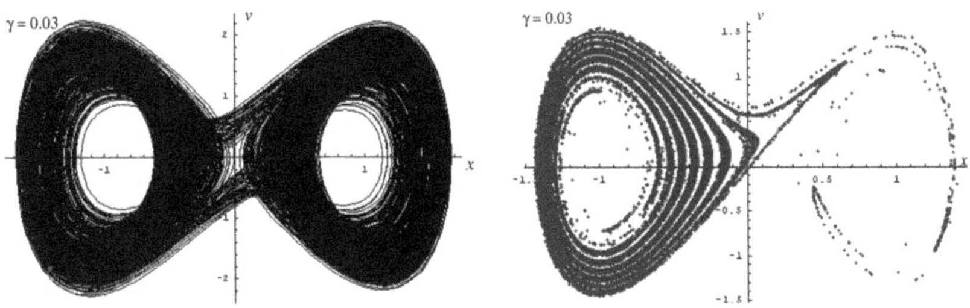

Figure 6.3.10. Chaotic attractor for $\gamma = 0.03$.

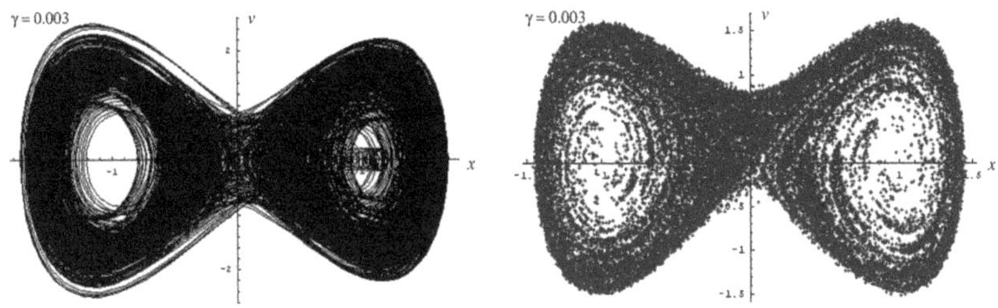

Figure 6.3.11. Attractor with a developed chaos for $\gamma = 0.003$ close to the dynamical chaos in conservative systems.

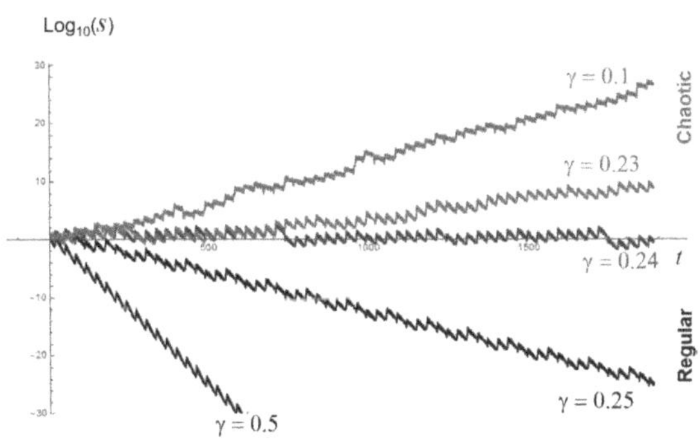

Figure 6.3.12. Trajectory deviations on the attractor for $\omega = 0.1$, $F_0 = 1.6$, and different values of γ. The borderline between the regular and chaotic behavior is $\gamma = 0.24$, where attractor becomes incommensurate with the periodic force.

that the borderline between the regular and chaotic behavior is $\gamma = 0.24$, where the attractor becomes incommensurate with the periodic force.

Summarizing, one can say that there are a lot of different types and scenarios of dynamical chaos, so that the latter is a big zoo.

Chapter 7

Microscopic Origin of Dissipation

One can read in textbooks on classical mechanics that *dissipation* that is usually taken into account by adding "phenomenological" *damping* terms in the equations of motion such as

$$m\ddot{\mathbf{r}}+2m\eta\dot{\mathbf{r}}+\partial U/\partial\mathbf{r}=0 \tag{7.0.1}$$

is itself of a non-mechanical origin because it is caused by the coupling of the system to the environment that is described statistically by temperature and other thermodynamic parameters. The origin of these statements goes back to the time when it was impossible to solve equations of motion for systems with a macroscopic number of degrees of freedom. However, with the growth of computing power, solving equations of motion for large systems has become a matter of fact, and a new area of physics, computer simulations, has emerged. Of course, simulating real macroscopic systems with an Avogadro number $N_A \sim 10^{23}$ particles is still prohibitive. Nevertheless, it turns out that using only $N \sim 10^3-10^5$ particles is sufficient to describe the macroscopic properties of many systems.

Already in the 19th century, Ludwig Boltzmann and other researchers argued about apparent irreversibility and dissipation in conservative many-body systems that obey deterministic equations of motion. Boltzmann himself obtained his famous kinetic equation under the assumption of the "molecular chaos" that introduced irreversibility "by hand". Some physicists disagreed with this, saying that the true equations are reversible and after making the Poincaré cycle, the system will return to the initial state, which is a rigorous mathematical result. Practically, however, for macroscopic systems, the Poincaré cycle is longer than the age of the Universe, so that a gas initially held in one half of the container will never completely return into this half after the separating wall is removed. Instead of completely returning into the initial state, the system will more likely make partial recurrences after recurrence times that are much shorter than the Poincaré cycle and also growing with the system's size.

Still, some of the 20th-century physicists believed that going beyond the conservative deterministic mechanics is needed to describe irreversibility and dissipation. However, computer simulations on conservative mechanical systems show that the recurrence phenomena, still visible for moderate-size systems, quickly disappear as the size increases, so that practically irreversible relaxation is observed. This means that dissipation is in fact a mechanical phenomenon. A recent review of irreversibility in conservative systems by Joel Lebowitz, "From Time-symmetric Microscopic Dynamics to Time-asymmetric Macroscopic Behavior: An Overview", can be recommended for further reading.

In the following, we will consider a particular conservative mechanical model, a "central" oscillator coupled to a bath of N oscillators. This model shows that the central oscillator is damped. Although the eigenfrequencies of this model are *real*, as is always the case for conservative systems, the many-body nature of the problem makes the solution for the central oscillator coincide with that of Eq. (7.0.1) for $N \to \infty$ in the weak-coupling limit. In this limit, one can replace the original problem having $N+1$ real eigenvalues by an effective one-particle problem that can be described by one complex eigenvalue.

On the other hand, for $N \to \infty$, an exact analytical solution with real eigenvalues for an arbitrarily strong coupling is possible, see Sec. 7.4.

7.1 The model

The Lagrange function of the model has the form

$$\mathcal{L} = \frac{m_0}{2}(\dot{X}^2 - \omega_0^2 X^2) + \sum_{i=1}^{N} \frac{m_i}{2}(\dot{x}_i^2 - \omega_i^2 x_i^2) - \frac{X}{\sqrt{N}} \sum_{i=1}^{N} \sqrt{m_0 m_i} C_i x_i. \tag{7.1.1}$$

The last term of this expression is the linear (more precisely, bilinear) coupling of the "central" oscillator X to the "environment" or "bath" of many oscillators x_i. The equations of motion for this model have the form

$$\begin{aligned} m_0(\ddot{X} + \omega_0^2 X) + \frac{1}{\sqrt{N}} \sum_{i=1}^{N} \sqrt{m_0 m_i} C_i x_i &= 0, \\ m_i(\ddot{x}_i + \omega_i^2 x_i) + \frac{1}{\sqrt{N}} \sqrt{m_0 m_i} C_i X &= 0. \end{aligned} \tag{7.1.2}$$

It is convenient to introduce renormalized coordinates Y and y_i as

$$Y \equiv \sqrt{m_0}X, \quad y_i \equiv \sqrt{m_i}x_i. \tag{7.1.3}$$

Then Eq. (7.1.2) simplifies to

$$\begin{aligned} \ddot{Y} + \omega_0^2 Y + \frac{1}{\sqrt{N}} \sum_{i=1}^{N} C_i y_i = 0, \\ \ddot{y}_i + \omega_i^2 y_i + \frac{1}{\sqrt{N}} C_i Y = 0. \end{aligned} \tag{7.1.4}$$

Searching for the solution in the form

$$Y(t) = a\cos(\Omega t - \Phi), \quad y_i(t) = a_i \cos(\Omega t - \varphi_i), \tag{7.1.5}$$

one obtains the system of linear equations

$$\begin{aligned} \left(-\Omega^2 + \omega_0^2\right) a + \frac{1}{\sqrt{N}} \sum_{i=1}^{N} C_i a_i = 0, \\ \left(-\Omega^2 + \omega_i^2\right) a_i + \frac{1}{\sqrt{N}} C_i a = 0, \quad i = 1 \ldots N, \end{aligned} \tag{7.1.6}$$

which defines the eigenfrequencies and normal modes of the system. There are $N+1$ eigenfrequencies Ω_α^2, all of which are positive.

7.2 General matrix solution and numerical results

To find the solution of the problem, one can write Eq. (7.1.4) in the vectorized form

$$\ddot{\mathbf{Y}} + \mathbb{K} \cdot \mathbf{Y} = 0, \tag{7.2.1}$$

where $\mathbf{Y}$ is the vector of oscillator displacements, $\mathbf{Y} = (y_0, y_1, y_2, \ldots, y_N)$ with $y_0 \equiv Y$ and the matrix $\mathbb{K}$ is given by

$$\begin{aligned} K_{ii} &= \omega_i^2, \quad i = 0, 1, \ldots, N, \\ K_{0i} &= K_{i0} = \frac{1}{\sqrt{N}} C_i, \quad i = 1, \ldots, N. \end{aligned} \tag{7.2.2}$$

The eigenvalue problem, Eq. (7.1.6), reads

$$\left(-\Omega^2 \mathbb{I} + \mathbb{K}\right) \cdot \mathbf{a} = \mathbf{0}, \tag{7.2.3}$$

where $\mathbb{I}$ is the unit matrix, $I_{ij} = \delta_{ij}$. There are $N+1$ eigenvalues Ω_α^2 and eigenvectors $\mathbf{a}_\alpha$, $\alpha = 0, 1, \ldots, N$ that satisfy the orthogonality condition

$$\mathbf{a}_\alpha^T \cdot \mathbf{a}_\beta = \delta_{\alpha\beta} \tag{7.2.4}$$

($\mathbf{a}_\beta$ are columns and $\mathbf{a}_\alpha^T$ are rows). One can stack all $\mathbf{a}_\alpha$ into the $(N+1)\times(N+1)$ matrix $\mathbb{A}$ that is orthogonal:

$$\mathbb{A}^T\cdot\mathbb{A}=\mathbb{I},\quad \mathbb{A}^{-1}=\mathbb{A}^T. \tag{7.2.5}$$

From Eq. (7.2.3) follows

$$\mathbf{a}_\alpha^T\cdot\mathbb{K}\cdot\mathbf{a}_\beta=\Omega_\alpha^2\delta_{\alpha\beta},\quad \mathbb{A}^T\cdot\mathbb{K}\cdot\mathbb{A}=\boldsymbol{\Omega}^2\equiv\mathrm{diag}\left(\Omega_\alpha^2\right). \tag{7.2.6}$$

Further, one can introduce the normal-coordinate vector $\boldsymbol{\zeta}$ by

$$\mathbf{Y}=\mathbb{A}\cdot\boldsymbol{\zeta},\quad \mathbf{Y}=(\mathbb{A}\cdot\boldsymbol{\zeta})^T=\boldsymbol{\zeta}^T\cdot\mathbb{A}^T. \tag{7.2.7}$$

These equations can be resolved for $\boldsymbol{\zeta}$ that yields the formula for ζ_α:

$$\boldsymbol{\zeta}=\mathbb{A}^{-1}\cdot\mathbf{Y}=\mathbb{A}^T\cdot\mathbf{Y},\quad \zeta_\alpha=\sum_{i=0}^{N}\left(\mathbf{A}^T\right)_{\alpha i}y_i=\sum_{i=0}^{N}a_{i\alpha}y_i. \tag{7.2.8}$$

Inserting $\mathbf{Y}=\mathbb{A}\cdot\boldsymbol{\zeta}$ into Eq. (7.2.1), left multiplying by $\mathbb{A}^T$, and using Eqs. (7.2.5) and (7.2.6), one obtains the equation of motion for the normal coordinates in the simple form

$$\ddot{\boldsymbol{\zeta}}+\boldsymbol{\Omega}\cdot\boldsymbol{\zeta}=\mathbf{0} \tag{7.2.9}$$

or

$$\ddot{\zeta}_\alpha+\Omega_\alpha^2\zeta_\alpha=0,\quad \alpha=0,1,\ldots,N. \tag{7.2.10}$$

The solution of these equations is

$$\begin{aligned}\zeta_\alpha(t)&=\zeta_\alpha(0)\cos\left(\Omega_\alpha t\right)+\dot{\zeta}_\alpha(0)\frac{\sin\left(\Omega_\alpha t\right)}{\Omega_\alpha}\\&=\sum_{i=0}^{N}a_{i\alpha}\left[y_i(0)\cos\left(\Omega_\alpha t\right)+\dot{y}_i(0)\frac{\sin\left(\Omega_\alpha t\right)}{\Omega_\alpha}\right].\end{aligned} \tag{7.2.11}$$

This result can also be brought into a vectorized form (excercize). Finally, from Eq. (7.2.7), one obtains

$$y_i(t)=\sum_{\alpha=0}^{N}a_{i\alpha}\zeta_\alpha(t)=\sum_{\alpha=0}^{N}\sum_{j=0}^{N}a_{i\alpha}a_{j\alpha}\left[y_j(0)\cos\left(\Omega_\alpha t\right)+\dot{y}_j(0)\frac{\sin\left(\Omega_\alpha t\right)}{\Omega_\alpha}\right]. \tag{7.2.12}$$

In particular, if in the initial state only the central oscillator $Y\equiv y_0$ was in a general state, whereas the bath oscillators were in their ground states, for the central oscillator, one obtains

$$Y(t) = \sum_{\alpha=0}^{N} a_{0\alpha}^2 \left[Y(0) \cos(\Omega_\alpha t) + \dot{Y}(0) \frac{\sin(\Omega_\alpha t)}{\Omega_\alpha} \right]. \tag{7.2.13}$$

The time-dependent energy of the central oscillator is given by

$$E(t) = \frac{m_0}{2} \left[\dot{X}^2(t) + \omega_0^2 X^2(t) \right] = \frac{1}{2} \left[\dot{Y}^2(t) + \omega_0^2 Y^2(t) \right]. \tag{7.2.14}$$

If, for instance, $\dot{Y}(0) = 0$, then one obtains

$$\frac{E(t)}{E(0)} = \left(\sum_{\alpha=0}^{N} a_{0\alpha}^2 \cos(\Omega_\alpha t) \right)^2 + \frac{1}{\omega_0^2} \left(\sum_{\alpha=0}^{N} a_{0\alpha}^2 \Omega_\alpha \sin(\Omega_\alpha t) \right)^2. \tag{7.2.15}$$

Let us now numerically calculate the time dependence of the energy for an equidistant spectrum of bath oscillators

$$\omega_i = \omega_{\min} + \frac{i-1}{N-1} \omega_{\max}, \qquad \omega_{\min} < \omega_0 < \omega_{\max}, \tag{7.2.16}$$

so that $\omega_1 = \omega_{\min}$ and $\omega_N = \omega_{\max}$. Such distribution of frequencies corresponds to the density of states

$$\rho(\omega) = \frac{1}{\omega_{\max} - \omega_{\min}}. \tag{7.2.17}$$

For the coupling, we use

$$C_i = C(\omega_i) = C_0 \left(\frac{\omega_i}{\omega_0} \right)^\xi. \tag{7.2.18}$$

It is not difficult to find eigenvalues and eigenvectors of the $\mathbb{K}$ matrix of Eq. (7.2.2) and to program Eq. (7.2.15) with the help Wolfram Mathematica or other software. The results in the case of a *weak* coupling to the bath are shown in Fig. 7.2.1 for $\omega_{\min} = 0$, $\omega_0 = 1$, $\omega_{\max} = 2$, $C_0 = 0.1$, and $\xi = 2$. One can see that up to a recurrence time that increases with N, the relaxation follows the exponential law of Eq. (7.3.16). The latter is compatible with Eq. (7.0.1).

The central-oscillator part of the normalized eigenvectors, $a_0^2(\Omega) \equiv a_{0\alpha}^2(\Omega_\alpha)$, is shown in Fig. 7.2.2. In the absence of the bath coupling, $C_i = 0$, the frequency of the central oscillator has a sharp value ω_0, so that $a_0^2(\Omega) = 1$ for $\Omega = \omega_0$ and zero otherwise. Coupling to the bath causes spreading and decreasing of the frequency of the central oscillator, as it can be seen in the figure. The values of $a_0^2(\Omega)$ obey the sum rule $\sum_\alpha a_{0\alpha}^2(\Omega_\alpha) = 1$, as the eigenvectors are normalized, see Eq. (7.2.4). These

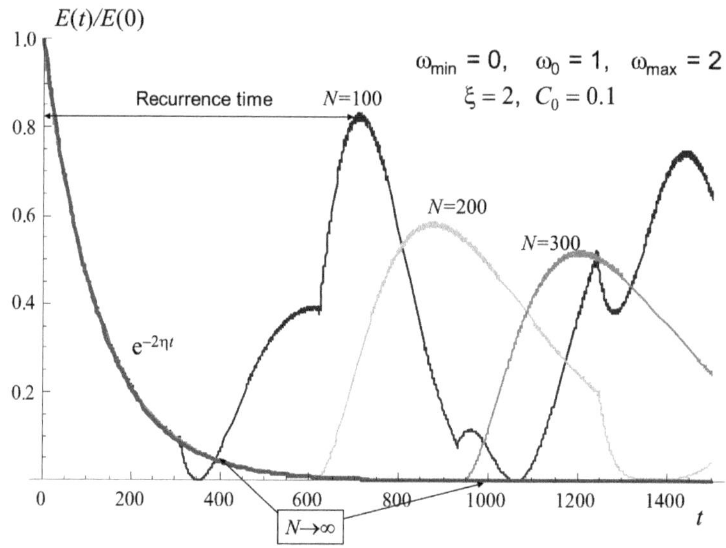

Figure 7.2.1. Time dependence of the energy of the central oscillator in the case of weak coupling to the bath. With increasing N, the numerical solution approaches the exponential curve and the recurrence time is increasing.

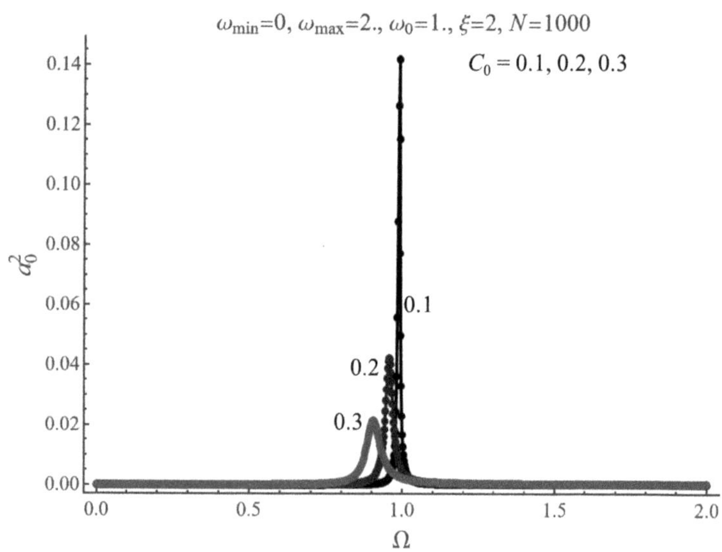

Figure 7.2.2. The central-oscillator part of the eigenvectors, $a_0^2(\Omega) \equiv a_{0\alpha}^2(\Omega_\alpha)$, for $N = 1000$ and different coupling strengths C_0.

results define the response of the central oscillator to a harmonic force, which has a similar form.

7.3 Reduced description in the weak-coupling limit

There is an analytical solution to our problem that has become standard in the literature. This solution concentrates on the central oscillator and integrates out the bath oscillators, which leads to a modification of the motion of the central oscillator. This motion is described by only one eigenvalue Ω that turns out to be complex. The imaginary part of the frequency eigenvalue Ω of the central oscillator is its damping η due to the coupling to the bath. It is not rigorous mathematically and should be considered only as a working recipe to obtain the damping η in the weak-coupling limit. An attractive feature of this method based on the *reduced description* of the system is that the complex value of Ω provides an immediate connection to Eq. (7.0.1).

Expressing in Eq. (7.1.6) a_i through a using the second equation, inserting the result into the first equation, and dividing the result by a, one obtains

$$\Omega^2 - \omega_0^2 - \frac{1}{N}\sum_{i=1}^{N}\frac{C_i^2}{\Omega^2 - \omega_i^2} = 0. \tag{7.3.1}$$

This is still a $(N+1)$th-order algebraic equation for Ω^2 that can be solved numerically as well as the whole eigenvalue problem of Eq. (7.1.6). In the case of weak coupling C_i, one can find the correction to the frequency of the central oscillator perturbatively. Replacing $\Omega \to \omega_0$ (unperturbed) in the denominator, for the latter, one obtains

$$\Omega^2 = \omega_0^2 - \frac{1}{N}\sum_{i=1}^{N}\frac{C_i^2}{\omega_i^2 - \omega_0^2}, \tag{7.3.2}$$

where the second term is the correction to ω_0^2. With the spectral-density function

$$\lambda(\omega) = \frac{1}{N}\sum_{i=1}^{N} C_i^2 \delta(\omega - \omega_i), \tag{7.3.3}$$

one can rewrite Eq. (7.3.2) as

$$\Omega^2 = \omega_0^2 - \int_{\omega_{\min}}^{\omega_{\max}} \frac{d\omega\lambda(\omega)}{\omega^2 - \omega_0^2}. \tag{7.3.4}$$

One can easily see that this expression is still equivalent to Eq. (7.3.2) since one can step back from Eq. (7.3.4) to Eq. (7.3.2). The crucial step is now to consider $\lambda(\omega)$ as a *continuous* function of ω rather than as a collection of δ-functions. Then Eq. (7.3.4) becomes a *true* integral that depends on how the singularity at $\omega = \omega_0$ is treated. In particular, adding a vanishingly small imaginary constant to ω, i.e., $\omega \to \omega + i\varepsilon$ with $\varepsilon \to 0$ (that makes no difference in the discrete case) is equivalent to shifting the integration contour up or down from the real axis into the complex plane that results in avoiding the singularity. Physically, $\varepsilon \to +0$ corresponds to adding an infinitesimally small damping to the bath oscillators. With $\omega \to \omega + i\varepsilon$, Eq. (7.3.4) becomes

$$\Omega^2 = \omega_0^2 - \int_{\omega_{\min}}^{\omega_{\max}} \frac{d\omega\lambda(\omega)}{(\omega + i\varepsilon)^2 - \omega_0^2} = \omega_0^2 - \int_{\omega_{\min}}^{\omega_{\max}} \frac{d\omega\lambda(\omega)}{(\omega + \omega_0)(\omega - \omega_0 + i\varepsilon)}. \tag{7.3.5}$$

Note that we kept $i\delta$ only in the singular factor containing $\omega - \omega_0$, while in all other places, $i\delta$ is irrelevant. Then using the formula

$$\frac{1}{x + i\varepsilon} = \mathcal{P}\frac{1}{x} - i\pi \mathrm{sign}(\varepsilon)\delta(x) \tag{7.3.6}$$

($\mathcal{P}$ stands for the principal value, while the second term stems from making a half-circle around the pole at $\omega = \omega_0$ in the complex plane), one obtains

$$\Omega^2 = \omega_0^2 + 2\omega_0\Sigma_{\omega_0} + 2i\omega_0\eta, \tag{7.3.7}$$

i.e.,

$$\Omega = \sqrt{\omega_0^2 + 2\omega_0\Sigma_{\omega_0} + 2i\omega_0\eta} \cong \tilde{\omega}_0 + i\eta, \quad \tilde{\omega}_0 \equiv \omega_0 + \Sigma_{\omega_0}. \tag{7.3.8}$$

Here, Σ_{ω_0} is a principal-value integral

$$\Sigma_{\omega_0} \equiv \frac{1}{2\omega_0}\mathcal{P}\int_{\omega_{\min}}^{\omega_{\max}} \frac{d\omega\lambda(\omega)}{\omega_0^2 - \omega^2}, \tag{7.3.9}$$

which gives a small real correction to ω_0. More interesting is the imaginary part defined by

$$\eta = \frac{\pi}{4}\frac{\lambda(\omega_0)}{\omega_0^2}\mathrm{sign}(\varepsilon), \quad \mathrm{sign}(\varepsilon) = \begin{cases} -1, & \varepsilon < 0, \\ 0, & \varepsilon = 0, \\ 1 & \varepsilon > 0. \end{cases} \tag{7.3.10}$$

If the dependence on i enters only through the frequency ω_i, i.e., $C_i = C(\omega_i)$, one can express $\lambda(\omega)$ through the density of states of the bath oscillators

$$\rho(\omega) = \frac{1}{N}\sum_{i=1}^{N} \delta\left(\omega - \omega_i\right), \tag{7.3.11}$$

which is normalized by $\int_{\omega_{\min}}^{\omega_{\max}} \rho(\omega) d\omega = 1$. This yields

$$\lambda(\omega) = C^2(\omega)\rho(\omega) \tag{7.3.12}$$

and

$$\eta = \frac{\pi}{4} C^2(\omega_0) \frac{\rho(\omega_0)}{\omega_0^2} \operatorname{sign}(\varepsilon). \tag{7.3.13}$$

The results obtained suggest that the central oscillator obeys the effective damped equation of motion

$$\ddot{Y} + 2\eta\dot{Y} + \tilde{\omega}_0^2 Y = 0, \tag{7.3.14}$$

which is a form of Eq. (7.0.1). However, the correction to the frequency Σ_{ω_0} is unrelated to the damping η. One can see that physically acceptable positive damping η requires $\varepsilon > 0$ in Eq. (7.3.10). This means that one has to ascribe an infinitesimal positive damping to the bath oscillators to obtain a finite positive damping of the central oscillator, as mentioned above. The solution of Eq. (7.3.14) in terms of the initial conditions at $t = 0$ has the form

$$Y(t) = \left[Y(0)\cos\left(\tilde{\omega}_0 t\right) + \dot{Y}(0)\frac{\sin\left(\tilde{\omega}_0 t\right)}{\tilde{\omega}_0}\right] e^{-\eta t}, \tag{7.3.15}$$

which is well approximating Eq. (7.2.13) for weak coupling. The energy of the central oscillator is given by

$$E(t) = \frac{1}{2}\left[\dot{Y}^2(t) + \omega_0^2(t) Y^2(t)\right] = E(0) e^{-2\eta t}, \tag{7.3.16}$$

where $E(0) = (1/2)\left[\dot{Y}^2(0) + \omega_0^2 Y^2(0)\right]$ and we haven't differentiated the slow function $e^{-\eta t}$.

The recipe-like nature of the approach described in this section is in the necessity of adding an infinitesimally small damping ε of bath oscillators to obtain the damping of the central oscillator η. On the other hand, neither numerical solution in Sec. 7.2 nor the rigorous analytical solution in Sec. 7.4 make use of this assumption. One can, for instance, add a small damping of the bath oscillators in the numerical solution and the result will be the same as without it. Adding the infinitesimal damping ε is a formal trick that, however, creates a room for deep physical speculations on the nature of dissipation, such as the idea that dissipation cannot emerge in purely conservative systems and it requires introducing irreversibility "by hand".

The method of this section cannot be used if the coupling to the bath is strong and should be considered non-perturbatively. The first idea to generalize the method for an arbitrary coupling is to take the continuous limit of Eq. (7.3.1) and consider it as a transcendental equation for the complex Ω:

$$\Omega^2 - \omega_0^2 - \int_{\omega_{\min}}^{\omega_{\max}} \frac{\lambda(\omega)}{\Omega^2 - \omega^2} d\omega = 0. \tag{7.3.17}$$

Including the imaginary part of Ω in the denominator leads to an inconsistency, however, since $\mathrm{Im}\Omega$ is *subtracted* from the infinitesimal ε that changes the location of the pole with respect to the real axis. Natural choice $\mathrm{Im}\Omega > 0$ in the denominator has the same effect as $\varepsilon < 0$ in the treatment above and renders the imaginary part of the integral *negative* that contradicts the initial assumption $\mathrm{Im}\Omega > 0$. On the other hand, the choice $\mathrm{Im}\Omega < 0$ in the denominator yields $\mathrm{Im}\Omega > 0$ after the integral is calculated. One has to conclude that Eq. (7.3.17) is naive and breaks down. In the following section, a rigorous analytical solution to the problem will be considered that does not require adding the infinitesimally small damping to the bath oscillators and work for strong coupling as well.

7.4 Exact analytical solution

The method applied in this section uses the idea of Fano[1] to find the eigenvectors of the problem in the continuous limit analytically using the fact that the eigenvalues Ω build a continuum, so that any value of Ω is good and one does not need to diagonalize a matrix to find them. One considers the model with parameters, depending on the frequency ω_i only and takes the continuous limit $\omega_i \Rightarrow \omega$, replacing summation by integration in equations as follows:

$$\frac{1}{N}\sum_{i=1}^{N} \ldots \Rightarrow \int_{\omega_{\min}}^{\omega_{\max}} \ldots \rho(\omega) d\omega, \tag{7.4.1}$$

[1] U. Fano, Phys. Rev. **124**, 1866 (1961).

where $\rho(\omega)$ is the density of states of bath oscillators, Eq. (7.3.11), considered as a continuous function of ω. The continuous limit of Eq. (7.1.4) is

$$\ddot{Y} + \omega_0^2 Y + \sqrt{N} \int_{\omega_{\min}}^{\omega_{\max}} C(\omega) y(\omega) \rho(\omega) d\omega = 0,$$
$$\ddot{y}(\omega) + \omega^2 y(\omega) + \frac{1}{\sqrt{N}} C(\omega) Y = 0. \tag{7.4.2}$$

The continuous limit of the eigenvalue problem, Eq. (7.1.6), becomes

$$\left(-\Omega^2 + \omega_0^2\right) a_\Omega + \sqrt{N} \int_{\omega_{\min}}^{\omega_{\max}} C(\omega) b_\Omega(\omega) \rho(\omega) d\omega = 0, \tag{7.4.3}$$

$$\left(-\Omega^2 + \omega^2\right) b_\Omega(\omega) + \frac{1}{\sqrt{N}} C(\omega) a_\Omega = 0, \tag{7.4.4}$$

where we use b for the bath oscillators for the sake of transparency. This problem differs from that of Eq. (7.1.6) since there is a continuum of eigenvalues in the interval $\omega_{\min} \leq \Omega \leq \omega_{\max}$ and there is no need to search for them. The only problem is to find corresponding eigenvectors

$$A_\Omega(\omega) \equiv \{a_\Omega, b_\Omega(\omega)\}, \tag{7.4.5}$$

which will be done analytically in the following. In addition, there may be discrete eigenvalues Ω_n beyond the interval $(\omega_{\min}, \omega_{\max})$. The orthonormality condition is

$$a_\Omega a_{\Omega'} + N \int_{\omega_{\min}}^{\omega_{\max}} b_\Omega(\omega) b_{\Omega'}(\omega) \rho(\omega) d\omega = \delta\left(\Omega - \Omega'\right) \tag{7.4.6}$$

for continuous eigenstates and

$$a_{\Omega_n} a_{\Omega_{n'}} + N \int_{\omega_{\min}}^{\omega_{\max}} b_{\Omega_n}(\omega) b_{\Omega_{n'}}(\omega) \rho(\omega) d\omega = \delta_{nn'} \tag{7.4.7}$$

for discrete eigenstates. Discrete and continuous eigenstates are, of course, orthogonal to each other.

Let us start with finding continuous eigenstates. The method is the following: One finds all $b_\Omega(\omega)$ from Eq. (7.4.4) in terms of a_Ω and then normalizes the eigenstates using Eq. (7.4.6). The first step yields

$$b_\Omega(\omega) = \left[\mathcal{P} \frac{C(\omega)}{\Omega^2 - \omega^2} + Z_\Omega \delta\left(\omega - \Omega\right)\right] \frac{a_\Omega}{\sqrt{N}}, \tag{7.4.8}$$

where $\mathcal{P}$ means principal value and $Z(\Omega)$ is a function to be determined. Note that this formula differs from Eq. (7.3.6). We will not shift the integration path away from the real axis into the complex plane, and all the quantities below will be real. The validity of the above solution can be checked by its substitution into Eq. (7.4.4), upon which the term with the δ-function vanishes. Substituting this expression into Eq. (7.4.3), one arrives at the eigenvalue problem in the form

$$\left[-\Omega^2+\omega_0^2+\omega_0\Sigma_\Omega+Z_\Omega C(\Omega)\rho(\Omega)\right]a_\Omega=0, \tag{7.4.9}$$

where

$$\omega_0\Sigma_\Omega\equiv\mathcal{P}\int_{\omega_{\min}}^{\omega_{\max}}\frac{C^2(\omega)}{\Omega^2-\omega^2}\rho(\omega)d\omega. \tag{7.4.10}$$

Now, the solution for Z_Ω that nullifies the expression in the square brackets becomes

$$Z_\Omega=\frac{\Omega^2-\omega_0^2-\omega_0\Sigma_\Omega}{\rho(\Omega)C(\Omega)}. \tag{7.4.11}$$

The last step of the calculation is to find a_Ω from Eq. (7.4.6) that takes the form

$$\begin{aligned}\delta\left(\Omega-\Omega'\right)=a_\Omega a_{\Omega'}&\left\{1+\int_{\omega_{\min}}^{\omega_{\max}}\left[-\frac{C(\omega)}{\omega^2-\Omega^2}+Z_\Omega\delta\left(\omega-\Omega\right)\right]\right.\\&\left.\times\left[-\frac{C(\omega)}{\omega^2-\Omega'^2}+Z_{\Omega'}\delta\left(\omega-\Omega'\right)\right]\rho(\omega)d\omega\right\}.\end{aligned} \tag{7.4.12}$$

In the following, we will be using mathematical identities

$$\delta(x-a)\delta(x-b)=\delta(a-b)\delta(x-a) \tag{7.4.13}$$

and

$$\frac{1}{(x-a)(x-b)}=\frac{1}{a-b}\mathcal{P}\left(\frac{1}{x-a}-\frac{1}{x-b}\right)+\pi^2\delta(a-b)\delta(x-a). \tag{7.4.14}$$

The former is obvious and the latter is derived in the Appendix. Using these formulas, one can rearrange Eq. (7.4.6) to

$$\begin{aligned}&\delta\left(\Omega-\Omega'\right)\\&=a_\Omega a_{\Omega'}\left\{1+\frac{1}{\Omega^2-\Omega'^2}\mathcal{P}\int_{\omega_{\min}}^{\omega_{\max}}C^2(\omega)\left(\frac{1}{\omega^2-\Omega^2}-\frac{1}{\omega^2-\Omega'^2}\right)\rho(\omega)d\omega\right.\\&\quad+C^2(\Omega)\pi^2\frac{\rho(\Omega)}{2\Omega}\delta\left(\Omega^2-\Omega'^2\right)+\frac{C(\Omega')\rho(\Omega')Z_{\Omega'}}{\Omega^2-\Omega'^2}-\frac{C(\Omega)\rho(\Omega)Z_\Omega}{\Omega^2-\Omega'^2}\\&\quad\left.+Z_\Omega^2\rho(\Omega)\delta\left(\Omega-\Omega'\right)\right\}.\end{aligned} \tag{7.4.15}$$

In the rhs, the terms without δ-functions with the help of Eqs. (7.4.10) and (7.4.11) can be shown to cancel:

$$a_\Omega a_{\Omega'}\left[1+\omega_0\frac{-\Sigma_\Omega+\Sigma_{\Omega'}}{\Omega^2-\Omega'^2}+\frac{\Omega'^2-\omega_0^2-\omega_0\Sigma_{\Omega'}}{\Omega^2-\Omega'^2}-\frac{\Omega^2-\omega_0^2-\omega_0\Sigma_\Omega}{\Omega^2-\Omega'^2}\right]=0. \tag{7.4.16}$$

Then Eq. (7.4.6) simplifies to

$$\delta\left(\Omega-\Omega'\right)=a_\Omega a_{\Omega'}\left[\frac{C^2(\Omega)}{(2\Omega)^2}\pi^2+Z_\Omega^2\right]\rho(\Omega)\delta\left(\Omega-\Omega'\right). \tag{7.4.17}$$

With the help of Eq. (7.4.11), this yields

$$a_\Omega^2=\frac{4}{\pi}\frac{\Omega^2\eta(\Omega)}{\left(\Omega^2-\omega_0^2-\omega_0\Sigma_\Omega\right)^2+4\Omega^2\eta^2(\Omega)}, \tag{7.4.18}$$

where

$$\eta(\Omega)\equiv\frac{\pi}{4}C^2(\Omega)\frac{\rho(\Omega)}{\Omega^2}. \tag{7.4.19}$$

After that, $b_\Omega(\omega)$ of Eq. (7.4.8) are found to be

$$b_\Omega(\omega)=\frac{C(\omega)C(\Omega)\rho(\Omega)/(\Omega^2-\omega^2)+\left(\Omega^2-\omega_0^2-\omega_0\Sigma_\Omega\right)\delta\left(\omega-\Omega\right)}{\sqrt{N\rho(\Omega)}\sqrt{\left(\Omega^2-\omega_0^2-\omega_0\Sigma_\Omega\right)^2+4\Omega^2\eta^2(\Omega)}}. \tag{7.4.20}$$

Let us now consider discrete eigenvalues Ω_n beyond the interval $(\omega_{\min},\omega_{\max})$. In this case, $Z_\Omega=0$, and instead of Eq. (7.4.9), one has the eigenvalue equation

$$\Omega^2-\omega_0^2-\int_{\omega_{\min}}^{\omega_{\max}}\frac{C^2(\omega)}{\Omega^2-\omega^2}\rho(\omega)d\omega=0, \tag{7.4.21}$$

whereas the eigenvectors can be obtained from Eq. (7.4.8) with $Z_\Omega=0$ and Eq. (7.4.7):

$$\begin{aligned}1&=a_{\Omega_n}^2+N\int_{\omega_{\min}}^{\omega_{\max}}b_{\Omega_n}(\omega)b_{\Omega_n}(\omega)\rho(\omega)d\omega\\&=a_{\Omega_n}^2\left(1+\int_{\omega_{\min}}^{\omega_{\max}}\frac{C^2(\omega)}{\left(\Omega_n^2-\omega^2\right)^2}\rho(\omega)d\omega\right).\end{aligned} \tag{7.4.22}$$

Equation (7.4.21) coincides with Eq. (7.3.17), but the solution is searched outside the continuum. Investigating this equation graphically, one can see that there can be only one discrete eigenvalue above the continuum and one below the continuum. If there is a solution $\Omega^2 < 0$, this means that the state with $X = x_i = 0$ is unstable and thus not the ground state of the system. This can happen if the coupling becomes too strong. We will exclude such situations. To properly deal with them, one has to include higher-order terms in the potential energy that prevent the system from going to infinity. We will number solutions of Eq. (7.4.21) as Ω_n with $n = 1, 2, \ldots$

Let us now consider the dynamics of the system. The relation between the original variables $Y(\omega, t) \equiv \{Y(t), y(\omega, t)\}$ and the normal modes $\zeta_\Omega(t)$ is

$$Y(\omega, t) = \sum_n A_{\Omega_n}(\omega)\zeta_{\Omega_n}(t) + \int_{\omega_{\min}}^{\omega_{\max}} A_\Omega(\omega)\zeta_\Omega(t)d\Omega \tag{7.4.23}$$

or, more explicitly,

$$Y(t) = \sum_n a_{\Omega_n}\zeta_{\Omega_n}(t) + \int_{\omega_{\min}}^{\omega_{\max}} a_\Omega\zeta_\Omega(t)d\Omega, \tag{7.4.24}$$

$$y(\omega, t) = \sum_n b_{\Omega_n}(\omega)\zeta_{\Omega_n}(t) + \int_{\omega_{\min}}^{\omega_{\max}} b_\Omega(\omega)\zeta_\Omega(t)d\Omega. \tag{7.4.25}$$

The inverse transformation has the form

$$\zeta_\Omega(t) = Y(t)a_\Omega + N\int_{\omega_{\min}}^{\omega_{\max}} y(\omega, t)b_\Omega(\omega)\rho(\omega)d\omega. \tag{7.4.26}$$

Consistency of the above definitions can be shown by combining Eqs. (7.4.24)–(7.4.26). One obtains

$$\begin{aligned}\zeta_\Omega(t) &= \sum_{n'} \zeta_{\Omega_{n'}}(t)\left(a_{\Omega_{n'}}a_\Omega + N\int_{\omega_{\min}}^{\omega_{\max}} b_{\Omega_{n'}}(\omega)b_\Omega(\omega)\rho(\omega)d\omega\right)\\ &\quad + \int_{\omega_{\min}}^{\omega_{\max}} d\Omega'\zeta_{\Omega'}(t)\left(a_{\Omega'}a_\Omega + N\int_{\omega_{\min}}^{\omega_{\max}} b_\Omega(\omega)b_\Omega(\omega)\rho(\omega)d\omega\right),\end{aligned} \tag{7.4.27}$$

which is again $\zeta_\Omega(t)$ for both discrete and continuous Ω, as follows from the orthonormality conditions (7.4.6) and (7.4.7). The normal modes satisfy the equations of motion

$$\ddot{\zeta}_\Omega + \Omega^2\zeta_\Omega = 0, \tag{7.4.28}$$

which has the solution

$$\zeta_\Omega(t) = \zeta_\Omega(0)\cos(\Omega t) + \dot{\zeta}_\Omega(0)\frac{\sin(\Omega t)}{\Omega}. \tag{7.4.29}$$

The initial conditions here can be expressed through $Y(0)$ and $y(\omega, 0)$ with the help of Eq. (7.4.26). If in the initial state, the central oscillator had

$Y(0) \neq 0$ and $\dot{Y}(0) = 0$, whereas all bath oscillators were not excited, one has $\zeta_\Omega(0) = a_\Omega Y(0)$ and $\dot{\zeta}_\Omega(0) = 0$. Taking into account both continuous and discrete eigenvalues and using Eqs. (7.4.18) and (7.4.22), one obtains

$$\frac{Y(t)}{Y(0)} = \int_{\omega_{\min}}^{\omega_{\max}} a_\Omega^2 \cos(\Omega t)\, d\Omega + \sum_n a_{\Omega_n}^2 \cos(\Omega_n t)$$

$$= Y(0)\frac{4}{\pi}\int_{\omega_{\min}}^{\omega_{\max}} \frac{\Omega^2 \eta(\Omega)}{\left(\Omega^2 - \omega_0^2 - \omega_0 \Sigma_\Omega\right)^2 + 4\Omega^2\eta^2(\Omega)} \cos(\Omega t)\, d\Omega$$

$$+ \sum_n \left(1 + \int_{\omega_{\min}}^{\omega_{\max}} \frac{C^2(\omega)}{\left(\Omega_n^2 - \omega^2\right)^2}\rho(\omega)d\omega\right)^{-1} \cos(\Omega_n t). \quad (7.4.30)$$

This result is similar to Eq. (7.2.13). Note that at $t = 0$, $Y(t) = Y(0)$ should be fulfilled. This implies the completeness relation

$$\int_{\omega_{\min}}^{\omega_{\max}} a_\Omega^2 d\Omega + \sum_n a_{\Omega_n}^2 = 1. \quad (7.4.31)$$

Proving this relation analytically should be a good exercise, especially as the integral is taken over a finite range and discrete eigenvalues are solutions of a transcendental equation, Eq. (7.4.21). Numerical check of Eq. (7.4.31) passes. Other completeness relations to check are

$$\int_{\omega_{\min}}^{\omega_{\max}} a_\Omega b_\Omega(\omega) d\Omega + \sum_n a_{\Omega_n} b_{\Omega_n}(\omega) = 0, \quad (7.4.32)$$

$$\int_{\omega_{\min}}^{\omega_{\max}} b_\Omega(\omega) b_\Omega(\omega') d\Omega + \sum_n b_{\Omega_n}(\omega) b_{\Omega_n}(\omega') = \delta(\omega - \omega'). \quad (7.4.33)$$

In the weak-coupling limit, discrete eigenvalues can be neglected, one has $\eta(\omega_0) \equiv \eta \ll \omega_0$ and $\omega_0 \Sigma_\Omega$ is negligible, the Lorentzian peak is very narrow and one can replace $\Omega \Rightarrow \omega_0$ in most places. Thus, Eq. (7.4.18) simplifies to

$$a_\Omega^2 \cong \frac{1}{\pi}\frac{\eta}{\left(\Omega - \omega_0\right)^2 + \eta^2}. \quad (7.4.34)$$

For $Y(t)$, one obtains

$$Y(t) = Y(0)\mathrm{Re}\frac{1}{\pi}\int_{-\infty}^{\infty} \frac{\eta}{\left(\Omega - \omega_0\right)^2 + \eta^2} e^{i\Omega t} d\Omega$$

$$= Y(0)\mathrm{Re}\left[e^{i\omega_0 t}\frac{1}{\pi}\int_{-\infty}^{\infty} \frac{\eta}{u^2 + \eta^2} e^{iut} du\right], \quad (7.4.35)$$

i.e.,

$$Y(t) = Y(0)e^{-\eta t}\cos(\omega_0 t). \tag{7.4.36}$$

The form of this solution suggests that $Y(t)$ satisfies Eq. (7.0.1) with a quadratic potential. To the contrary, the general solution of Eq. (7.4.30) cannot generate a closed differential equation for $Y(t)$.

The energy of the central oscillator, Eq. (7.2.14), is given by

$$\frac{E(t)}{E(0)} = \left(\int_{\omega_{\min}}^{\omega_{\max}} a_\Omega^2 \cos(\Omega t)\, d\Omega + \sum_n a_{\Omega_n}^2 \cos(\Omega_n t)\right)^2 + \frac{1}{\omega_0^2}\left(\int_{\omega_{\min}}^{\omega_{\max}} a_\Omega^2 \Omega \sin(\Omega t)\, d\Omega + \sum_n a_{\Omega_n}^2 \Omega_n \sin(\Omega_n t)\right)^2, \tag{7.4.37}$$

which is similar to Eq. (7.2.15). In the weak-coupling limit, this yields

$$E(t) = E(0)e^{-2\eta t}, \tag{7.4.38}$$

which coincides with Eq. (7.3.16). At $t \to \infty$, the integral terms in Eq. (7.4.37) disappear, while the discrete terms continue to oscillate with different frequencies around the asymptotic level

$$\left.\frac{E(t)}{E(0)}\right|_{t\to\infty} = \frac{1}{2}\sum_n a_{\Omega_n}^4 \left(1 + \frac{\Omega_n^2}{\omega_0^2}\right). \tag{7.4.39}$$

To compare the analytical result of this section with the numerical results of Sec. 7.2, one should note that the equidistant distribution of the bath-oscillator frequencies of Eq. (7.2.16) corresponds to the density of states

$$\rho(\omega) = \frac{1}{\omega_{\max} - \omega_{\min}} \tag{7.4.40}$$

and change from summation over Ω_α to integration over Ω using Eq. (7.4.1). Effectively, one should multiply a_Ω^2 obtained numerically by the factor $N\rho(\omega)$.

Let us illustrate how the theory works using the constant density of states ρ of Eq. (7.4.40), with $\omega_{\min} = 0$, $\omega_0 = 1$, $\omega_{\max} = 2$ and $\xi = 2$, $C_0 = 0.5$. Figure 7.4.1 shows the results for the eigenvalue spread for the central oscillator obtained from the numerical solution of the discrete problem with $N = 300$ in Sec. 7.2 and the analytical solution of the

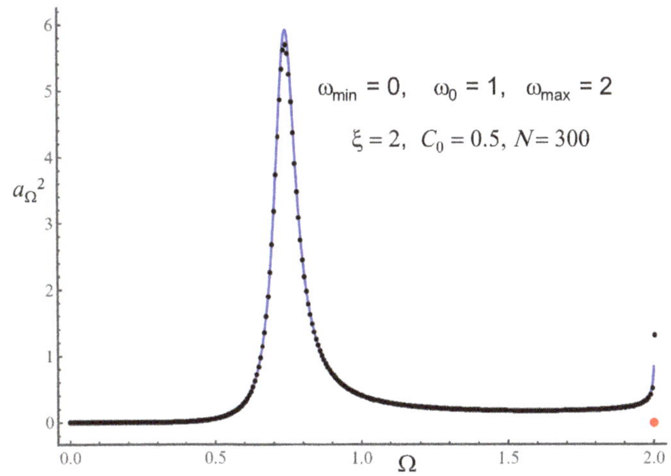

Figure 7.4.1. Spread of the state of the central oscillator over exact eigenstates of the system as a result of its medium-strong coupling to the bath. Solid circles is the numerical solution of the discrete model with $N = 300$ and the blue solid line is the analytical solution of the continuous problem. Note the shift of the maximum from $\Omega = \omega_0$ in the weak-coupling limit to $\Omega \simeq 0.75\omega_0$ in the present case. The red circle is the isolated state.

continuous problem in this section, Eq. (7.4.18). Apart from the continuous spectrum, the analytical solution yields one discrete eigenvalue above the continuum, $\Omega_1 = 2.00069$ with $a_{\Omega_1}^2 = 0.00553$ (the red circle on the right) that makes a small but still noticeable contribution into observables. Deviations from the Lorentzian are indeed quite large, and the peak is shifted to the left from $\omega_0 = 1$ because of the coupling to the bath. There is a good accordance between the solutions of the discrete and continuous problems. Figure 7.4.2 shows the relaxation of the energy of the central oscillator for different values of the bath coupling C_0. For a large coupling, the relaxation becomes non-exponential. Small regular oscillations seen for $C_0 = 0.1$ and 0.2 are due to a finite value of C_0. Again, the curves for the discrete model with $N = 300$ and the continuous model coincide. For strong coupling, relaxation is non-exponential and it cannot be described with Eq. (7.2.1). With the increasing of the coupling C_0 and gradual approaching to the instability limit $\Upsilon = 1$ (see the following), the gap between Ω_1 and the continuum increases and the contribution of the discrete eigenvalue to the physical quantities becomes more noticeable.

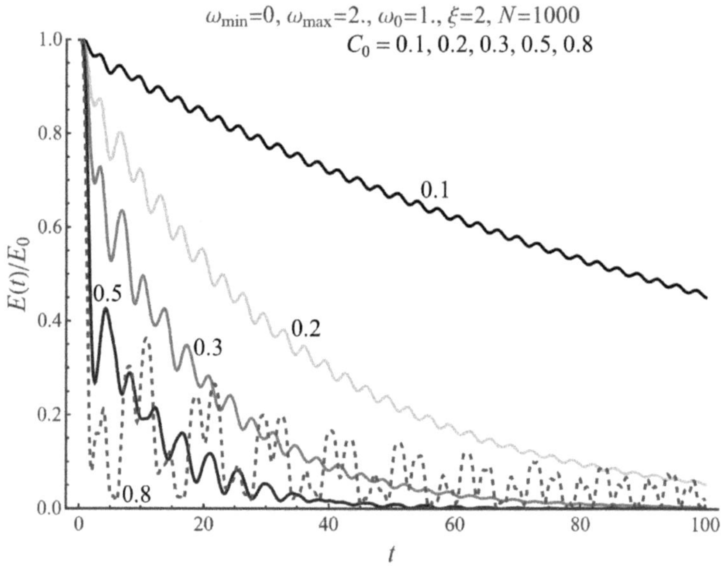

Figure 7.4.2. Relaxation of the energy of the central oscillator becomes non-exponential in the case of a strong coupling to the bath. These results were obtained numerically, but the analytical solution in this section yields the same curves.

Let us now investigate the applicability of the weak-coupling approximation in more detail. The width of the maximum at $\Omega = \omega_0$ is of order η, thus one of the applicability conditions is $\eta \ll \omega_0$. Another condition is that $\eta(\Omega)$ does not change much when Ω changes by η, i.e.,

$$\left.\frac{\partial \eta(\Omega)}{\partial \Omega}\right|_{\Omega=\omega_0} \ll 1. \tag{7.4.41}$$

Typically, $\eta(\Omega) \propto \Omega^\alpha$, so that $\partial\eta(\Omega)/\partial\Omega = \alpha\eta(\Omega)/\Omega$ and both criteria are equivalent. Using Eq. (7.4.19), one can rewrite the weak-coupling criterion in the form

$$\frac{C^2(\omega_0)}{\omega_0^3}\rho(\omega_0) \ll 1. \tag{7.4.42}$$

More stringent, however, is the criterion of smallness of Σ_Ω. If the coupling $C(\omega)$ significantly grows with ω and $\omega_{\max}$ is large, Σ_Ω has a large negative value. The effect of this is that the maximum of a_Ω^2 can be shifted from $\Omega^2 = \omega_0^2$ to smaller Ω^2. For a sufficiently strong coupling, a discrete eigenvalue $\Omega^2 < 0$ can arise that implies instability of the state $X = x_i = 0$.

The instability threshold corresponding to $\Omega = 0$ follows from Eq. (7.4.21):

$$\Upsilon \equiv \int_{\omega_{\min}}^{\omega_{\max}} \frac{C^2(\omega)}{\omega_0^2\omega^2}\rho(\omega)d\omega = 1. \tag{7.4.43}$$

If the integrand grows toward $\omega_{\max}$, which is a typical case, the weak-coupling criterion becomes

$$\Upsilon \sim \frac{C^2(\omega_{\max})}{\omega_0^2\omega_{\max}}\rho(\omega_{\max}) \ll 1, \tag{7.4.44}$$

which is more stringent than Eq. (7.4.42). As an illustration, consider the case $\omega_{\min} = 0$ and

$$C(\omega) = C_0\left(\frac{\omega}{\omega_0}\right)^{\xi}, \quad \rho(\omega) = \rho_0\left(\frac{\omega}{\omega_0}\right)^{\nu}. \tag{7.4.45}$$

Then, for $2\xi + \nu > 1$,

$$\begin{aligned}\Upsilon &= \int_0^{\omega_{\max}} \frac{C^2(\omega)}{\omega_0^2\omega^2}\rho(\omega)d\omega = \frac{C_0^2\rho_0}{\omega_0^{2\xi+\nu+2}}\int_0^{\omega_{\max}} \omega^{2\xi+\nu-2}d\omega \\ &= \frac{C_0^2\rho_0}{\omega_0^3}\frac{(\omega_{\max}/\omega_0)^{2\xi+\nu-1}}{2\xi+\nu-1} = \frac{C^2(\omega_{\max})}{\omega_0^2\omega_{\max}}\frac{\rho(\omega_{\max})}{2\xi+\nu-1}.\end{aligned} \tag{7.4.46}$$

For $\nu = 0$ and $\rho = 1/\omega_{\max}$, setting $\Upsilon = 1$, one obtains the instability threshold for the bath coupling

$$C_0^* = \omega_0\omega_{\max}\left(\frac{\omega_0}{\omega_{\max}}\right)^{\xi}\sqrt{2\xi - 1}. \tag{7.4.47}$$

For $\omega_0 = 1$, $\omega_{\max} = 2$, and $\xi = 2$, it becomes $C_0^* = \sqrt{3}/2 \simeq 0.866$. Thus, in the numerical examples shown in Fig. 7.4.2, the system is stable. The highest value $C_0 = 0.8$ is close to the instability threshold and this is why the energy relaxation curve is strongly fluctuating.

Appendix

To obtain the nontrivial δ-function term in Eq. (7.4.14), consider the regularized difference

$$\begin{aligned}F(x) = &\frac{x-a}{(x-a)^2+\epsilon^2}\frac{x-b}{(x-b)^2+\epsilon^2} \\ &-\frac{1}{a-b}\left(\frac{x-a}{(x-a)^2+\epsilon^2} - \frac{x-b}{(x-b)^2+\epsilon^2}\right),\end{aligned} \tag{7.4.48}$$

where $\epsilon \to 0$. This function simplifies to

$$F(x) = \frac{\epsilon}{(x-a)^2 + \epsilon^2} \frac{\epsilon}{(x-b)^2 + \epsilon^2}. \tag{7.4.49}$$

One can already see that in the limit $\epsilon \to 0$, it becomes negligible if x, a, and b are not close to each other. Thus, it is a product of two δ-functions: $F(x) \propto \delta(a-b)\delta(x-a)$. To find the coefficient, one can integrate $F(x)$ over x and over a (or over b). The first integration yields

$$\int_{-\infty}^{\infty} F(x)dx = \frac{2\pi\varepsilon}{(a-b)^2 + 4\epsilon^2}. \tag{7.4.50}$$

The second integration, over a or over b, yields π^2, so that $F(x) = \pi^2\delta(a-b)\delta(x-a)$.

Chapter 8

Disk on a Plane

A wheel in the form of a disk or a ring rolling on a plane without skidding is an example of a non-holonomic system. Due to the latter, one needs to use Euler equations to describe its dynamics.

8.1 Equations of motion

Let us consider a disk of radius R and mass M with moments of inertia

$$I_1 = I_2 = I/2, \quad I_3 = I. \tag{8.1.1}$$

All the following results can be generalized for a non-flat body with a sharp round edge, for which $I_1 = I_2 \neq I_3/2$. The disk is rolling without skidding on a plane, see Fig. 8.0.1, under the action of the gravity force and a constant horizontal force:

$$\mathbf{F} = -Mg\mathbf{e}_z + F_x\mathbf{e}_x. \tag{8.1.2}$$

The force F_x may be, for instance, due to the inclination of the plane, $F_x = Mg\sin\alpha$, or due to the wind. The rolling constraint, in our case,

$$\mathbf{V} = R\left[\boldsymbol{\omega} \times \mathbf{e}^{(2)}\right] \tag{8.1.3}$$

is non-holonomic, i.e., it depends on velocities in such a way that it cannot be integrated and expressed as a relation between the coordinates. Thus, the Lagrange approach, at least in its standard form, is inapplicable and one needs to use Euler equations.

For a body rolling without skidding, one can use the dynamical equation for the angular momentum with respect to the contact point (CP) of the disk and the ground O', Eq. (4.2.22), which is decoupled from the translational dynamics. The latter can be determined from Eq. (8.1.3) after

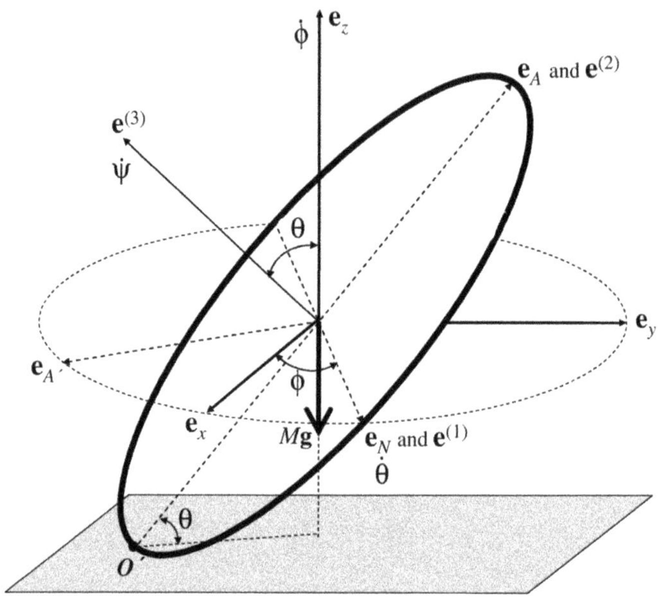

Figure 8.0.1. Disk on a plane.

the rotational problem has been solved. As the principal axes of the rolling disk, one should use the sliding vectors $\mathbf{e}^{(1)} = \mathbf{e}_N$ and $\mathbf{e}^{(2)} = \mathbf{e}_A$ introduced in Sec. 4.2.2, as shown in Fig. 8.0.1. The moments of inertia with respect to these axes, shifted from the disk center to the CP, are given by

$$I_1' = I/2 + MR^2, \quad I_2 = I/2, \quad I_3' = I + MR^2. \tag{8.1.4}$$

Since the disk is axially symmetric, they do not change as it is rolling. Note that, although the disk is rotationally symmetric, the shifted moments of inertia are all different, so that the problem is of the same grade of difficulty as that of a fully asymmetric body.

The Euler equations with respect to the CP are Eqs. (4.4.13) with primes added and the torque augmented by the rolling-constraint term according to Eq. (4.2.22). The torque due to the forces is given by

$$\begin{aligned}\boldsymbol{\tau}' &= [R\mathbf{e}^{(2)} \times \mathbf{F}] = -MgR[\mathbf{e}^{(2)} \times \mathbf{e}_z] + F_x R[\mathbf{e}^{(2)} \times \mathbf{e}_x] \\ &= -MgR\cos\theta\mathbf{e}^{(1)} + F_x R[\mathbf{e}^{(2)} \times (\cos\phi\mathbf{e}^{(1)} + \sin\phi\mathbf{e}_{A'})] \\ &= -MgR\cos\theta\mathbf{e}^{(1)} - F_x R\cos\phi\mathbf{e}^{(3)} + F_x R\sin\theta\sin\phi\mathbf{e}^{(1)},\end{aligned} \tag{8.1.5}$$

where $\mathbf{e}_{A'} = \mathbf{e}^{(3)} \sin\theta - \mathbf{e}^{(2)} \cos\theta$ is another antinode vector. The part of the effective torque due to the constraint with $\mathbf{b} = R\mathbf{e}^{(2)}$ in Eq. (4.2.22) can be rewritten as

$$\begin{aligned}\boldsymbol{\tau}_{\text{constr}} &= MR^2\dot{\mathbf{e}}^{(2)} \times [\boldsymbol{\omega} \times \mathbf{e}^{(2)}] = -MR^2\mathbf{e}^{(2)}(\dot{\mathbf{e}}^{(2)} \cdot \boldsymbol{\omega}) \\ &= MR^2\mathbf{e}^{(2)}(\dot{\phi}\cos\theta - \omega_3)\dot{\theta},\end{aligned} \tag{8.1.6}$$

where we used $\left(\dot{\mathbf{e}}^{(2)} \cdot \mathbf{e}^{(2)}\right) = 1/2\partial_t\left(\mathbf{e}^{(2)} \cdot \mathbf{e}^{(2)}\right) = 0$ and Eqs. (4.4.9) and (4.2.12). Now, the components of the total effective torque $\tilde{\boldsymbol{\tau}}' = \boldsymbol{\tau}' + \boldsymbol{\tau}_{\text{constr}}$ become

$$\begin{aligned}\tilde{\tau}_1' &= -MgR\cos\theta + F_x R\sin\theta\sin\phi, \\ \tilde{\tau}_2' &= MR^2(\dot{\phi}\cos\theta - \omega_3)\dot{\theta}, \\ \tilde{\tau}_3' &= -F_x R\cos\phi.\end{aligned} \tag{8.1.7}$$

Adding the effective torque to Eqs. (4.4.13) with primes and simplifying the second line yield the system of nonlinear differential equations

$$\begin{aligned}I_1'\ddot{\theta} &= (I_2\dot{\phi}\cos\theta - I_3'\omega_3)\dot{\phi}\sin\theta - MgR\cos\theta + F_x R\sin\theta\sin\phi, \\ I_2\ddot{\phi}\sin\theta &= [-(I_1' + I_2)\,\dot{\phi}\cos\theta + I_3'\omega_3]\dot{\theta} + MR^2(\dot{\phi}\cos\theta - \omega_3)\dot{\theta} \\ &= [-(I_1 + I_2)\dot{\phi}\cos\theta + I_3\omega_3]\dot{\theta}, \\ I_3'\dot{\omega}_3 &= MR^2\dot{\theta}\dot{\phi}\sin\theta - F_x R\cos\phi,\end{aligned} \tag{8.1.8}$$

where the moments of inertia I_1, I_1', etc., are given by Eqs. (8.1.1) and (8.1.4). Note that adding $\boldsymbol{\tau}_{\text{constr}}$ only modifies the coefficients in the second equation. However, this term is needed for the conservation of the total energy

$$E = E_{\text{rot}} + U = \frac{I_1'}{2}\dot{\theta}^2 + \frac{I_2}{2}(\dot{\phi}\sin\theta)^2 + \frac{I_3'\omega_3^2}{2} + MgR\sin\theta - F_x X, \tag{8.1.9}$$

where X is the coordinate of the CM. The total order of this system of equations is $2 + 2 + 1 = 5$.

After the solution of this system of equations is found, one can obtain the velocity of the center of mass from Eq. (8.1.3):

$$\mathbf{V} = [\boldsymbol{\omega} \times R\mathbf{e}^{(2)}] = R(\omega_1\mathbf{e}^{(3)} - \omega_3\mathbf{e}^{(1)}). \tag{8.1.10}$$

Using Eqs. (4.2.11) and (4.2.12), for the components of $\mathbf{V} \equiv (\dot{X}, \dot{Y}, \dot{Z})$, one finally obtains

$$\dot{X} = R\left(\dot{\theta}\sin\theta\sin\phi - \frac{L_3}{I_3'}\cos\phi\right),$$

$$\dot{Y} = -R\left(\dot{\theta}\sin\theta\cos\phi + \frac{L_3}{I_3'}\sin\phi\right), \tag{8.1.11}$$

as well as $\dot{Z} = R\dot{\theta}\cos\theta$. The latter can be immediately integrated, leading to the obvious result:

$$Z = R\sin\theta. \tag{8.1.12}$$

Note that the z-components of the non-holonomic constraint, Eq. (8.1.3), is holonomic since it can be integrated! From Eq. (8.1.11), one can see that for $\phi = 0$ and $\omega_3 < 0$, as well as for $\phi = \pi$ and $\omega_3 > 0$, the disk rolls in the positive X-direction.

The position of the contact point $\boldsymbol{\xi}$ is related to that of the CM:

$$\mathbf{R}_{\mathrm{CM}} = X\mathbf{e}_x + Y\mathbf{e}_y + Z\mathbf{e}_z \tag{8.1.13}$$

by the formula

$$\boldsymbol{\xi} = \mathbf{R}_{\mathrm{CM}} - R\mathbf{e}^{(2)}. \tag{8.1.14}$$

The components of this formula are

$$\xi_x = X + R\cos\theta\sin\phi,$$

$$\xi_y = Y - R\cos\theta\cos\phi. \tag{8.1.15}$$

In the case of the forces that depend on the disk's horizontal position X,Y, the problem cannot be solved in two steps. In this case, one has to solve Eq. (8.1.8) with appropriately modified torques together with Eq. (8.1.11).

8.2 Rotation with respect to the center of mass

Rotational equation of motion for the disk (8.1.8) can also be obtained by considering rotations with respect to the CM. Here, the only torque is that from the reaction force $\mathbf{F}^{(R)}$ applied to the CP. The whole system of Newtonian equations of motion is given by

$$M\dot{\mathbf{V}} = \mathbf{F} + \mathbf{F}^{(R)},$$

$$\dot{\mathbf{L}} = \boldsymbol{\tau} = [-R\mathbf{e}^{(2)} \times \mathbf{F}^{(R)}], \tag{8.2.1}$$

plus the rolling-constraint relation, Eq. (8.1.3). In Eq. (1.3.11), $\mathbf{v}_O = \mathbf{V} = \mathbf{P}/M$ and the kinematic term vanishes. The reaction force can be eliminated from the first equation and the constraint:

$$\mathbf{F}^{(R)} = M\dot{\mathbf{V}} - \mathbf{F} = MR\frac{d}{dt}[\boldsymbol{\omega} \times \mathbf{e}^{(2)}] - \mathbf{F}, \tag{8.2.2}$$

where $\left[\boldsymbol{\omega} \times \mathbf{e}^{(2)}\right] = \mathbf{e}^{(3)}\omega_1 - \mathbf{e}^{(1)}\omega_3$. The torque then becomes

$$\boldsymbol{\tau} = [R\mathbf{e}^{(2)} \times \mathbf{F}] - MR^2[\mathbf{e}^{(2)} \times (\mathbf{e}^{(3)}\dot{\omega}_1 - \mathbf{e}^{(1)}\dot{\omega}_3 + \dot{\mathbf{e}}^{(3)}\omega_1 - \dot{\mathbf{e}}^{(1)}\omega_3)]. \tag{8.2.3}$$

It is convenient to use the sliding principal axes, for which the time derivatives are given by Eqs. (4.4.9). This yields

$$\boldsymbol{\tau} = [R\mathbf{e}^{(2)} \times \mathbf{F}] - MR^2[\mathbf{e}^{(1)}(\dot{\omega}_1 + \dot{\phi}\sin\theta\omega_3) + \mathbf{e}^{(3)}(\dot{\omega}_3 - \dot{\phi}\sin\theta\omega_1)]. \tag{8.2.4}$$

Using Eq. (8.1.5) for the first term of this expression and Eq. (4.2.12) for the components of the angular velocity ω_1 and ω_2, one finally obtains the components of the torque:

$$\begin{aligned}
\tau_1 &= -MgR\cos\theta + F_x R\sin\theta\sin\phi - MR^2(\ddot{\theta} + \dot{\phi}\sin\theta\omega_3),\\
\tau_2 &= 0,\\
\tau_3 &= -F_x R\cos\phi - MR^2(\dot{\omega}_3 - \dot{\theta}\dot{\phi}\sin\theta).
\end{aligned} \tag{8.2.5}$$

Substituting this torque into Euler's equation (4.4.13) with $I_1 = I_2$ and combining the terms, one obtains

$$\begin{aligned}
I_1'\ddot{\theta} &= (I_2\dot{\phi}\cos\theta - I_3'\omega_3)\dot{\phi}\sin\theta - MgR\cos\theta + F_x R\sin\theta\sin\phi,\\
I_2\ddot{\phi}\sin\theta &= [-(I_1 + I_2)\dot{\phi}\cos\theta + I_3\omega_3]\dot{\theta},\\
I_3'\dot{\omega}_3 &= MR^2\dot{\theta}\dot{\phi}\sin\theta - F_x R\cos\phi,
\end{aligned} \tag{8.2.6}$$

which coincides with Eq. (8.1.8). This equation for the rotational motion, decoupled from the translational motion, could be obtained because the rotation of the disk can be represented as the rotation with respect to the instantaneous axis that goes through the contact point.

8.3 Possibility of an analytical solution for $F_x = 0$

For the pure-gravity torque, $F_x = 0$, the rolling-disk problem can be solved analytically. For this, it is convenient to write Euler's equations in terms of the angular velocities, see Eq. (4.4.12). This yields

$$\begin{aligned} I_1'\dot{\omega}_1 &= (I_2\omega_2 \cot\theta - I_3'\omega_3)\,\omega_2 - MgR\cos\theta, \\ I_2\dot{\omega}_2 &= (-I_1\omega_2 \cot\theta + I_3\omega_3)\,\omega_1, \\ I_3'\dot{\omega}_3 &= MR^2\omega_1\omega_2. \end{aligned} \tag{8.3.1}$$

Here, one can make a trick and express ω_2 and ω_3 as the functions of θ. Now, $\dot{\omega}_2 = (d\omega_2/d\theta)\,\dot{\theta} = (d\omega_2/d\theta)\,\omega_1$ and similar for ω_2. Substituting this into the second and third equations leads to their linearization:

$$\begin{aligned} I_2\frac{d\omega_2}{d\theta} &= -I_1\omega_2 \cot\theta + I_3\omega_3, \\ I_3'\frac{d\omega_3}{d\theta} &= MR^2\omega_2. \end{aligned} \tag{8.3.2}$$

This is a system of two linear differential equations with a variable coefficient. Differentiating the second equation and substituting $d\omega_2/d\theta$ from the first equation, one obtains one second-order differential equation

$$\frac{d^2\omega_3}{d\theta^2} + \cot\theta\frac{d\omega_3}{d\theta} - \frac{MR^2I_3}{I_2I_3'}\omega_3 = 0, \tag{8.3.3}$$

where $I_1 = I_2$ was taken into account. With the substitution $\cos\theta = 1-2x$, this equation can be rewritten as a Gauss hypergeometrical equation. Thus, the problem of a disk rolling without skidding on a plane can be solved in terms of the Gauss hypergeometrical functions. With the functional form of $\omega_2(\theta)$ and $\omega_3(\theta)$ known, one can find $\dot{\theta}$ from the energy integral, Eq. (8.1.9), as

$$\dot{\theta} = \pm\sqrt{\frac{2E - I_2\omega_2^2(\theta) - I_3'\omega_3^2(\theta) - 2MgR\sin\theta}{I_1'}}. \tag{8.3.4}$$

Integrating this equation, $\int d\theta/\sqrt{\ldots} = t$, one obtains the solution of the problem in quadratures. However, this way is difficult, and many particular interesting cases can be treated analytically with much simpler methods, while in general, these equations can be solved numerically.

As the problem without a horizontal force is integrable, it does not show any dynamical chaos, although Euler's equations are nonlinear. With the horizontal force, chaotic motion is, in principle, possible.

8.4 Steadily rolling disk

For $F_x = 0$, Eqs. (8.1.8) have a stationary solution with θ, L_3, and $\dot{\phi}$ constant. In this case, the second and third equations are satisfied automatically, while the first equation becomes

$$\left(I_2\dot{\phi}\cos\theta - I_3'\omega_3\right)\dot{\phi}\sin\theta - MgR\cos\theta = 0. \tag{8.4.1}$$

It has two solutions for $\dot{\phi}$:

$$\dot{\phi}_\pm = \frac{I_3'\omega_3 \pm \sqrt{\left(I_3'\omega_3\right)^2 + 4MgRI_2\cos^2\theta/\sin\theta}}{2I_2\cos\theta}. \tag{8.4.2}$$

Then, one can find the radius of the circle circumscribed by the CM of the disk by integrating Eqs. (8.1.11). The result reads

$$\mathcal{R} = \frac{\omega_3}{\dot{\phi}}R. \tag{8.4.3}$$

In the weak-gravity/fast rotating limit, $MgRI_2\cos^2\theta/\sin\theta \ll \left(I_3'\omega_3\right)^2$, Eq. (8.4.2) simplifies and one obtains

$$\dot{\phi}_+ \cong \frac{I_3'\omega_3}{I_2\cos\theta} + \frac{MgR}{I_3'\omega_3}\cot\theta \cong \frac{I_3'\omega_3}{I_2\cos\theta}, \quad \mathcal{R} \cong \frac{I_2\cos\theta}{I_3'}R. \tag{8.4.4}$$

This is the fast-precessing solution that does not require gravity. For the second solution, one obtains

$$\dot{\phi}_- \cong -\frac{MgR}{I_3'\omega_3}\cot\theta, \quad \mathcal{R} \cong \frac{I_3'\omega_3^2\tan\theta}{Mg}, \tag{8.4.5}$$

the slow-precessing solution due to gravity. The fast precession has a limit $\omega_3 \to 0$ and $\theta \to \pi/2$ with $\dot{\phi}_+ = \text{const}$. This is a pure precession with $\mathcal{R} \to 0$. The slow precession case has the limit of the straight upright rolling, $\theta \to \pi/2$ with $\dot{\phi}_- \to 0$ and $\mathcal{R} \to \infty$.

In the opposite limit, $MgRI_2\cos^2\theta/\sin\theta \gg \left(I_3'\omega_3\right)^2$, which is realized if the disk's CM is close to the surface, $\theta \ll 1$, one obtains

$$\dot{\phi}_\pm \cong \pm\sqrt{\frac{MgR}{I_2\sin\theta}}. \tag{8.4.6}$$

In this fast-wobbling regime, the spin of the disk is non-essential. If the energy of the system is slowly decreasing due to dissipation, θ decreases, and the wobbling frequency increases. This is the working principle of the "Euler's disk" toy. A heavy polished disk is wobbling on a very smooth slightly concave surface, which resonates to the motion and produces the sound of the frequency $\omega = \dot{\phi}$. The dissipation is very low, so that the disk wobbles for a long time. As its energy gradually decreases, the frequency of the sound increases.

The energy in the fast-wobbling regime can be obtained from Eq. (8.1.9) by discarding everything except for the wobbling and gravity:

$$E = \frac{I_2}{2}(\dot{\phi}\sin\theta)^2 + MgR\sin\theta. \tag{8.4.7}$$

Expressing $\sin\theta$ with the use of Eq. (8.4.6), $\sin\theta = MgR/(I_2\dot{\phi}^2)$, and substituting it into the formula above, one obtains $E = 3/2(MgR)^2/(I_2\dot{\phi}^2)$, and hence,

$$\dot{\phi} = \sqrt{\frac{3}{2}}\frac{MgR}{\sqrt{I_2 E}} \tag{8.4.8}$$

for the frequency of the generated sound.

8.5 Stability of the upright rolling and precessing

For $F_x = 0$, let us investigate the stability of the straight upright rolling with $\theta = \pi/2$, $\dot{\phi} = 0$, and $\omega_3 =$ const. Introducing the small deviation $\delta\theta$ from the upright orientation of the disk, $\theta = \pi/2 + \delta\theta$ and discarding quadratic terms, from the third of Eqs. (8.1.8), one obtains $\dot{L}_3 = 0$ in the linear approximation, whereas the first two equations become

$$\begin{aligned} I_1'\delta\ddot{\theta} &= -I_3'\omega_3\dot{\phi} + MgR\delta\theta, \\ I_2\ddot{\phi} &= \omega_3 I_3\delta\dot{\theta}. \end{aligned} \tag{8.5.1}$$

Integrating the second equation, one obtains $I_2\dot{\phi} = \omega_3 I_3\delta\theta$. Substituting this into the first equation yields

$$I_1'\delta\ddot{\theta} = \left(-\frac{\omega_3^2 I_3 I_3'}{I_2} + MgR\right)\delta\theta. \tag{8.5.2}$$

One can see that the solution of this equation exponentially decreases, and thus, the motion is stable for the fast-rotating disk if

$$\omega_3^2 > \omega_{3,\text{crit}}^2 \equiv \frac{MgRI_2}{I_3 I_3'} = \frac{MgR}{2\,(I + MR^2)}. \tag{8.5.3}$$

This stability condition is similar to that for the "sleeping" top, Eq. (4.3.13).

Let us now investigate the stability of the upright precessing. In this regime (with $F_x = 0$), one has $\theta = \pi/2$, $\omega_3 = 0$, and $\dot{\phi} =$ const. Euler's equations should be linearized in the small deviations from this solution. One has $\theta = \pi/2 + \delta\theta$, so that $\cos\theta \cong -\sin\theta\delta\theta \cong -\delta\theta$, while corrections to $\dot{\phi}$ can be neglected. The second equation contains only quadratic terms on the right side and can be discarded. The third equation, after linearization, can be immediately integrated and yields $\omega_3 = \left(MR^2/I_3'\right)\dot{\phi}\delta\theta$. Substituting this into the first equation, linearized, yields

$$I_1'\delta\ddot{\theta} = -\left(I_2 + MR^2\right)\dot{\phi}^2\delta\theta + MgR\delta\theta.$$

One can see that the gravity term causes instability, while the precession favors stability. The condition for the stable upright precession reads

$$\dot{\phi}^2 > \dot{\phi}^2_{\text{crit}} \equiv \frac{MgR}{I_2 + MR^2} = \frac{MgR}{I/2 + MR^2}. \tag{8.5.4}$$

8.6 Rapidly rolling disk

If the wheel is spinning with a large angular frequency ω_3, Euler's equations (8.1.8) simplify. As the second equation contains the term $\omega_3\dot{\theta}$, one has $\dot{\theta} \propto 1/\omega_3$ and the time derivatives of θ can be neglected. Also, in the first equation, one can discard $I_2\dot{\phi}\cos\theta$ that is small in comparison to $I_3'\omega_3$. As a result, the first and third equations form a system of two first-order differential equations

$$\begin{aligned}
\dot{\phi} &= -\frac{R}{I_3'\omega_3}\left(Mg\cot\theta - F_x\sin\phi\right) = -\frac{RF_x}{I_3'\omega_3}(\kappa - \sin\phi),\\
\dot{\omega}_3 &= -\frac{RF_x}{I_3'}\cos\phi,
\end{aligned} \tag{8.6.1}$$

where

$$\kappa \equiv \frac{Mg}{F_x}\cot\theta \tag{8.6.2}$$

is the parameter that defines the weak-force/strong-gravity regime for $|\kappa| > 1$ and the strong-force/weak-gravity regime for $|\kappa| < 1$. For $|\kappa| > 1$, the angle ϕ monotonically increases or decreases with time, and thus, $\sin\phi$, $\cos\phi$, and ω_3 are oscillating. For $|\kappa| < 1$, there is a stationary solution with $\sin\phi_\phi = \kappa$ and L_3 growing linearly with time, which causes the wheel to accelerate in the direction, making an angle $\phi_\infty = \arcsin\kappa$ or $\pi - \arcsin\kappa$ with the horizontal force. In the sequel, we assume $F_x > 0$ without a limitation of generality.

The system of differential equations above can be solved by representing L_3 as a function of ϕ. Dividing the second equation by the first one yields

$$\frac{d\omega_3}{d\phi} = \frac{\omega_3 \cos\phi}{\kappa - \sin\phi}. \tag{8.6.3}$$

This can be easily integrated to yield

$$\omega_3 = \frac{\kappa\bar{\omega}_3}{\kappa - \sin\phi}, \tag{8.6.4}$$

where $\bar{\omega}_3 = \text{const.}$, depending on the initial conditions. In the limit $\kappa \to \infty$ $(F_x \to 0)$, one has $\omega_3 = \bar{\omega}_3 = \text{const.}$

Let us now consider the motion of the center of the wheel. From Eq. (8.1.11), with the use of Eqs. (8.6.1), one obtains

$$X = R\int dt\omega_3(t)\cos\phi(t) = -\frac{I_3'}{F_x}\int dt\omega_3(t)\dot{\omega}_3(t) = -\frac{I_3'}{2F_x}\int dt\frac{d}{dt}\omega_3^2(t), \tag{8.6.5}$$

which results in

$$X = -\frac{I_3'\omega_3^2}{2F_x} + \text{const.} = -\frac{I_3'\kappa^2\bar{\omega}_3^2}{2F_x\left(\kappa - \sin\phi\right)^2} + \text{const.} \tag{8.6.6}$$

Now, we consider the case $|\kappa| > 1$. Here, X is oscillating without growing on average. Then

$$\begin{aligned} Y &= -R\int dt\omega_3(t)\sin\phi(t) = -R\int\frac{d\phi}{\dot{\phi}}\omega_3(\phi)\sin\phi \\ &= \frac{I_3'\bar{\omega}_3^2\kappa^2}{F_x}\int d\phi\frac{\sin\phi}{\left(\kappa - \sin\phi\right)^3}, \end{aligned} \tag{8.6.7}$$

which after integration yields

$$Y = \frac{I_3'\bar{\omega}_3^2\kappa^2}{F_x}f(\phi) + \text{const.}, \tag{8.6.8}$$

where

$$\begin{aligned} f(\phi) &= \frac{3\kappa}{\left(\kappa^2 - 1\right)^{5/2}}\arctan\left[\frac{\kappa\tan(\phi/2) - 1}{\sqrt{\kappa^2 - 1}}\right] \\ &\quad - \frac{\left[\kappa\left(1 + 2\kappa^2\right) - \left(2 + \kappa^2\right)\sin\phi\right]\cos\phi}{2\left(\kappa^2 - 1\right)^2\left(\kappa - \sin\phi\right)^2}. \end{aligned} \tag{8.6.9}$$

As arctan is a multivalued function, the result should be defined properly by setting

$$\arctan\left[\frac{\kappa\tan(\phi/2)-1}{\sqrt{\kappa^2-1}}\right] \Rightarrow \arctan\left[\frac{\kappa\tan(\phi/2)-1}{\sqrt{\kappa^2-1}}\right] + \pi\left\lfloor\frac{\phi+\pi}{2\pi}\right\rfloor\frac{\kappa}{|\kappa|}, \tag{8.6.10}$$

where in the rhs, arctan is confined to the interval $(-\pi/2, \pi/2)$ and $\lfloor x\rfloor$ means the integer part of floor of x. With this refinement, $f(\phi)$ has no jumps and increases on average for both signs of κ. Its increase over the period of motion defines the lateral displacement

$$\Delta Y = \frac{I_3'\bar{\omega}_3^2\kappa^2}{F_x}\left[f(2\pi) - f(0)\right] = \frac{I_3'\bar{\omega}_3^2\kappa^2}{F_x}\frac{3\pi\,|\kappa|}{(\kappa^2-1)^{5/2}}. \tag{8.6.11}$$

This result is valid when ϕ increases with time, i.e., in Eq. (8.6.1), $\cot\theta/L_3 < 0$. If ϕ decreases with time, the factor -1 should be introduced to this formula. One can see that ΔY diverges for $\kappa^2 \to 1$. In the opposite limit, $\kappa^2 \gg 1$, one obtains

$$\Delta Y = \frac{I_3'\bar{\omega}_3^2}{F_x}\frac{3\pi}{\kappa^2} = \frac{I_3'\bar{\omega}_3^2}{F_x}\frac{3\pi}{(\frac{Mg}{F_x}\cot\theta)^2} = \frac{3\pi I_3'\omega_3^2 F_x}{(Mg)^2}\tan^2\theta, \tag{8.6.12}$$

i.e., the lateral displacement is proportional to F_x and increases with the spin. As X is oscillating, while Y has an additional monotonically growing component, the trajectory (X, Y) is of a trochoidal type.

The time dependences follow from the solution of the differential equation

$$\dot{\phi} = -\frac{RF_x}{\kappa I_3'\bar{\omega}_3}(\kappa - \sin\phi)^2, \tag{8.6.13}$$

where Eq. (8.6.4) was used. Separating variables and integrating, one obtains the implicit solution

$$\int\frac{d\phi}{(\kappa-\sin\phi)^2} = g(\phi) = -\frac{RF_x}{\kappa I_3'\bar{\omega}_3}t + \text{const.}, \tag{8.6.14}$$

where

$$g(\phi) \equiv \frac{2\kappa}{(\kappa^2-1)^{3/2}}\arctan\left[\frac{\kappa\tan(\phi/2)-1}{\sqrt{\kappa^2-1}}\right] - \frac{\cos\phi}{\sqrt{\kappa^2-1}(\kappa-\sin\phi)}, \tag{8.6.15}$$

where arctan should be treated as above. The fundamental frequency of this anharmonic motion, Ω_0, is defined by

$$g(2\pi) - g(0) = \frac{2\pi\,|\kappa|}{\left(\kappa^2 - 1\right)^{3/2}} = -\frac{RF_x}{\kappa I_3'\bar{\omega}_3}\frac{2\pi}{\Omega_0}, \tag{8.6.16}$$

where one finds

$$\Omega_0 = -\frac{RF_x}{I_3'\bar{\omega}_3}\frac{\left(\kappa^2 - 1\right)^{3/2}}{\kappa\,|\kappa|}. \tag{8.6.17}$$

This goes to zero for $\kappa \to 0$ as expected. For $|\kappa| \to \infty$ ($F_x \to 0$), the result simplifies to $\Omega_0 = -RF_x\kappa/(I_3'\bar{\omega}_3) = -MgR\cot\theta/(I_3'\bar{\omega}_3)$, which coincides with $\dot{\phi}$ in Eq. (8.4.5). The average speed of the motion in the y direction is given by

$$\langle \dot{Y} \rangle = \Delta Y \frac{|\Omega_0|}{2\pi} = \frac{3}{2}\frac{\kappa\bar{\omega}_3 R}{\kappa^2 - 1}. \tag{8.6.18}$$

For $\kappa^2 \gg 1$, this simplifies to

$$\langle \dot{Y} \rangle = \frac{3}{2}\frac{\bar{\omega}_3 RF_x}{Mg}\tan\theta. \tag{8.6.19}$$

In the strong-force/weak-gravity regime, $|\kappa| < 1$, the solution of Eq. (8.6.1) asymptotically approaches $\sin\phi = \kappa$, while, according to Eq. (8.6.4), $\omega_3 \to \pm\infty$. In this region, the equation for ϕ with $\phi = \phi_\infty + \delta\phi$ can be linearized in $\delta\phi$. Then this equation becomes

$$I_3'\omega_3\delta\dot{\phi} = F_x R\cos\phi_\infty\delta\phi, \tag{8.6.20}$$

where $\cos\phi_\infty = \pm\sqrt{1-\kappa^2}$. Since ω_3 is large, one can assume that its relative change during the characteristic stability/instability time is small, so that it can be considered as a constant. Searching for the solution in the exponential form, $\delta\phi \propto e^{\lambda t}$, one obtains $\lambda = F_x R\cos\phi/(I_3'\omega_3)$. In the case, $\cos\phi < 0$ with $\omega_3 > 0$ or $\cos\phi > 0$ with $\omega_3 < 0$ (rolling in the direction of the horizontal force in both cases), one has $\lambda < 0$ and the motion is stable. For the rolling against the horizontal force, $\cos\phi < 0$ and $\omega_3 < 0$ or $\cos\phi > 0$ and $\omega_3 > 0$ (which is physically the same), one has $\lambda > 0$ and the motion is unstable.

As an example, for $\theta = \pi/4$, $Mg = 1$, and $F_x = 3$, one has $\kappa = 1/3$ and for a large negative ω_3, the wheel is asymptotically rolling with $\phi_\infty = \arcsin(1/3) = 19.47°$ in the positive x and y directions closer to the x-axis. Another solution is $\phi_\infty = \pi - \arcsin(1/3) = 160.5°$ and large positive ω_3 in which the wheel is rolling in the positive x- and negative y-directions, closer to the x-axis. These two solutions differ by the tilt of the wheel. In the first case, it is tilted to the left (if one looks along the x-axis in Fig. 8.0.1), and in the second case, it is tilted to the right.

8.7 Rapid rolling via angular momentum

If the wheel is rapidly rotating around its symmetry axis, the problem simplifies and can be treated by a method similar to that of Sec. 4.2.9. Again, L_3 dominates in the angular momentum and one can use the approximate relation $\mathbf{L} = L\mathbf{e}^{(3)}$ with L positive or negative. This implies $\mathbf{e}^{(3)} = \mathbf{L}/L$. Using the sliding axes $\mathbf{e}^{(1)}$ and $\mathbf{e}^{(2)}$, as above in this chapter, one obtains the rotational dynamical equation in the form

$$\dot{\mathbf{L}} = \boldsymbol{\tau}' = [R\mathbf{e}^{(2)} \times \mathbf{F}], \tag{8.7.1}$$

where the torque $\boldsymbol{\tau}'$ is the same as in Eq. (8.1.5). The constraint contribution to the torque, Eq. (8.1.6), can be discarded because, as we shall see, for the fast-spinning disk, $\theta = \text{const}$. Here, we will express the torque via $\mathbf{L}$ and $\mathbf{F}$, bypassing Euler's angles. For this, we define

$$\mathbf{e}^{(1)} = \frac{[\mathbf{e}_z \times \mathbf{e}^{(3)}]}{\sqrt{[\mathbf{e}_z \times \mathbf{e}^{(3)}]^2}} = \eta \frac{[\mathbf{e}_z \times \mathbf{L}]}{\sqrt{[\mathbf{e}_z \times \mathbf{L}]^2}}, \tag{8.7.2}$$

where $\eta \equiv |L|/L = L/|L| = \pm 1$ stands for the direction of the spin with respect to $\mathbf{e}^{(3)}$. As in our approximation, L is large and does not go through zero, η does not change dynamically. Further,

$$\mathbf{e}^{(2)} = [\mathbf{e}^{(3)} \times \mathbf{e}^{(1)}] = \frac{\eta}{L} \frac{[\mathbf{L} \times [\mathbf{e}_z \times \mathbf{L}]]}{\sqrt{[\mathbf{e}_z \times \mathbf{L}]^2}} = \frac{1}{|L|} \frac{\mathbf{e}_z L^2 - \mathbf{L}(\mathbf{e}_z \cdot \mathbf{L})}{\sqrt{[\mathbf{e}_z \times \mathbf{L}]^2}}. \tag{8.7.3}$$

Now, the rotational dynamical equation becomes

$$\dot{\mathbf{L}} = \frac{L^2 [\mathbf{e}_z \times R\mathbf{F}] - (\mathbf{e}_z \cdot \mathbf{L}) [\mathbf{L} \times R\mathbf{F}]}{|L| \sqrt{[\mathbf{e}_z \times \mathbf{L}]^2}}. \tag{8.7.4}$$

This rapid-wheel equation is similar to the rapid-top equation (4.2.55), although it is more complicated.

If there is only the gravity force, $\mathbf{F} = -Mg\mathbf{e}_z$, the dynamics simplify to a pure precession described by

$$\dot{\mathbf{L}} = \frac{MgR}{|L|} \frac{(\mathbf{e}_z \cdot \mathbf{L})}{\sqrt{[\mathbf{e}_z \times \mathbf{L}]^2}} [\mathbf{L} \times \mathbf{e}_z]. \tag{8.7.5}$$

Note that the direction of $\dot{\mathbf{L}}$ does not depend on the sign of L. This means that the wheel precesses in different directions for different signs of L. Due to the prefactor in this equation, the precession rate depends on the angle θ. One has $\mathbf{e}_z \cdot \mathbf{L} = L_z = L\cos\theta$ and

$$\eta\sqrt{[\mathbf{e}_z \times \mathbf{L}]^2} = \eta\sqrt{L^2 - (\mathbf{e}_z \cdot \mathbf{L})^2} = \eta\sqrt{L^2 \sin^2\theta} = L\sin\theta, \tag{8.7.6}$$

which is correct for both signs of L. Thus, the precession rate is given by

$$\dot{\phi} = -\frac{MgR}{L}\cot\theta. \tag{8.7.7}$$

For $L > 0$, vector $\mathbf{L}$ precesses clockwise around $\boldsymbol{\Omega} \parallel \mathbf{e}_z$, see Fig. 8.0.1, thus $\dot{\phi} < 0$. This is the case of a slow precession of the disk, Eq. (8.4.5) with $I_3'\omega_3 = L$.

As we have seen, for $F_x = 0$, the length of the vector $\mathbf{L}$ is conserved. In general, $L \neq$ const., and to account for its change, one can write $\mathbf{L} \equiv L\mathbf{n}_L$, where $\mathbf{n}_L \equiv \mathbf{e}^{(3)}$ shows the direction of the precession axis. From this follows $\dot{\mathbf{L}} = \dot{L}\mathbf{n}_L + L\dot{\mathbf{n}}_L$. Here, one can split the equation for $\dot{L}$ and $\dot{\mathbf{n}}_L$ by projecting $\dot{\mathbf{L}}$ onto the directions collinear and perpendicular to $\mathbf{L}$. The latter has the form $\dot{\mathbf{n}}_L \propto [\mathbf{n}_L \times \mathbf{e}_z]$ (with an orientation-dependent coefficient), which entails $\theta =$ const. Thus, the dynamics of $\mathbf{n}_L$ is described by the angle ϕ only. The corresponding equation reduces to the first equation in Eqs. (8.6.1). The equation for $\dot{L}$ becomes the second of Eqs. (8.6.1).

8.8 Numerical results

The dynamics of a wheel (disk) rolling on a plane is very rich, having a number of different regimes. In this section, we consider some types of behavior, illustrated by the numerical solution of Eqs. (8.1.8), implemented in Wolfram Mathematica (supplementary download). The parameters of the wheel are set to $M = R = I = 1$ for simplicity.

8.8.1 *Wheel with gravity and no horizontal force*

If the wheel's initial spin is very slow, the wheel nearly falls flat because of gravity, but, as θ approaches 0 or π, the rotation dramatically increases so that both $\dot{\phi}$ and L_3 become large and the sign of $\dot{\theta}$ reverses (in real situations, the wheel skids and falls flat, but here skidding is ruled out). During the short nearly-fall time interval, both the CM and the contact point assume large velocities and the contact point makes a bow around the CM. This regime is illustrated for $Mg = 1$ and $F_x = 0$ and the initial

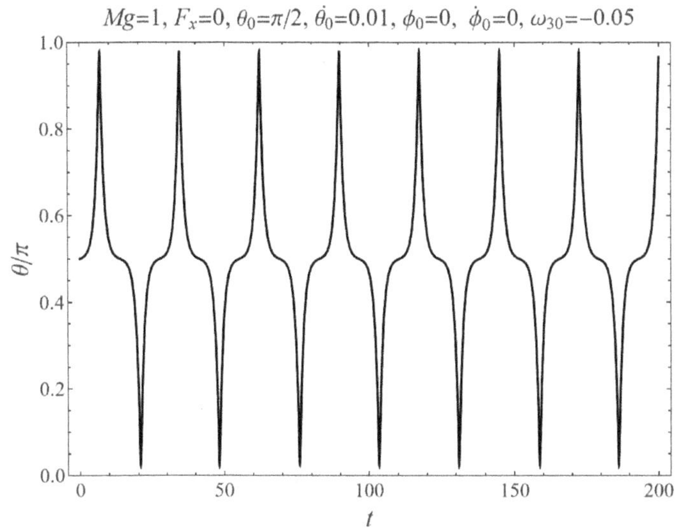

Figure 8.8.1. Wheel on a plane: $\theta(t)$, unstable motion, nearly falling on both sides.

conditions $\theta(0) = \pi/2$, $\dot{\theta}(0) = 0.01$, and $\phi(0) = \dot{\phi}(0) = 0$ with $\omega_3(0) = -0.05$ (slow spin). That is, in the initial state, the wheel is upright, its plane is parallel to the x-axis and it begins rolling very slowly in the positive x-direction. A small side push, $\dot{\theta}(0) = 0.01$, together with the gravity torque cause the wheel to nearly fall flat. However, at the lowest point, $\dot{\theta}$ reverses and the wheel stands up again because the energy is conserved. Now, $\dot{\theta}$ has a different sign, and the wheel falls onto the other side. The process is repeating infinitely. Figure 8.8.1 shows the dependence $\theta(t)$ nearly falling on both sides ($\theta = 0, \pi$) and recovering. Figure 8.8.2 shows that the wheel is rapidly precessing in alternating directions and its spin sharply increases when the wheel nearly hits the ground. The trajectories of the wheel's center and the contact point obtained by the integration of Eq. (8.1.11) are shown in Fig. 8.8.3. Since ϕ takes nearly all possible values, the wheel rolls in all directions. Still, there is no full symmetry that can be seen from the dependence $\phi(t)$ in the top panel of Fig. 8.8.2. As a result, on average, the wheel is moving in the negative directions along x and y. For twice as high spin, $\omega_3(0) = -0.1$, the trajectories of CM and CP are nearly close, and the flat CP trajectory makes a figure of eight, as illustrated in Fig. 8.8.4.

The critical value of ω_3 above which the upright rolling of the wheel becomes stable is given by Eq. (8.5.3). For our parameters, $|\omega_{3,\mathrm{crit}}| = 0.5$.

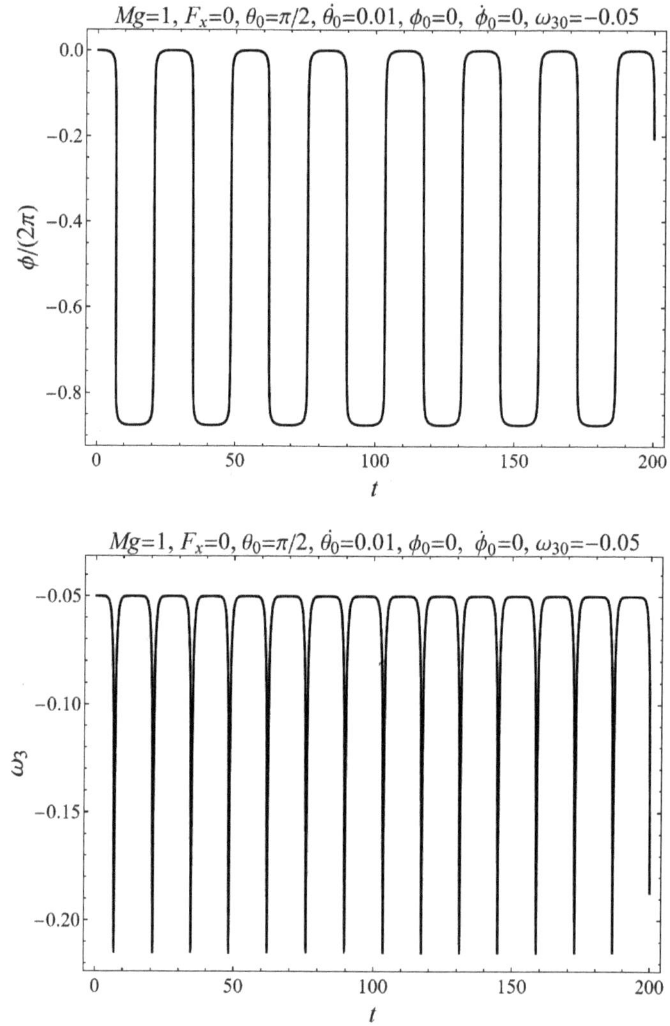

Figure 8.8.2. Time dependences of $\phi(t)$ and $\omega_3(t)$. When the wheel nearly falls on its side, both $|\dot{\phi}|$ and $|\omega_3|$ become large.

The solution for $\theta(t)$ for the subcritical value $\omega_{30} = -0.49$ is shown in Fig. 8.8.5. One can see that the motion is unstable, although the wheel's center does not come close to the ground. There is no drastic increase in ω_3 near the lowest point in contrast to the results for the slow rolling, $\omega_{30} = -0.05$ and -0.1. The biggest effect is in $\phi(t)$, which shows the values $|\phi| > \pi/2$,

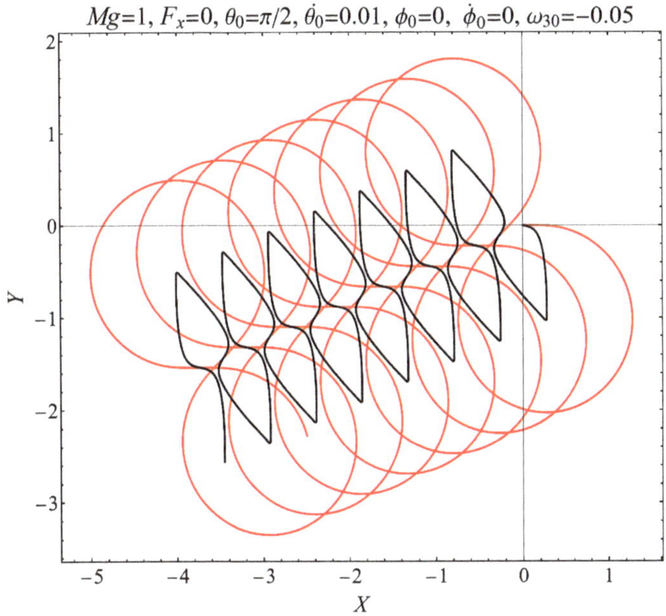

Figure 8.8.3. Trajectory of the CM (black) and CP (red) for the wheel on a plane with a low spin for $\omega_{30} = 0.05$ (starting point $X(0) = Y(0) = 0$).

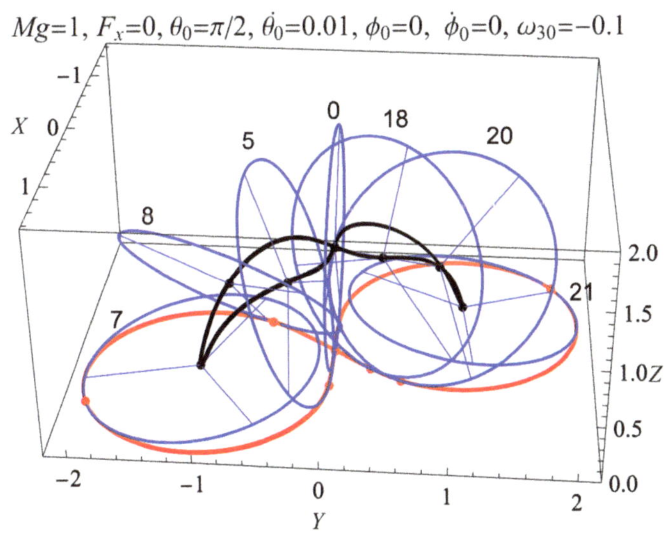

Figure 8.8.4. Wheel in the unstable motion with $\omega_{30} = -0.1$ at the moments of time $t = 0$, 5, 7, 8, 18, 20, and 21.

similar to Fig. 8.8.2. Due to this, the wheel is rolling in different directions, as shown in Fig. 8.8.5 (bottom). Similar to Fig. 8.8.3, the wheel is rolling on average in the negative directions, although the effect is smaller. The same trajectories in 3D with the wheel shown at different moments of time are shown in Fig. 8.8.4.

Figure 8.8.6 shows stable rolling in the supercritical (stable) regime with $\omega_{30} = -0.51$. Here, $\phi(t)$ remains close to zero. As follows from Eq. (8.1.11), in this case, the wheel is rolling in the positive x-direction (the slower motion in the y-direction is due to the initial side push).

Initial tilt of the wheel is dynamically preserved because the wheel cannot get upright due to the energy conservation. So, it remains tilted and precesses in the same direction, its CM making rosettes. An example is Fig. 8.8.7 (top), obtained with the initial conditions $\theta(0) = \pi/4$, $\dot{\theta}(0) = \phi(0) = \dot{\phi}(0) = 0$, and $\omega_3(0) = -0.5$. If the initial condition for $\dot{\phi}(0)$ is chosen according to Eq. (8.4.2), the wheel makes perfect circles with $\theta =$ const. and $\dot{\phi} =$ const. Initial precession makes the motion more sophisticated, as shown in Fig. 8.8.7 (bottom), obtained with the initial conditions $\theta(0) = \pi/2$, $\dot{\theta}(0) = \phi(0) = 0$, $\dot{\phi}(0) = 1$, and $\omega_3(0) = -0.5$.

8.8.2 *Wheel with a gravity and a horizontal force*

Switching on the horizontal force F_x causes several effects, depending on the relations between F_x, the gravity force, initial spin $\omega_3(0)$, and the initial precession $\dot{\phi}(0)$. For a rapid spin, the problem was solved analytically in Sec. 8.6. Numerical calculations confirm these solutions. For a slow spin, the motion of the wheel becomes unstable, and the numerical solution shows a rather complicated behavior, especially in the case of a significant initial precession.

If a small horizontal force $F_x = 0.3$ is added to the problem with the initial conditions $\theta(0) = \pi/2$, $\dot{\theta}(0) = \phi(0) = 0$, $\dot{\phi}(0) = 1$, and $\omega_3(0) = -0.5$, as in Fig. 8.8.7 (bottom), then the rosette structure gets destroyed, the CM and CP shifting on average in the positive y-direction, perpendicular to the force, as shown in Fig. 8.8.8 (top). Increasing the force leads to the wheel turning over and moving in the negative y-direction after some time, as shown in other panels of Fig. 8.8.8. Finally, the force wins the competition with the precession, and the trajectory becomes an imperfect cycloid. Increasing the spin works in the same direction as increasing the force. The bottom panel of 8.8.8 shows a nearly perfect cycloid for $F_x = 0.3$ and $\omega_3(0) = -1.5$. Note that in these figures, the axes X and Y are

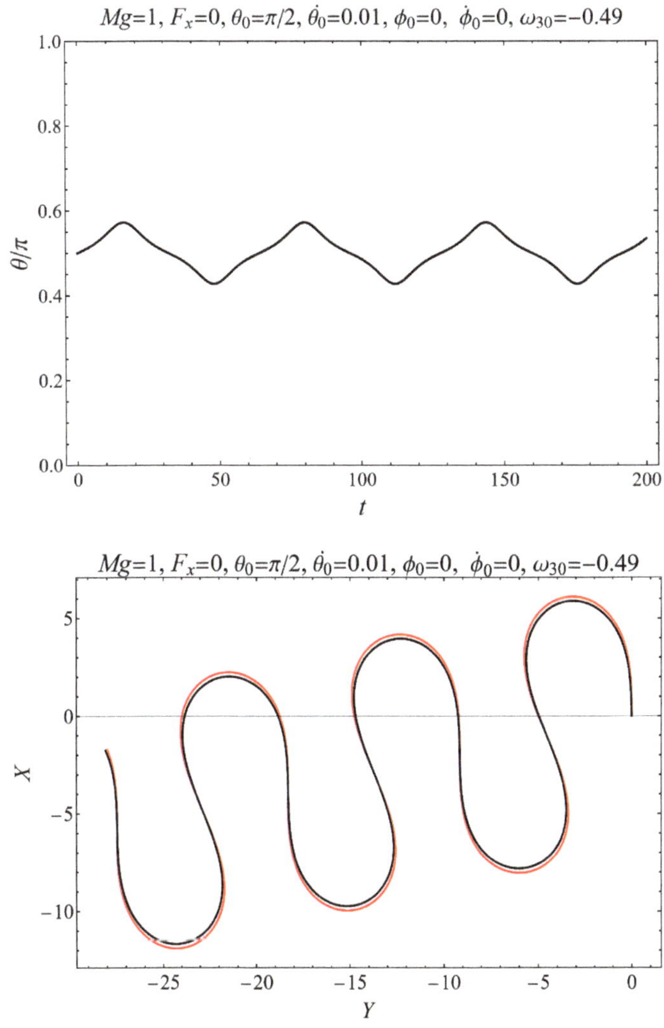

Figure 8.8.5. Unstable rolling at the subcritical value $\omega_{30} = -0.49$, with time dependence of $\theta(t)$ (top) and trajectory of the CM (black) and CP (red) (bottom). The motion is still in the negative direction on average.

interchanged in comparison to the preceding figures with trajectories. In the first of these figures, the maximal time is $t_{\max} = 600$, while in all others, $t_{\max} = 200$.

Computations over longer time intervals show that, for a sufficiently large F_x, after a period of initial wandering the wheel finally accelerates

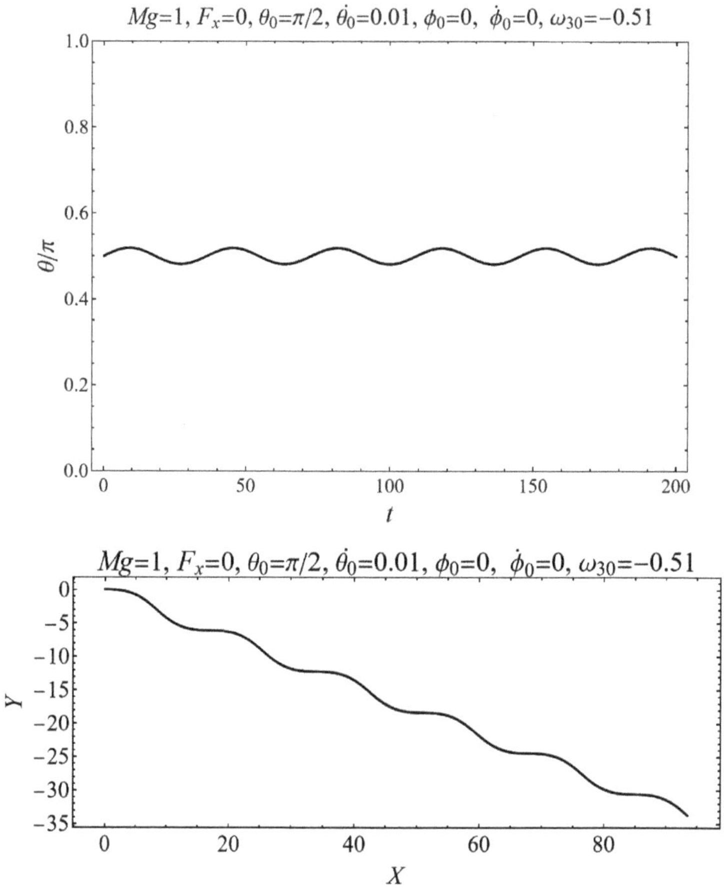

Figure 8.8.6. Rolling in the supercritical (stable) regime with $\omega_{30} = -0.51$, now in the positive x-direction.

by the horizontal force, so that its spin becomes large and the motion simplifies to the accelerating unidirectional rolling or to the trochoidal motion considered in Sec. 8.6. Which of these regimes is realized depends on the parameter $\kappa = Mg \cot\theta / F_x$, whereas θ is constant for a fast spin. What value of θ gets chosen by the system depends on the prehistory — the slow-spin unstable motion. Thus, it is difficult to predict the asymptotic behavior of the wheel from the initial condition, leading to unstable motion.

For a smaller F_x, with a significant initial precession, the unstable regime persists for an infinite time, as in the first and second panels of Fig. 8.8.8.

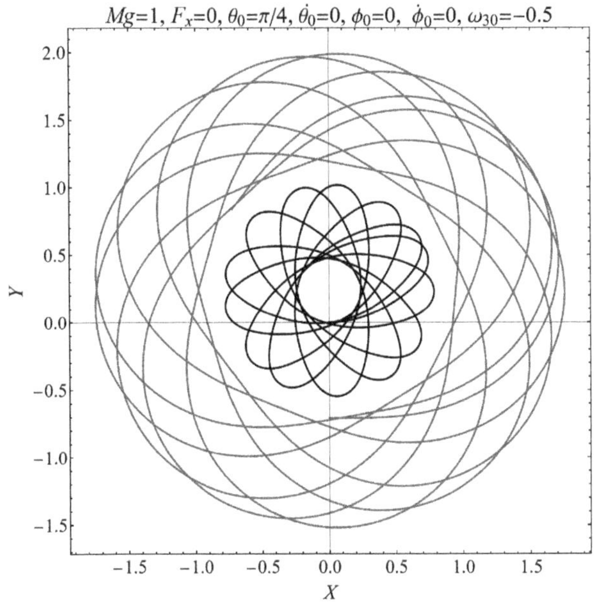

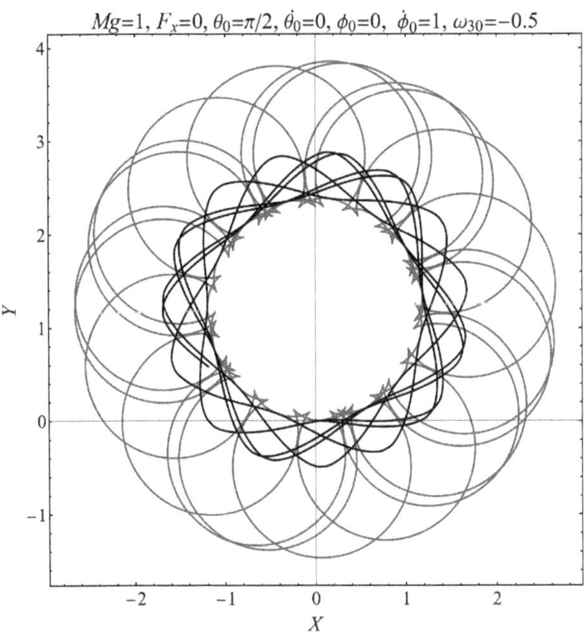

Figure 8.8.7. Trajectories of the wheel rolling with the initial tilt, $\theta(0) = \pi/4$ (top) and the initial precession, $\dot{\phi}(0) = 1$ (bottom).

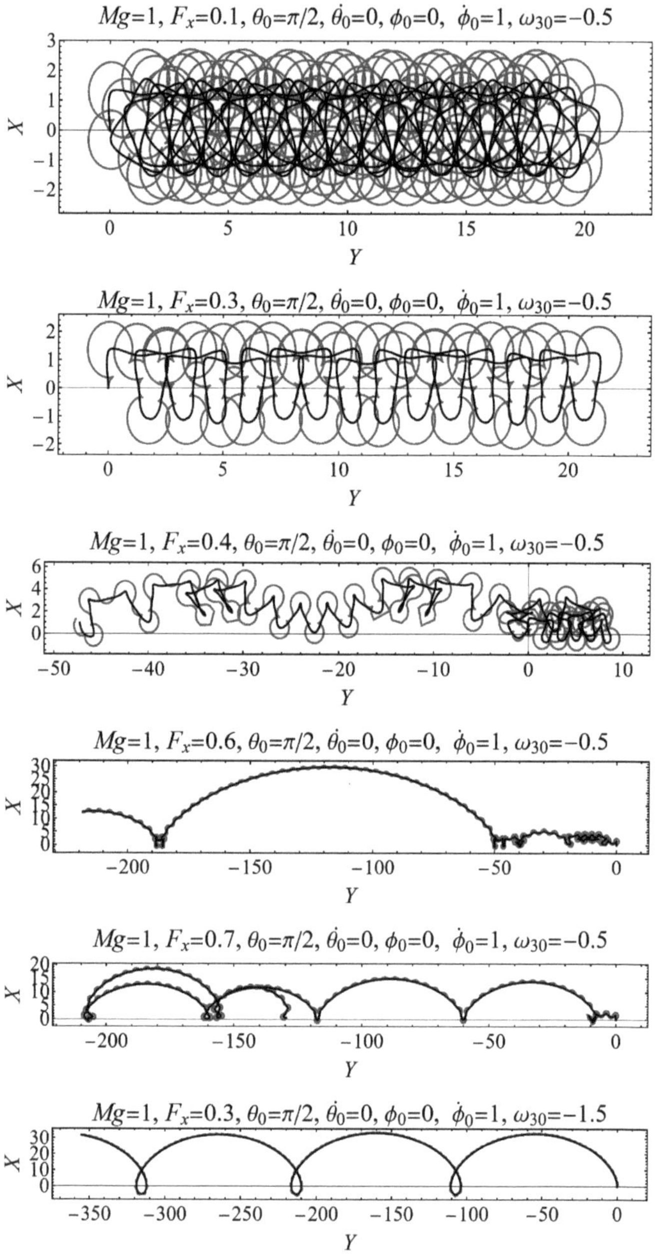

Figure 8.8.8. Trajectory of the wheel in the presence of the initial precession and a horizontal force.

In these cases, the average motion is in the direction perpendicular to the horizontal force.

8.9 Problems

8.9.1 *Disk on a plane without friction*

Set up Euler equations in terms of Euler's angles for the disk on a plane with no friction, taking into account gravity and a horizontal force. Use sliding principal axes. Prove that the energy of the system and the angular momentum components L_3 and L_z are integrals of motion.

Solution: Consider the rotation of the disk with respect to the CM and use the constraint equation (4.1.17). The set of equations of motion reads

$$\begin{aligned} M\dot{\mathbf{V}} &= \mathbf{F} + F^{(R)}\mathbf{e}_z, \\ \dot{\mathbf{L}} &= \boldsymbol{\tau} = [-R\mathbf{e}^{(2)} \times F^{(R)}\mathbf{e}_z] = -RF^{(R)}\cos\theta\mathbf{e}^{(1)}, \\ 0 &= (\mathbf{V} - R[\boldsymbol{\omega} \times \mathbf{e}^{(2)}]) \cdot \mathbf{e}_z, \end{aligned} \tag{8.9.1}$$

where we have taken into account that the reaction force $\mathbf{F}^{(R)}$ is directed vertically and that the kinematic term in Eq. (1.3.11) is absent if the rotation around the CM is considered. Expressing $F^{(R)}$ from the first equation and using the constraint, one obtains

$$\begin{aligned} F^{(R)} &= -F_z + M\dot{V}_z = -F_z + MR\frac{d}{dt}[\boldsymbol{\omega} \times \mathbf{e}^{(2)}] \cdot \mathbf{e}_z \\ &= -F_z + MR\frac{d}{dt}[\mathbf{e}^{(2)} \times \mathbf{e}_z] \cdot \boldsymbol{\omega} = -F_z + MR\frac{d}{dt}(\omega_1\cos\theta)\,. \end{aligned} \tag{8.9.2}$$

Now, τ_1 becomes

$$\tau_1 = -RF^{(R)}\cos\theta = R\cos\theta\left[F_z - MR\frac{d}{dt}(\omega_1\cos\theta)\right], \tag{8.9.3}$$

while other components of the torque are zero. Substituting the torque into Euler's equation (4.4.12), using $F_z = -Mg$ and $I_1 = I_2$, one obtains

$$\begin{aligned} I_1\ddot{\theta} + MR^2\cos\theta\frac{d}{dt}\left(\dot{\theta}\cos\theta\right) &= (I_2\omega_2\cot\theta - I_3\omega_3)\,\omega_2 - MgR\cos\theta, \\ I_2\dot{\omega}_2 &= (-I_1\omega_2\cot\theta + I_3\omega_3)\,\dot{\theta}, \\ I_3\dot{\omega}_3 &= 0, \end{aligned} \tag{8.9.4}$$

where $\omega_1 = \dot{\theta}$. This is a closed system of differential equations for the rotation of the disk. Its translational motion is described by the two autonomous equations

$$M\dot{V}_x = F_x, \quad M\dot{V}_y = 0, \tag{8.9.5}$$

while the vertical coordinate of the disk's center is defined from the constraint condition in Eqs. (8.9.1):

$$V_z = \dot{Z} = R\dot{\theta}\cos\theta = R\frac{d}{dt}\sin\theta, \tag{8.9.6}$$

and thus, $Z = R\sin\theta$.

The energy of the system is given by

$$\begin{aligned} E &= \frac{M\left(V_x^2 + V_y^2 + V_z^2\right)}{2} + \frac{I_1\dot{\theta}^2}{2} + \frac{I_2\omega_2^2}{2} + \frac{I_3\omega_3^2}{2} + MgR\sin\theta - F_x X \\ &= \frac{M\left(V_x^2 + V_y^2\right)}{2} - F_x X + \frac{MR^2}{2}\left(\dot{\theta}\cos\theta\right)^2 + \frac{I_1\dot{\theta}^2}{2} + \frac{I_2\omega_2^2}{2} + \frac{I_3\omega_3^2}{2} \\ &\quad + MgR\sin\theta. \end{aligned} \tag{8.9.7}$$

The translational part of the energy, the first two terms, is obviously conserved. The remainder yields

$$\dot{E} = \left[MR^2\cos\theta\frac{d}{dt}\left(\dot{\theta}\cos\theta\right) + I_1\ddot{\theta} + MgR\cos\theta\right]\dot{\theta} + I_2\omega_2\dot{\omega}_2 + I_3\omega_3\dot{\omega}_3. \tag{8.9.8}$$

Using Euler's equations, one obtains

$$\dot{E} = \left(I_2\omega_2\cot\theta - I_3\omega_3\right)\omega_2\dot{\theta} + \left(-I_1\omega_2\cot\theta + I_3\omega_3\right)\dot{\theta}\omega_2 = 0. \tag{8.9.9}$$

Apart from the energy, this problem has an obvious integral of motion $L_3 = I_3\omega_3$. Let us check if

$$L_z = I_2\omega_2\sin\theta + I_3\omega_3\cos\theta \tag{8.9.10}$$

is conserved. This is likely since there is only the torque τ_1 that rotates the disk in the direction perpendicular to the z-axis.

Using the second Euler equation, one obtains

$$\dot{L}_z = \left(-I_1\omega_2\cot\theta + I_3\omega_3\right)\dot{\theta}\sin\theta + I_2\omega_2\dot{\theta}\cos\theta - I_3\omega_3\dot{\theta}\sin\theta = 0, \tag{8.9.11}$$

as it should be. Thus, the problem of the disk on a frictionless plane is integrable and resembling that of a heavy symmetric top considered in Sec. 4.3.3. With

$$I_2\omega_2 = \frac{L_z - L_3\cos\theta}{\sin\theta}, \tag{8.9.12}$$

one can rewrite the conserved rotational part of the energy as

$$E_{\text{rot}} = \frac{\left(I_1 + MR^2\cos^2\theta\right)\dot{\theta}^2}{2} + \frac{\left(L_z - L_3\cos\theta\right)^2}{2I_2\sin^2\theta} + \frac{L_3^2}{2I_3} + MgR\sin\theta. \tag{8.9.13}$$

This can be resolved for $\dot{\theta}$ and integrated, which yields an implicit solution for $\theta(t)$. As E_{rot} diverges for $\theta \to 0$ and $\theta \to \pi$, the rotating disk cannot fall onto the plane.

8.9.2 *Disk on ice*

Set up Euler equations in terms of Euler's angles for the disk on ice, taking into account gravity and a horizontal force F_x. Use sliding principal axes. Prove that the energy of the system and the angular momentum components L_3 and L_z are integrals of motion. For $F_x = 0$, find the solution of the equations with $\theta = \text{const}$.

Solution: Disk on ice can freely spin around its symmetry axis, but its contact point cannot move sideways. This constraint condition is Eq. (4.1.18), in which $\mathbf{n} = \mathbf{e}_z$ and $\mathbf{e}_O = \mathbf{e}^{(2)}$. It is convenient to express $\mathbf{e}_z$ as $\mathbf{e}_z = \mathbf{e}^{(2)}\sin\theta + \mathbf{e}^{(3)}\cos\theta$, see Fig. 8.0.1. Then the constraint equations become:

$$\begin{aligned}
&\mathbf{V}\cdot\mathbf{e}^{(2)} = V_2 = 0,\\
&\mathbf{V}\cdot\mathbf{e}^{(3)} = V_3 = R\left[\boldsymbol{\omega}\times\mathbf{e}^{(2)}\right]\cdot\mathbf{e}^{(3)} = R\left[\mathbf{e}^{(2)}\times\mathbf{e}^{(3)}\right]\cdot\boldsymbol{\omega} = R\omega_1,
\end{aligned} \tag{8.9.14}$$

whereas the component V_1 remains unconstrained. The disk can freely slide on the ice in the direction of $\mathbf{e}^{(1)}$, thus V_1 is the sliding velocity. The equations of motion for the translational and rotational motion around the CM are, as usual,

$$\begin{aligned}
M\dot{\mathbf{V}} &= \mathbf{F} + \mathbf{F}^{(R)},\\
\dot{\mathbf{L}} &= \boldsymbol{\tau} = \left[-R\mathbf{e}^{(2)}\times\mathbf{F}^{(R)}\right],
\end{aligned} \tag{8.9.15}$$

where $\mathbf{F} = -Mg\mathbf{e}_z + F_x\mathbf{e}_x$ and $\mathbf{F}^{(R)}$ is the reaction force. Similar to Eq. (4.4.3), these equations can be represented in the form

$$M\left(\frac{\tilde{d}\mathbf{V}}{dt} + [\boldsymbol{\omega} \times \mathbf{V}]\right) = \mathbf{F} + \mathbf{F}^{(R)}$$

$$\frac{\tilde{d}\mathbf{L}}{dt} + [\boldsymbol{\omega} \times \mathbf{L}] = \boldsymbol{\tau} = [-R\mathbf{e}^{(2)} \times \mathbf{F}^{(R)}]. \tag{8.9.16}$$

There is no reaction force in the direction $\mathbf{e}^{(1)}$, so $\mathbf{F}^{(R)}$ can be searched for in the form

$$\mathbf{F}^{(R)} = F_2^{(R)}\mathbf{e}^{(2)} + F_3^{(R)}\mathbf{e}^{(3)}. \tag{8.9.17}$$

Expressing the reaction force from the first equation as $\mathbf{F}^{(R)} = -\mathbf{F} + M\dot{\mathbf{V}}$, projecting it onto $\mathbf{e}^{(2)}$ and $\mathbf{e}^{(3)}$, and using the constraint relations yield

$$\begin{aligned} F_2^{(R)} &= -F_2 + M(\dot{V}_2 + [\boldsymbol{\omega} \times \mathbf{V}]_2) = -F_2 + M\left(-\omega_1 V_3 + \omega_3 V_1\right) \\ &= -F_2 - MR\omega_1^2 + M\omega_3 V_1 \end{aligned} \tag{8.9.18}$$

and

$$\begin{aligned} F_3^{(R)} &= -F_3 + M(\dot{V}_3 + [\boldsymbol{\omega} \times \mathbf{V}]_3) = -F_3 + MR\dot{\omega}_1 + M\left(\omega_1 V_2 - \omega_2 V_1\right) \\ &= -F_3 + MR\dot{\omega}_1 - M\omega_2 V_1. \end{aligned} \tag{8.9.19}$$

Here, $F_2^{(R)}$ does not contribute to the torque:

$$\boldsymbol{\tau} = [-R\mathbf{e}^{(2)} \times \mathbf{F}^{(R)}] = -RF_3^{(R)}\mathbf{e}^{(1)}. \tag{8.9.20}$$

With $F_3 = -Mg\cos\theta + F_x\sin\theta\sin\phi$, one finally obtains

$$\tau_1 = -Mg\cos\theta + F_x\sin\theta\sin\phi - MR^2\dot{\omega}_1 + MR\omega_2 V_1, \tag{8.9.21}$$

whereas the two other components of the torque are zero. Substituting the torque into Euler's equations (4.4.12) yields

$$\begin{aligned} I_1'\dot{\omega}_1 &= (I_2\omega_2\cot\theta - I_3\omega_3)\,\omega_2 - MgR\cos\theta + F_xR\sin\theta\sin\phi + MR\omega_2 V_1, \\ I_2\dot{\omega}_2 &= (-I_1\omega_2\cot\theta + I_3\omega_3)\,\omega_1, \\ I_3\dot{\omega}_3 &= 0, \end{aligned} \tag{8.9.22}$$

where $I_1' = I_1 + MR^2$ and we used $I_1 = I_2$. One can see that the system of Euler's equations for the rotation of the disk is not closed and is coupled to the sliding velocity V_1 which is not constrained. From the third equation, one immediately obtains the integral of motion $L_3 = \text{const}$.

The equation of motion for V_1 has the form

$$M(\dot{V}_1 + [\boldsymbol{\omega} \times \mathbf{V}]_1) = F_1, \tag{8.9.23}$$

where $F_3^{(R)} = 0$ was taken into account. Calculating

$$[\boldsymbol{\omega} \times \mathbf{V}]_1 = \omega_2 V_3 - \omega_3 V_2 = \omega_2 R \omega_1, \tag{8.9.24}$$

and with the use of Eq. (4.2.11),

$$F_1 = F_x(\mathbf{e}_x \cdot \mathbf{e}^{(1)}) - Mg(\mathbf{e}_z \cdot \mathbf{e}^{(1)}) = F_x \cos\phi, \tag{8.9.25}$$

one obtains the equation of motion

$$M\dot{V}_1 = F_x \cos\phi - MR\omega_1\omega_2. \tag{8.9.26}$$

This equation is coupled to the rotational motion, so that Eqs. (8.9.22) and (8.9.26) form a closed system of equations.

After V_1 has been found, one can find the total vector $\mathbf{V}$ from the constraints. First, the vertical coordinate of the disk's center is defined by the constraint condition

$$Z = R\sin\theta, \tag{8.9.27}$$

which is holonomic. From this follows

$$V_z = \dot{Z} = R\dot{\theta}\cos\theta. \tag{8.9.28}$$

The in-plane components of the velocity are given by

$$\begin{aligned} V_x &= V_1(\mathbf{e}^{(1)} \cdot \mathbf{e}_x) + V_2(\mathbf{e}^{(2)} \cdot \mathbf{e}_x) + V_3(\mathbf{e}^{(3)} \cdot \mathbf{e}_x) \\ &= V_1 \cos\phi + R\omega_1 \sin\theta \sin\phi, \\ V_y &= V_1(\mathbf{e}^{(1)} \cdot \mathbf{e}_y) + V_2(\mathbf{e}^{(2)} \cdot \mathbf{e}_y) + V_3(\mathbf{e}^{(3)} \cdot \mathbf{e}_y) \\ &= V_1 \sin\phi - R\omega_1 \sin\theta \cos\phi. \end{aligned} \tag{8.9.29}$$

After the system of rotation-translational equations of motion has been solved, one can integrate V_x and V_y to find the position of the CM in the xy-plane.

The energy of the system can be written as

$$E = \frac{M\left(V_1^2 + V_2^2 + V_3^2\right)}{2} + \frac{I_1\dot{\theta}^2}{2} + \frac{I_2\omega_2^2}{2} + \frac{I_3\omega_3^2}{2} + MgR\sin\theta - F_x X. \tag{8.9.30}$$

Using $V_2 = 0$ and $V_3 = R\omega_1 = R\dot{\theta}$ fixed by Eq. (8.9.14), one can simplify this to

$$E = \frac{MV_1^2}{2} + \frac{I_1'\dot{\theta}^2}{2} + \frac{I_2\omega_2^2}{2} + \frac{I_3\omega_3^2}{2} + MgR\sin\theta - F_x X. \quad (8.9.31)$$

Let us check if E is conserved by differentiating E and substituting the equations of motion and other formulas obtained above:

$$\begin{aligned}\dot{E} &= MV_1\dot{V}_1 - F_x V_x + (I_1'\ddot{\theta} + MgR\cos\theta)\dot{\theta} + I_2\omega_2\dot{\omega}_2 + I_3\omega_3\dot{\omega}_3 \\ &= V_1(F_x\cos\phi - MR\dot{\theta}\omega_2) - F_x(V_1\cos\phi + R\dot{\theta}\sin\theta\sin\phi) \\ &\quad + \left[(I_2\omega_2\cot\theta - I_3\omega_3)\,\omega_2 + F_x R\sin\theta\sin\phi + MR\omega_2 V_1\right]\dot{\theta} \\ &\quad + (-I_1\omega_2\cot\theta + I_3\omega_3)\,\omega_2\dot{\theta} = (I_2 - I_1)\,\omega_2^2\dot{\theta}\cot\theta. \end{aligned} \quad (8.9.32)$$

For $I_1 = I_2$, this turns to zero, thus the energy is conserved.

Let us check if

$$L_z = I_2\omega_2\sin\theta + I_3\omega_3\cos\theta \quad (8.9.33)$$

is conserved. This is likely since there is only the torque τ_1 that rotates the disk in the direction perpendicular to the z-axis.

Using the second Euler equation, one obtains

$$\dot{L}_z = (-I_1\omega_2\cot\theta + I_3\omega_3)\,\dot{\theta}\sin\theta + I_2\omega_2\dot{\theta}\cos\theta - I_3\omega_3\dot{\theta}\sin\theta = 0, \quad (8.9.34)$$

as it should be. With

$$I_2\omega_2 = \frac{L_z - L_3\cos\theta}{\sin\theta}, \quad (8.9.35)$$

one can rewrite the energy as

$$E = \frac{MV_1^2}{2} + \frac{I_1'\dot{\theta}^2}{2} + \frac{(L_z - L_3\cos\theta)^2}{2I_2\sin^2\theta} + \frac{L_3^2}{2I_3} + MgR\sin\theta - F_x X. \quad (8.9.36)$$

As E_{rot} diverges for $\theta \to 0$ and $\theta \to \pi$, the rotating disk cannot fall onto the plane.

Let us write down the system of equations for the disk on ice, taking into account the integrals of motion L_3 and L_z. In the first Euler equation,

$$\begin{aligned}(I_2\omega_2\cot\theta - I_3\omega_3)\,\omega_2 &= \left(\frac{L_z - L_3\cos\theta}{\sin\theta}\cot\theta - L_3\right)\frac{L_z - L_3\cos\theta}{I_2\sin\theta} \\ &= \frac{(L_z\cos\theta - L_3)\,(L_z - L_3\cos\theta)}{I_2\sin^3\theta}, \end{aligned} \quad (8.9.37)$$

and it simplifies. With $\omega_2 = \dot{\phi}\sin\theta$, Eq. (8.9.35) becomes an equation of motion for ϕ. Then, there is Eq. (8.9.26) for V_1. Collecting all equations, one obtains the system

$$\begin{aligned} I_1'\ddot{\theta} &= \frac{(L_z\cos\theta - L_3)(L_z - L_3\cos\theta)}{I_2\sin^3\theta} - MgR\cos\theta \\ &\quad + F_x R\sin\theta\sin\phi + MRV_1\frac{L_z - L_3\cos\theta}{I_2\sin\theta}, \\ \dot{\phi} &= \frac{L_z - L_3\cos\theta}{I_2\sin^2\theta}, \\ M\dot{V}_1 &= F_x\cos\phi - MR\frac{L_z - L_3\cos\theta}{I_2\sin\theta}\dot{\theta}. \end{aligned} \tag{8.9.38}$$

This system of equations is still rather complicated. For $F_x = 0$, it simplifies because the first and second equations become independent of ϕ. In this case, one can search for V_1 as a function of θ and write $M\dot{V}_1 = M\,(dV_1/d\theta)\,\dot{\theta}$, which results in the equation

$$\frac{dV_1}{d\theta} = -R\frac{L_z - L_3\cos\theta}{I_2\sin\theta}. \tag{8.9.39}$$

Integrating it, one obtains

$$V_1(\theta) = \frac{R}{I_2}\left[L_z\ln\left(\cot\frac{\theta}{2}\right) + L_3\ln(\sin\theta)\right] + \text{const.} \tag{8.9.40}$$

After that, one can substitute it into the first equation in Eqs. (8.9.38), introduce the effective potential energy $U_{\text{eff}}(\theta)$, so that rhs of the equation for θ is $-dU_{\text{eff}}(\theta)/d\theta$, and after that, obtain the implicit solution for $\theta(t)$ as a quadrature. Or, shortcutting this, one can substitute $V_1(\theta)$ into the energy, Eq. (8.9.36), resolve it for $\dot{\theta}$ and obtain the solution for $\theta(t)$ as a quadrature.

There is a solution of Eqs. (8.9.38) with $\theta = \text{const.}$ and

$$MRV_1 = \frac{MgI_2\sin\theta\cos\theta}{L_z - L_3\cos\theta} - \frac{L_z\cos\theta - L_3}{\sin^2\theta} = \text{const.} \tag{8.9.41}$$

The constant precession frequency $\dot{\phi}$ is a free parameter that defines the frequency of the motion, and V_1 is the velocity of the motion along the trajectory, see Fig. 8.0.1. The trajectory of the CM is a circle with the radius

$$\mathcal{R} = \frac{V_1}{\dot{\phi}}. \tag{8.9.42}$$

A particular case is $\theta = \pi/2$, for which

$$MRV_1 = L_3, \quad \mathcal{R} = \frac{I_2}{MR}\frac{L_3}{L_z}. \tag{8.9.43}$$

Note that the disk rolling without skidding for $\theta = \pi/2$ rolls straight and cannot make a circle. Numerical solution shows that this motion becomes unstable below some threshold frequency $\dot{\phi}$.

An interesting feature of the solution found above is the unusual direction of the spin. For a body rolling on the surface, the cross-product $\boldsymbol{\omega} \times \mathbf{V}$ is directed down. Here, for $\theta = \pi/2$, both vectors $\mathbf{e}^{(1)}$ and $\mathbf{e}^{(3)}$ are horizontal. One has

$$\boldsymbol{\omega} \times \mathbf{V} = \left(\omega_3 \mathbf{e}^{(3)} + \omega_2 \mathbf{e}^{(2)}\right) \times V_1 \mathbf{e}^{(1)} = \omega_3 V_1 \mathbf{e}^{(2)} - \omega_2 V_1 \mathbf{e}^{(3)}, \tag{8.9.44}$$

where the first term is a vector directed up since $L_3 V_1 = I_3 \omega_3 V_1 > 0$. Thus, the disk is spinning in the "wrong" direction, while sliding on the surface.

Index

www.ingramcontent.com/pod-product-compliance
Lightning Source LLC
LaVergne TN
LVHW051056210926
848486LV00002B/128